LIMNOLOGICA

Ecology of Inland Waters Vol. 20, No. 2

1990

Nordic Limnology

Akademie-Verlag Berlin

LIMNOLOGICA

Ecology of Inland Waters

Akademie der Wissenschaften der DDR
Forschungsbereich Biowissenschaften und Medizin
Zentralinstitut für Mikrobiologie und Experimentelle Therapie Jena
Redaktion LIMNOLOGICA
Beutenbergstraße 11
Jena DDR-6900

GENERAL INFORMATION

Scope. LIMNOLOGICA were founded by THEODOR SCHRÄDER in 1962. They will publish papers which fall in the scope of limnology, i.e. papers dealing with any aspects of ecology and hydrobiology of inland waters and adjacent biotopes, as well as reviews, discussions, and essays that develop limnological-ecological theory (including all qualitative and quantitative aspects of planktonic, periphytic, benthonic, neustonic, and nektonic organisms, nutrient and energy flow, bioindication, biomonitoring, water intern control mechanisms, ecological modelling and simulation, biological-ecotechnological water quality management). Proceedings of specialized symposia may also be accepted for publication, if the scientific standard set by the journal is met. Not acceptable is purely hydrological, hydrophysical, biochemical, physiological, genetical, or marine biological work, that, while using inland water test organisms, is unrelated to the scope of LIMNOLOGICA.

Submission of articles. Manuscripts should be submitted to the editor-in-chief, the members of the editorial board, and the secretary.

Publication schedule. LIMNOLOGICA is published in one volume of nearly 480 pages per year. Each volume comprises two issues.

Subscriptions. All issues are available upon your usual supplier all over the world. If you can not get LIMNOLOGICA on this way, please directly send subscription orders to:
Akademie-Verlag Berlin
Leipziger Straße 3—4
DDR-1086 Berlin

Please send orders to:
Kunst und Wissen
Erich Bieber GmbH
Ein Unternehmen der Faxon-Gruppe
Wilhelmstraße 4—6
D-7000 Stuttgart 1

Cover Photo:

Lake Stora Idegölen, a typical Swedish forest lake, in the national park of Norra Kvill (Photo T. WILLÉN).

Nordic Limnology

Editor: BIRGER PEJLER

LIMNOLOGICA
Band/Volume 20 · 1990

LIMNOLOGICA

Ecology of Inland Waters

BAND/VOLUME 20

AKADEMIE-VERLAG BERLIN
1990

Band/Volume 20

Mit 224 Abbildungen und 109 Tabellen/With 224 Figures and 109 tables

Heft/Number 1, April/April 1990, Seite/Page 1—212
Heft/Number 2, Juni/June 1990, Seite/Page 213—374

ISBN 3-05-500643-7
ISSN 0075-9511

Erschienen im Akademie-Verlag Berlin, Leipziger Straße 3—4, Berlin, DDR-1086

Printed in the German Democratic Republic
Gesamtherstellung: Druckhaus „Thomas Müntzer“, GmbH, DDR-5820 Bad Langensalza

Inhalt/Contents

Heft 1

Originalarbeiten

Proceedings of the Estuary Symposium, Kühlungsborn, German Democratic Republic, November 16—21, 1987

ÁBARZÚA, S., & SCHIEWER, U.: NH_4rhythm of Cyanobacteria in Eutrophic Shallow Waters and in Laboratory Cultures. 113

ADAMKIEWICZ-CHOJNACKA, B., & RÓŻAŃSKA, Z.: Seasonal Variations of the Biomass of Copepoda and Rotatoria: Permanent Components of the Zooplankton in the Vistula Lagoon. 71

ARNDT, E. A.: Ecological, Physiological and Historical Aspects of the Brackish Water Fauna Distribution 61

ARNDT, H., SCHIEWER, U., JOST, G., WASMUND, N., WALTER, T., HEERKLOSS, R., ARLT, G., & ARNDT, E. A.: The Importance of Pelagic and Benthic Microfauna in a Shallow-Water Community of the Darss-Zingst Estuary, Southern Baltic, During Mesocosm Experiments. 101

BACHOR, A.: On the Variations of Some Chemical Criteria in in the Unterwarnow — an Estuary of the Western Baltic Sea. 29

BEHRENDT, H., & KROCKER, M.: A Method for the Separation of Zooplankton in High Eutrophic Water Bodies. 79

BERGER, U., & HEYER, J.: Distribution and Activity of Methanotrophic Bacteria in a Brackish Water Ecosystem. 141

BOIKOVA, E.: Protozoans as a Component of Biological Monitoring of the Baltic Sea. 119

BRÜGMANN, L., & LANGE, D.: Metal Distribution in Sediments of the Baltic Sea. 15

CISZEWSKI, P., & STYCZYNSKA-JUREWICZ, E.: Degradation and Restoration of Puck Bay (A Project). 191

DAHLKE, S.: Denitrification in Sediments of North Rugian Estuaries, Preliminary Communication. 145

ERDMANN, N., HAGEMANN, M., BERG, C., & FULDA, S.: Basis of Salt Adaptation in Cyanobacteria. 57

HEERKLOSS, R., BRENNING, U., & RING, M.: Secondary Production of Calanoids (Copepoda, Crustacea) in Brackish Waters of the Southern Baltic. 65

HEYER, J., BERGER, U., & SUCKOW, R.: Methanogenesis in Different Parts of a Brackish Water Ecosystem. 135

HÜBEL, H., & HÜBEL, M.: Cyanobacterial Blooms and Nitrogen Fixation in Coastal Waters of the Arkona Sea, 1974—1986. 37

KAISER, W., IRMISCH, A., NEHRING, D., GEORGI, F., & BREUEL, G.: Ecological Investigations in the Onshore Pelagic Zone Near Warnemünde from March 1985 to March 1986. 33

KHLEBOVICH, V. V.: Some Physico-chemical and Biological Phenomena in the Salinity Gradient. 5

KLINKENBERG, G., SCHLUNGBAUM, G., & BRÜGMANN, L.: Der Einfluß von Schadstoffen auf den Denitrifikationsprozeß in flachen Küstengewässern — Untersuchungen zur Mikrobiologie und Stoffbilanz.
Influence of Pollution Factors on the Denitrification Process in Shallow Coastal Waters — Experiments to Microbiology and Nutrient Balance. 149

KRÜGER, B., MIKKAT, S., & FULDA, S.: Mischkulturen und die Analyse biotischer Wechselwirkungen.
Mixed Cultures: Their Use for Analysing Interspecific Relationships. 107

KÜNNIS, K., & SAAVA, A.: Some Aspects of Microbiology of the Gulf of Finland. 127

MESSNER, U., & VON OERTZEN, J.-A.: Recent Changes in the Phytal Zone of Greifswald Bay. 183

MUDRYK, Z.: Bacteriological Studies of Lake Łebsko. 131

NIEDERMEYER, R.-G., & LANGE, D.: An Actualistic Model of Mud Deposition and Diagnesis for the Western Baltic Sea. 9

NIXDORF, B., BEHRENDT, H., & STELLMACHER, R.: Comparison of Methods for Estimation of Integral Primary Production in Shallow Aquatic Ecosystems with Special Regard to Turbulent Mixing. 53

PLIŃSKI, M.: Important Ecological Features of the Polish Coastal Zone of the Baltic Sea. 39

RADZIEJEWSKA, T., & DRZYCIMSKI, I.: Dynamics of Meiobenthic Communities in a Eutrophic and Polluted Estuary . . . 83

Round Table Discussion on "Perspectives in Estuary Research and Management". 199

SCHIEWER, U., ARNDT, E. A., & SCHLUNGBAUM, G.: Estuary Symposium, Kühlungsborn, GDR, Nov. 16—21, 1987. Preface. 3

— ARNDT, H., JOST, G., HEERKLOSS, R., & WASMUND, N.: Carbon Flux Dynamics in a Shallow Eutrophic Estuary. 95

SCHMIDT, I.: Untersuchungen über das Phytoplankton der Unterwarnow 1979/1980.
Investigations on the Phytoplankton of the Unterwarnow in 1979/1980. 165

SCHULZ, S., BREUEL, G., & IRMISCH, A.: Results of Ecological Enclosure Experiments with Pelagic Communities of the Arkona Sea (Baltic Sea). 89

VIETINGHOFF, H., PUTA, H., KLAPPER, H., STENDER, M., HUBERT, M. L., & KUNTZE, B.: Ecosystem Management in Estuaries: Costs and Benefits. 157

VON OERTZEN, J.-A., & SCHARF, E.-M.: Energy Requirement of Some Brackish Water Invertebrates. 73

WASMUND, N.: Characteristics of Phytoplankton in Brackish Waters of Different Trophic Levels. 47

WIKTOR, K.: Biomass and Abundance of the Zooplankton from the Gulf of Gdańsk. 75

— The Role of Common Mussel *Mytilus edulis* L. in the Biocoenosis of the Gulf of Gdańsk. 187

WINKLER, H. M.: Fishery in Shallow Estuaries of the GDR, Possibilities and Limitations. 195

WOLFF, W. J.: Anthropogenic Influences and Management of Estuaries . . . 153

Rezensionen

Avifauna der Deutschen Demokratischen Republik. 5 Bände. Band 1: Die Vogelwelt Mecklenburgs (D. FLÖSSNER) 210
Avifauna der Deutschen Demokratischen Republik. 5 Bände. Band 3: Die Vogelwelt Thüringens (D. FLÖSSNER) 210
BARCLAY, W. R., & MC INTOSH, R. P. (ed.): Algal Biomass Technologies (L. KRIENITZ) . . . 205
BARTH, W.-E.: Praktischer Umwelt- und Naturschutz (W. SCHÖNBORN) . . . 204
Beiträge zur Meereskunde, Heft 57 (L. KREY) . . . 207
Berichte des Geobotanischen Instituts der ETH, Stiftung Rübel. Heft 53 (H.-D. KRAUSCH) . . . 211
Biologia Acuatica, Heft 8 (A. LANGE) . . . 209
BOURNAUD, M., & TACHET, H. (eds.): Proceedings of the 5th International Symposium on Trichoptera, Lyon (France) 21.—26. July 1986 (Series Entomologica 39) (W. ZIMMERMANN) . . . 210
Decheniana 140. Verhandlungen des Naturhistorischen Vereins der Rheinlande und Westfalens (A. LANGE) . . . 203
DE DEKKER, P., & WILLIAMS, W. D. (eds.): Limnology in Australia (Monogr. Biol. 61) (R. KOSCHEL) . . . 201
Der Personalcomputer PC — 1715. Hrsgeg. von einem Autorenkollektiv des Kombinates ROBOTRON unter Ltng. von R. ZETH (P. KASPRZAK) . . . 212
EPPLEY, R. W. (ed.): Plankton Dynamics of the Southern California Bight (Lecture Notes on Coastal and Estuarine Studies Vol. 15) (J.-A. VON OERTZEN) . . . 206
FEDOROV, K. N.: The Physical Nature and Structure of Oceanic Fronts (Lecture Notes on Coastal and Estuarine Studies Vol. 19) (H. U. LASS) . . . 206
FENCHEL, T.: Ecology of Protozoa (W. SCHÖNBORN) . . . 209
FITTER, A.: Blumen - Wildblühende Pflanzen; Biologie + Bestimmen + Ökologie (S. J. CASPER) . . . 212
GOVI, G.: Introduzione alla Micologia (G. HIRSCH) . . . 211
HEGI, G.: Illustrierte Flora von Mitteleuropa, Band I. (S. J. CASPER, Jena) . . . 211
Isotope Techniques in Water Resources Development (R. KOSCHEL) . . . 207
KINDL, H.: Biochemie der Pflanzen (R. KOSCHEL) . . . 209
KLEE, R., & STEINBERG, C.: Kieselalgen bayerischer Gewässer (S. J. CASPER) . . . 204
KOHL, J.-G., & NICKLISCH, A.: Ökophysiologie der Algen (L. KRIENITZ) . . . 204
LENHART, B.: Limnologische Studien am Ammersee (1984—1986) (R. KOSCHEL) . . . 202
MABBERLEY, D. J.: The Plant-Book (S. J. CASPER) . . . 211
MC INTOSH, R. P.: The Background of Ecology (P. KASPRZAK) 202
MÜLLER, H.: Hydrocarbons in the Freshwater Environment — A Literature Review (P. CASPER) . . . 208
MUNAWAR, M. (ed.): Proceedings of an International Symposium on the Phycology of Large Lakes of the World (L. KRIENITZ) . . . 205
NACHTIGALL, W.: Lebensräume (A. LANGE) . . . 203
PETERS, R. H.: The Ecological Implication of Body Size (P. KASPRZAK) . . . 203
PROSSER, J. I. (ed.): Nitrification (Special Publ. of the Society for General Microbiology, Vol. 20) (H.-D. BABENZIEN) . . . 208
REN, M.-E. (ed.): Modern Sedimentation in the Coastal and Nearshore Zones of China (W. LEMKE, D. LANGE) . . . 207
ROBERTS, M. J.: The Spiders of Great Britain and Ireland (D. FLÖSSNER) . . . 211
SMIRNOV, N. N. (ed.): Lake Glubokoe (Developm. in Hydrobiol. 36) (P. KASPRZAK) . . . 201
THOMPSON, G., COLDREY, J., & BERNARD, G.: Der Teich (W. SCHEFFLER) . . . 201
TSCHÄPE, H.: Plasmide (H.-D. BABENZIEN) . . . 209
Veröffentlichungen des Geobotanischen Institutes der ETH, Stiftung Rübel, Zürich (Heft 99): Genetisch-ökologische Untersuchungen an alpinen Pflanzenarten auf verschiedenen Gesteinsunterlagen: Keimungs- und Aussaatversuche (H.-D. KRAUSCH) . . . 211
VINER, A. B. (ed.): Inland Waters of New Zealand (DSIR Bulletin 241) (R. KOSCHEL) . . . 207
WORBES, M.: Lebensbedingungen und Holzwachstum in zentralamazonischen Überschwemmungswäldern (H.-D. KRAUSCH) . . . 203

Heft 2

Originalarbeiten

Nordic Limnology

ANDERSSON, I., PARKMAN, H., & JERNELÖV, A.: The Role of Sediments as Sink or Source for Environmental Contaminants — a Case Study of Mercury and Chlorinated Hydrocarbons. . . . 347
APPELBERG, M.: Population Regulation of the Crayfish *Astacus astacus* L. after Liming an Oligotrophic, Low-alkaline Forest Lake. . . . 319
ARVOLA, L., SALONEN, K., & RASK, M.: Chemical Budgets for a Small Dystrophic Lake in Southern Finland. . . . 243
BLOMQVIST, P., & HEYMAN, U.: Effects of Arsenate Additions to Pelagic Mesocosms on Phytoplankton Biomass and Composition. . . . 329
BONSDORFF, E., & STORBERG, K.-E.: Ecological Changes in a Formerly Meromictic Coastal Lake. . . . 279
HÅKANSON, L.: Environmental Hazard Analysis — Contamination of Nutrients, Mercury and Cesium-137 in Natural Waters. . . . 339
HOLTAN, H.: Lake Mjøsa, Recipient and Ecosystem. . . . 285
ILMAVIRTA, V., & HUTTUNEN, P.: Characterization of 151 Small Lakes in Eastern Finland Based on their Water Chemistry, Land-use Catchment and Phytoplankton Communities. . . . 229
JÓNASSON, P. M., LINDEGAARD, C., DALL, P. C., HAMBURGER, K., & ADALSTEINSSON, H.: Ecosystem Studies on Temperate Lake Esrom and the Subarctic Lakes Mývatn and Thingvallavatn . . . 259
LINDHOLM, T., & ERIKSSON, J.: Limnological Changes in Two Recently Isolated Lakes. . . . 273
MORLING, G., & PEJLER, B.: Acidification and Zooplankton Development in Some West-Swedish Lakes 1966—1983. 307
— & WILLÉN, T.: Acidification and Phytoplankton Development in Some West-Swedish Lakes 1966—1983. . . . 291
PEJLER, B.: Preface . . . 215
STENSON, J. A. E.: Interspecific and Intertrophic Relations in Oligotrophic Water Systems. . . . 267
WALLS, M., RAJASILTA, M., SARVALA, J., & SALO, J.: Diel Changes in Horizontal Microdistribution of Littoral Cladocera . . . 253
WILLÉN, E., HADJU, S., & PEJLER, Y.: Summer Phytoplankton in 73 Nutrient-poor Swedish Lakes. Classification, Ordination and Choice of Long-term Monitoring Objects. . . . 217

Rezensionen

Archiv für Hydrobiologie. Suppl.-Band 76, Heft 1/2. Monographische Beiträge (S. J. CASPER) 365

Archiv für Hydrobiologie. Suppl.-Band 76, Heft 3 (H.-D. KRAUSCH) 371

ARNDT, U., NOBEL, W., & SCHWEIZER, B.: Bioindikatoren (H.-D. KRAUSCH) 361

Beiträge zur Meereskunde. Heft 58, Heft 59 (L. KREY) 365

Biologia Acuatica, Heft 13 (A. LANGE) 366

CAPPENBERG, TH., & STEENBERGEN, C. L. M. (eds.): Proceedings of the Third International Workshop on the Measurement of Microbial Activities in the Carbon Cycle in Aquatic Ecosystems (Arch. Hydrobiol., Beih. Ergebn. Limnol. 31) (H.-D. BABENZIEN) 363

DÖRFELT, H.: Die Erdsterne. Geastraceae und Astraeaceae (S. J. CASPER) 371

Decheniana 141. Verhandlungen des Naturhistorischen Vereins der Rheinlande und Westfalens. (A. LANGE) 365

ENNET, D.: Heilpflanzen und Drogen (S. J. CASPER) 371

ETTL, H., & GÄRTNER, G.: Chlorophyta II. Tetrasporales, Chlorococcales, Gloeodendrales (L. KRIENITZ) 369

Exkursionsflora für die Gebiete der DDR und BRD. Begründet von W. ROTHMALER, Bd. 3: Atlas der Gefäßpflanzen (S. J. CASPER) 367

FRESENIUS, W., QUENTIN, K. E., & SCHNEIDER, W. (eds.): Water Analysis. A Practical Guide to Physico-chemical, Chemical and Microbiological Water Examination and Quality Assurance (P. CASPER) 364

HÜTTER, L. A.: Wasser und Wasseruntersuchung. Methodik, Theorie und Praxis chemischer, chemisch-physikalischer und bakteriologischer Untersuchungsverfahren (G. PROFT) 364

JONES, R. I., & ILMAVIRTA, V.: Flagellates in Freshwater Ecosystems (Developm. in Hydrobiol. 45) (S. J. CASPER) 362

KERFOOT, W. C., & SIH, A. (eds.): Predation. Direct and Indirect Impacts on Aquatic Communities (P. KASPRZAK) . . . 363

KRAMMER, K., & LANGE-BERTALOT, H.: Bacillariophyceae. 2. Teil. Bacillariaceae, Epithemiaceae, Surirellaceae (H. HORN) 369

KRAUSS, G. J., & KRAUSS, G.: Hochleistungs-Flüssigchromatographie. SERVA-Praxishefte, Band 2 (P. CASPER) 364

KREMER, B. P.: Welche Heilpflanze ist das? Kennen — Verwenden — Anbauen (A. LANGE) 371

LAWSON, G. W., & JOHN, D. M.: The Marine Algae and Coastal Environment of Tropical West Africa (2. edition) (Beiheft zur Nova Hedwigia, Heft 93) (S. J. CASPER) 372

LEINS, P., TUCKER, S. C., & ENDRESS, P. K. (eds.): Aspects in Floral Development (S. J. CASPER) 372

MANIAK, U.: Hydrologie und Wasserwirtschaft. Eine Einführung für Ingenieure (L. KREY) 364

MATTHES, D.: Tierische Parasiten — Biologie und Ökologie (G. SCHÄLLER) 366

MC ALLISTER, D., & KOTT, E. (eds.): On Lampreys and Fishes: A Memorial Anthology in Honor of VADIM D. VLADYKOV (Developm. in Environm. Biol. of Fish 8) (D. RONNEBERGER) 366

MUNAWAR, M. (ed.): Limnology and Fisheries of Georgian Bay and the North Channel Ecosystems (Developm. in Hydrobiol. 46) (D. RONNEBERGER) 362

PLATT, T., & LI, W. K. W. (eds.): Photosynthetic Picoplankton (Can. Bull. Fish. Aquat. Sci. 214) (S. J. CASPER) 368

POKORNY, J., LHOTSKY, O., DENNY, P., & TURNER, G. (eds.): Waterplants and Wetland Processes (Arch. Hydrobiol., Beih. Ergebn. Limnol. 27) (H.-D. KRAUSCH) 368

RAGAN, M. A., & BIRD, C. (eds.): Twelfth International Seaweed Symposium. Proceeding of the Twelfth International Seaweed Symposium held in Sao Paulo, Brazil, July 27—August 1, 1986 (Developm. in Hydrobiol. 41) (S. J. CASPER) . . 373

SKRESLET, S. (ed.): The Role of Freshwater Outflow in Coastal Marine Ecosystems (H. HÜBEL) 365

SOULE, D. F., & KLEPPEL, G. S. (ed.): Marine Organisms as Indicators (S. J. CASPER) 361

STOLP, H.: Microbial Ecology: Organisms, Habitats, Activities (H.-D. BABENZIEN) 367

SYMOENS, J. J. (ed.): Vegetation of Inland Waters (Handbook of Vegetation Science 15/1) (H.-D. KRAUSCH) 368

THOMAS, R., EVANS, R., HAMILTON, A., MUNAWAR, M., REYNOLDSON, T., & SADAR, H.: Effects of In-Situ Contaminants (Developm. in Hydrobiol. 39) (G. PROFT) 361

Umweltprobenbank, Bericht und Bewertung der Pilotphase. — Hrsgeg. vom Bundesministerium für Forschung und Technologie (P. KASPRZAK) 373

Verhandlungen der Gesellschaft für Ökologie. Band XVI (Jahrestagung Gießen 1986) (H.-D. KRAUSCH) 363

Veröffentlichungen des Geobotanischen Instituts der ETH, Stiftung Rübel (Heft 95) (S. J. CASPER) 370

Veröffentlichungen des Geobotanischen Instituts der ETH, Stiftung Rübel (Heft 96) (H.-D. KRAUSCH) 370

Veröffentlichungen des Geobotanischen Instituts der ETH, Stiftung Rübel (Heft 98) (S. J. CASPER) 370

Veröffentlichungen des Geobotanischen Instituts der ETH, Stiftung Rübel (Heft 100) (H.-D. KRAUSCH) 370

VIETS, K. O.: Die Milben des Süßwassers (Hydrachnellae und Halacaridae [part.], Acari). Teil 2: Katalog (Sonderbände des Naturwissenschaftlichen Vereins in Hamburg, Band 8) (M. MORITZ) 366

VINCENT, W. F. (ed.): Dominance of Bloom-forming Cyanobacteria (Blue-green Algae) (New Zealand Journal of Marine and Freshwater Research Vol. 21, Issue 3) (G. DUDEL) . . . 374

Wasser und Abwasser, Supplementband I/88. KALMAN BIRO: Kleiner Bestimmungsschlüssel für Zuckmückenlarven (Diptera, Chironomide) (R. SAMIETZ) 373

ZEHNDER, A. J. B. (ed.): Biology of Anaerobic Microorganisms (Wiley Series in Ecological and Applied Microbiology) (H.-D. BABENZIEN) 367

Limnologica (Berlin) **20** (2), 215—216, Juni 1990

Preface

The idea behind the present volume was to give a cross-section of the limnological work being conducted in the Nordic countries. For this purpose a circular was sent to the directors of institutes where such research is in progress, with a request to forward the message to their collaborators and other scientists in their surroundings. My hope was to get a representative picture of current limnological research in the countries in question.

However, because the deadline for sending in manuscripts could not be set too far in advance, many limnologists were not prepared to contribute during the period available. Thus, several important research groups are not represented in this issue, and the low number of contributions from Denmark and Norway is especially regrettable, because much good limnological research is being conducted in those countries. Iceland is not represented at all, which is perhaps not so surprising, considering the fairly few scientists working there within the limnological field. However, two Icelandic lakes are intensely discussed in the paper by JÓNASSON et al. PÉTUR JÓNASSON is in fact from the beginning an Icelander, and during his time as professor at Hillerød, Denmark, he has been the leader of several projects dealing with Icelandic lakes. The highest number of contributions to the volume comes from Sweden, which ought to be natural because this is the most populous Nordic country. However, especially Uppsala is overrepresented, which should not be connected with the fact that this is the home town of the editor.

In spite of the uneven and partly arbitrary composition, the present volume still ought to contain information from most fields of current limnological research (some exceptions being touched upon below). It is interesting in this context to follow the historical development of our science in northern Europe. Because of the wealth of lakes and watercourses of different types in this area, regional limnology has a long and solid tradition and has played from early on a central role for Nordic limnologists [see, e. g., E. NAUMANN (1932): Grundzüge der regionalen Limnologie. Die Binnengewässer, Band 11]. In the beginning, these regional investigations were mainly descriptive, but later on became more and more analytical, as in W. RODHE's work on primary production where the ^{14}C method was used for the first time in freshwater ecosystems. Nowadays regional limnology is performed by aid of modern techniques such as computer treatment, and the present volume begins with two papers within this field.

The ecology of separate organisms and communities has also been in focus for a long time, dealing mainly with categories which are more easily available for sampling and analysis, such as macrophytes, fishes, benthic macrofauna of the profundal zone, metazoan zooplankton and net phytoplankton. Concerning the smallest organisms, as well as the littoral zone, pioneering research largely remains for present and future activities. One investigation within this field is reported in the paper by M. WALLS et al., concerned with Lake Pyhäjärvi. The littoral community has been more comprehensively analysed in another Finnish lake, viz. L. Pääjärvi, close to the ecological station at Lammi.

Based upon detailed studies of the separate communities, the pathways of energy within the ecosystem become available for analysis, which is done in the paper by JÓNASSON et al. The three lakes compared here (in Denmark and Iceland) cover together a very broad spectrum, though they are all situated in Nordic countries. Trophic relationships are also discussed in the paper by J. STENSON, who gives a survey of the work performed by his research group in Gothenburg, western Sweden.

The research on trophic relations may be applied for biomanipulation purposes, which was the theme of a previous special issue in *Limnologica*, Vol. **19** (1), 1988, with one Nordic contribution (H. OLOFSSON et al.). Intensive studies within this field are being conducted in different parts of Scandinavia and Finland, and in September 1988 a Nordic symposium was held in Lahti, Finland, about ecological lake management. Some contributions from this symposium are published in Aqua Fennica, Vol. **19** (1), 1989.

In order to get a complete picture of the carbon flux within the ecosystem, the role of the most minute organisms (bacteria, flagellates, ciliates) and of detritus must be analysed. This is the theme of projects in progress in Denmark (Hillerød), Finland (Lammi), Norway (Oslo) and Sweden (Uppsala). A book dealing with this problem has recently been edited by the Danish authors, B. RIEMANN & M. SØNDERGAARD (1986): Carbon dynamics in eutrophic, temperate lakes.

The watercourses in the Nordic countries have to a large extent been regulated for production of hydroelectric power. Because there is not so much more to gain in terms of energy, and because of environmental protection considerations, these activities have now decreased. However, there is still some research being conducted as concerns the consequences of such regulation. Close to the sea other types of water constructions are sometimes performed, and the effects of such human activities are discussed in the papers by LINDHOLM et al. and BONSDORFF et al.

The role of nutrients (especially nitrogen and phosphorus) has been in focus for a long time in Nordic limnology. This is reflected, e.g., in the volume dedicated to WILHELM RODHE when he retired from his professorship [Arch. Hydrobiol., Vol. **89** (1—2), 1980], with contributions from RODHE's pupils at that time and previously, and with the present editor as coordinator. Since that time, a more refined analysis has been in progress, e.g. as concerns the separate fractions of phosphorus. A symposium on the last-named theme was held in Uppsala during September 1985, resulting in a volume recently published (Hydrobiologia, Vol. **170**, 1988). In the present volume, the paper by H. HOLTAN is centred around problems concerning nutrient pollution and recovery. It deals with a very special case, which has been paid much attention, viz. the largest lake of Norway, L. Mjøsa. Much work has been devoted to finding measures for reducing the nutrient load to our waters and for studying the consequences obtained after such reductions, e.g. by the teams directed by T. AHL and C. FORSBERG in Uppsala. At the limnological institute of Lund, under the leadership of S. BJÖRK, lake restoration has been in focus for a long time.

During the latest decades, acidification has been realized as a very serious problem for the waters of northern Europe. Much work has been devoted in these countries to determining causal connections behind acidification, as well as to disentangling its consequences and finding remedies [see, e.g., two special issues on this matter, Hydrobiologia **101** (1–2), 1983, and Rep. Inst. Freshw. Res., Drottningholm, **61**, 1984]. Almost all research on this matter began, for natural reasons, after the first signs of acidification appeared. The long-term follow-up reported in the two papers by G. MORLING and her co-authors in the present volume is unique in this respect, because it started at a much earlier point of time, initiated for quite other reasons. In order to restore acidified lakes, liming has occurred on a large scale, and the consequences of this measure have also been studied. Here this problem area is represented in the paper by M. APPELBERG.

Finally, ecotoxicology accounts for a great part of Nordic water research today, and the three last papers of the present volume belong to this field. The role of heavy metals and chlorinated hydrocarbons is intensely studied in our countries, as well as the impact of the catastrophe at Chernobyl.

Some of the contributions to this volume are of review character, while others mainly report primary results from original research. Such a mixture ought to be appropriate in a context like this. The papers have been read critically by one or more referees, and the undersigned editor has examined them all.

Uppsala, February 1989 BIRGER PEJLER

Limnologica (Berlin) **20** (2), 217—227, Juni 1990

National Environmental Protection Board, Environmental Quality Laboratory, Uppsala, Sweden

Summer Phytoplankton in 73 Nutrient-poor Swedish Lakes. Classification, Ordination and Choice of Long-term Monitoring Objects

Eva Willén, Susanna Hajdu & Ylva Pejler

With 8 Figures

Key words: Phytoplankton; Life-form pattern; Oligotrophic lakes; TWINSPAN classification; CANOCO ordination; Long-term monitoring; Acidification.

Summary

A summer study of phytoplankton in 73 mainly oligotrophic lakes, distributed from latitude 56 to 68 °N in Sweden, was performed in two steps, 1983 and 1986. The aim was to make an evaluation of objects for long-term monitoring. Lakes, species' biomasses and 9 environmental variables were tested by multivariate techniques using the TWINSPAN classification, a discriminant analysis and the CANOCO ordination. The ordination technique is considered superior in this context due to its integrated analysis of both chemical and biological variables. In this way, some evident groups of lake-types could be defined. The very acidified (median pH 4:5) clearwater lakes in western Sweden were especially discerned and the humic, acid (median pH 5.5) lakes containing *Gonyostomum*, situated south of limes norrlandicus — a climatic transition zone in Sweden. High mountain lakes in the northern latitudes were also easily distinguished. The large group of dystrophic lakes, however, occurring in the whole country, formed transitions along gradients of pH (5.3—7), conductivity (2—7.4 $mS \cdot m^{-1}$), total phosphorus (7—17 $\mu g \cdot l^{-1}$) and light climate (Secchi depth 2.3—5.1 m). The phytoplankton life-form pattern is discussed as a reflection of the growth conditions in the lakes. A final choice of 25 monitoring lakes is presented. The choice is based partly on an ecoregional lake classification using chemical variables from a large (160) number of lakes, and partly on results from the classified and ordered biological material, which also included the bottom fauna.

Introduction

The work presented here was undertaken as part of an introductory study preceding the final choice of lakes suitable for limnological long-term monitoring in Sweden commencing in 1988. On the basis of biological and chemical information from lakes distributed over the whole country (73 lakes sampled for phytoplankton and bottom fauna, 160 lakes sampled for water-chemistry), 25 representative objects were chosen for combined biological and chemical monitoring under the responsibility of the Swedish Environmental Protection Board. This monitoring is primarily aimed at:

1. detecting long-term changes and evaluate the range of cyclicity due to climatic variations,
2. collecting background information from relatively unaffected regions as a reference for local affected waters,
3. quantifying the flux of different substances and their influence on the biota.

It is important that lakes representative of the most prominent physical-geographical regions and vegetation zones are included (Nordiska ministerrådet 1984). Another prerequisite is to reflect the state and modifications in those parts of Sweden which are especially exposed to acidification and other air-borne pollution. Further factors to consider when choosing characteristic objects are an appropriate size of the lake and its drainage area to reflect changes within a reasonable time-span. Too large and too small lakes are unsuitable. The former may react slowly or represent numerous subsystems, the latter may be too influenced by the litoral, the bottom or the drainage area, causing interpretation difficulties. The most limiting selection criterion is, however, the possibility to protect a lake from serious human disturbance. Apart from the fact that interference by the owner (e.g. liming, clear-cutting in the drainage area) must be minimized, also intense recreational use is inconvenient but hard to prevent in a country where the ancient right to common access to land and water is so well established. Water-chemical analyses from the 160 lakes have been performed since 1983. This more extensive material formed the basis of a water quality classification of the lakes, where median values of pH, alkalinity, total phosphorus (TP) and $KMnO_4$-consumption were used. In this classification lakes representative of 16 different physical-geographical regions (ecoregions) were distinguished, when densely populated and especially affected areas were omitted.

Phytoplankton from 73 lakes (Fig. 1), which is discussed in this paper, was classified and ordered with the use of multivariate techniques involving species biomasses, lakes and nine environmental variables. The aim was to distinguish a phytoplankton composition representative of a number of lakes with a specific water quality and, if possible, also of the various ecoregions. Similar methods were used in the treatment of the bottom fauna from these lakes (Johnson & Wiederholm 1989; Johnson 1989).

Sampling, analysis and statistical treatment

This introductory study was undertaken in two steps; 22 lakes were investigated in August 1983 and 51 in August 1986. August was chosen because it is a climatically stable period in Swedish lakes with high temperatures, stratified water-columns and with little influence from drainage areas. In spite of the large latitudinal difference in Sweden, 55°—70 °N, most lakes are in a comparable successional stage at that time.

Phytoplankton samples were collected with a tube from the epilimnion (stated by temperature measurements) of five well separated localities in each lake. These samples were all pooled. In this way, one lake-representative sample was finally obtained, from which 300 ml was subsampled. This subsample, aimed at a quantitative evaluation of included species, was preserved with an iodine solution supplemented with acetic acid (Throndsen 1978). Net samples (mesh-size 25 µm) were also collected to facilitate species identification. These samples were preserved with a formaldehyde solution. The counting of phytoplankton was performed after sedimentation in 50 and 100 ml settling chambers according to the Utermöhl method (Utermöhl 1958) with the practical application described in WILLÉN (1976) and HOBRO & WILLÉN (1977).

Water which was sampled simultaneously for chemical analyses represented the surface layer (0—1 m depth). The following variables were analysed: N (NO_2—N, NO_3—N; NH_4—N, organic N), P (PO_4—P, organic P), Si, Mg^{2+}, Na^+, K^+, Ca^{2+}, SO_4^{2-}, Cl^-, pH, alkalinity (HCO_3^-), $KMnO_4$-consumption, conductivity and turbidity. Water temperature and Secchi depth were measured directly in each lake. All analyses mentioned were made according to Swedish Standards and to procedures described in AHL (1972).

After a logarithmic transformation (except for pH), making allowance for skewed distributions, the results were classified and ordered using multivariate techniques. To classify the lakes on the basis of their phytoplankton species biomasses a two-way indicator analysis (TWINSPAN) was chosen (HILL 1979). TWINSPAN gives an integrated dichotomized analysis of both species and sites (= lakes) in an ordered data matrix (as illustrated in Fig. 6), so that each group of sites is characterized by a group of differential species (so-called indicator species and preferential species). This ordination is based on reciprocal averaging (HILL 1973). The ordination axis is broken near its middle to get a negative and positive group of sites and species. After this division, further refinements are made by reclassifications as described by HOGEWEG (1976). A discriminant analysis was then used to compare the agreement between selected (usually with a small covariation) environmental variables (water temperature, light climate, Secchi depth, pH, conductivity, $KMnO_4$-consumption, TP, NO_3—N and NH_4—N) and the TWINSPAN groups of phytoplankton. In this analysis the probability for the lakes to belong to a certain group is given. Altogether 70% of the lakes had consistent classifications between chemical and biological variables.

Besides, an ordination technique, CANOCO — canonical correspondence analysis (TER BRAAK 1986), was carried out based on a simultaneous treatment of species, lakes and environmental variables. For this purpose, the same nine environmental variables used in the discriminant analysis were chosen. The number of quantified species was 160. In this analysis lakes and species are separated along compositional axes which are optimally related to linear combinations of the chosen environmental variables. This technique has been successfully tested on parts of this material (FÄNGSTRÖM & WILLÉN 1987) and also when dealing with diatoms in sediments (STEVENSON et al. 1989) and in macrophyte investigations (CRAMER & HYTTEBORN 1987; ROSÉN & SJÖGREN 1988). Further insights into the multivariate techniques described above are given in GAUCH (1982) and JONGMAN et al. (1987).

Lake characteristics

The lakes chosen for this study (Fig. 1) have a median size of 0.3 km² (varying between 0.02—2 km², except 5 larger lakes) and a median ratio of 15 between drainage area and lake area. 65% of the lakes are of an oligotrophic character (TP < 15 µg · l⁻¹), while the rest are mesotrophic. Only one lake is truly eutrophic (TP > 40 µg · l⁻¹). Trophic classifications are made according to FORSBERG & RYDING (1980) and OECD (1982). The lakes

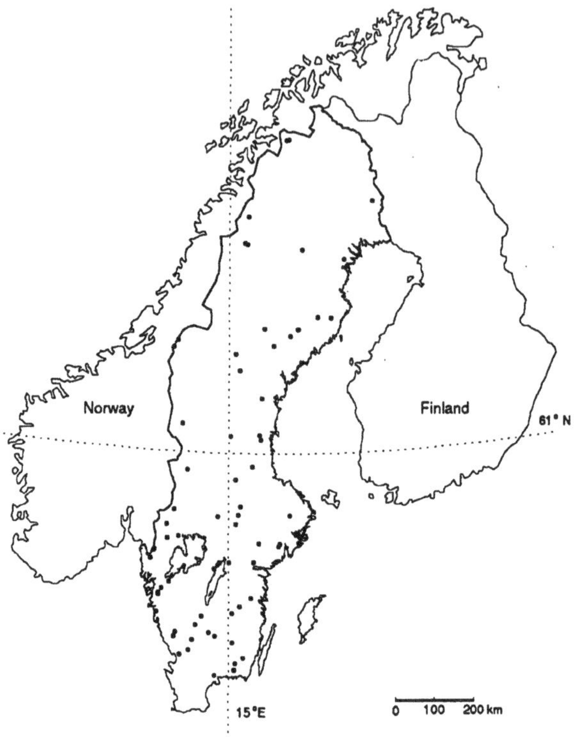

Fig. 1. Distribution of 73 oligotrophic Swedish lakes investigated in August 1983 and 1986

Table 1. Positions and some lake characteristics of investigated lakes in Sweden (August 1983, 1986) preceding a choice for long-term limnological monitoring. Lakes indicated by an asterisk are the chosen monitoring objects.

	CANOCO ordination	Lat.	Long.	pH	$KMnO_4$ (mg · l⁻¹)	Tot.-P. (µg · l⁻¹)
1. Lillesjö	C	5612	1432	4.5	6.8	5
2. Sännen*	D	5619	1521	5.5	20.0	15
3. Hallsjön	A1	5620	1521	6.7	34.1	10
4. Fersjön	A2	5628	1524	6.0	35.7	15
5. Brunnsjön*	A2	5635	1543	5.2	78.2	16
6. St. Skärsjön*	A1	5640	1304	6.9	12.8	10
7. Rammsjön	D	5645	1325	4.9	41.7	35
8. Storasjö*	D	5656	1516	5.6	33.6	14
9. Harasjön*	D	5700	1334	5.4	59.3	13
10. St. Pipsjön	A2	5704	1246	4.7	12.9	18
11. Fiolen*	A1	5705	1431	6.5	13.9	18
12. Tjärnesjön	C	5710	1251	5.0	10.9	6
13. Älgarydssjön*	A2	5710	1416	5.7	34.1	18
14. Hagasjön*	A1	5720	1342	6.5	29.4	8
15. Mossjön	D	5732	1356	5.5	44.7	21
16. Fisjön	C	5736	1159	4.5	6.0	4
17. Fjärasjö	A1	5736	1515	7.0	36.2	17
18. St. Idgölen	D	5745	1535	5.4	58.6	8
19. Allgjuttern*	A1	5756	1605	6.6	24.1	6
20. Kvarnsjön	C	5800	1203	5.6	3.8	4
21. Härsvatten*	C	5801	1201	4.6	0.2	3
22. Torrgårdsvattnet	C	5803	1203	4.5	3.5	2

23. Fräcksjön*	A1	5808	1211	6.6	30.4	8
24. Alsjön	D	5818	1230	5.5	51.8	13
25. Humsjön*	A1	5837	1428	6.8	18.5	33
26. Lommasjön	C	5842	1438	4.3	37.5	5
27. Storesjön	C	5842	1438	4.4	39.0	21
28. Fagertärn	D	5845	1442	6.1	38.2	17
29. Grissjön*	A2	5845	1508	5.6	29.9	5
30. Skärgölen*	A1	5845	1614	6.7	25.8	9
31. Råsjön	D	5846	1713	6.6	40.4	18
32. Rotehogstjärnen*	D	5848	1136	5.6	48.2	20
33. St. Abborrtjärnet	C	5900	1145	5.1	3.5	4
34. St. Tresticklan	C	5901	1147	4.7	3.3	3
35. St. Envättern*	A1	5906	1721	6.7	27.9	4
36. N. Yngern	A1	5910	1724	7.4	20.5	13
37. Bylsjön	D	5910	1815	4.7	28.2	8
38. St. Fjällsjön	D	5911	1628	4.6	68.4	23
39. Bysjön*	A1	5918	1220	6.8	17.9	6
40. Sandsjön	D	5921	1251	6.7	40.5	39
41. Ulvsjön	A2	5936	1217	6.2	25.2	8
42. Ämten S-län	A2	5937	1526	5.7	31.9	7
43. St. Gloppsjön	A1	5948	1437	6.9	38.8	13
44. Edasjön	A1	5948	1754	7.6	46.5	47
45. Övre skärsjön*	A2	5950	1533	5.3	26.3	5
46. Ämten	D	5956	1236	6.0	94.3	33
47. Mäsen*	A1	6001	1539	7.0	20.0	10
48. Spjutsjön	A2	6038	1526	6.6	13.7	24
49. Bockskärstjärn	A2	6051	1309	4.7	25.5	6
50. Källsjö-Holmsjön	A2	6056	1614	5.8	48.9	18
51. Laxtjärn, Laxtj.B.	A2	6132	1638	6.4	29.8	8
52. Hundsjön	A2	6138	1511	6.9	36.1	9
53. Stensjön*	A2	6138	1634	6.2	23.4	7
54. Gränjåstjärnen	A2	6154	1250	5.7	33.3	8
55. Gransjön	A1	6228	1642	6.9	39.9	24
56. Stugu-Sittsjön	A1	6306	1539	7.5	25.6	9
57. Fyrsjön	A1	6328	1525	7.9	21.1	9
58. Stor-Björsjön	A1	6336	1214	7.3	19.1	5
59. Bergsjön S.	A2	6338	1723	6.4	51.2	15
60. Lilltjärnen	B	6346	1226	6.7	27.9	7
61. Remmarsjön*	A1	6351	1816	6.6	32.4	32
62. Lång Viskasjön	A2	6400	1842	6.7	30.1	8
63. V. Stenbitsjön	A2	6402	1655	5.9	30.0	7
64. Bjännsjön	A1	6414	2026	6.5	29.1	12
65. Stortjärnen	A2	6415	1945	5.3	71.6	15
66. Brännträsket*	A1	6531	2125	6.8	42.4	10
67. Laxtjärnen, Reivo*	B	6547	1906	7.0	23.2	10
68. Stor-Tjulträsket*	B	6557	1604	7.3	7.5	7
69. Raurejaure	B	6559	1556	7.3	9.6	10
70. Guoletsijaure	B	6634	1608	7.3	8.6	4
71. Pahajärvi	B	6646	2321	6.9	19.7	14
72. Påtjujaure	B	6817	1831	7.5	3.4	4
73. Abiskojaure*	B	6818	1839	7.3	12.1	6

are listed in Table 1 and their geographical positions are indicated from south to north in Sweden. Some characteristic environmental variables are also given (TP, $KMnO_4$, pH).

About 55% of the lakes are situated south of what is called limes norrlandicus (indicated in Fig. 5) which is a climatic transition zone in Sweden, approximately coinciding with latitude 61 °N. The growing period (i.e. where the mean diurnal temperature exceeds 6 °C) north of limes is shorter than 160 days, the period of snowcover lasts for more than 130 days and the mean temperature in January falls below −5 °C (Fransson 1965; Sjörs 1965). These features have a decisive impact on the vegetation and limes norrlandicus is a dividing zone between the boreo-nemoral and the boreal zone. This means that many deciduous trees like oak, elm, hazel and ash are not present in the northern part of the country.

The areas especially influenced by acidifying substances are mainly concentrated to the western and south-central regions of Sweden (Fig. 2). It is in these regions that about 30% of the investigated lakes are situated.

To get an overall view of the water-chemical characteristics of the investigated lakes a classification is made in Fig. 3. The typical lake of this material is oligotrophic, mesohumic, slightly acid with comparatively warm water (> 18 °C) and a very low buffering capacity (median: TP $10\,\mu g \cdot l^{-1}$, TN $400\,\mu g \cdot l^{-1}$, $KMnO_4$ $29\,mg \cdot l^{-1}$, pH 6.4, alkalinity $0.035\,mmoles \cdot l^{-1}$).

Phytoplankton — classification and ordination

A survey of the total phytoplankton volumes in all lakes (Fig. 4) clearly indicates their oligo- and mesotrophic character. Several lakes containing the very large raphidophycean flagellate *Gonyostomum semen* (Ehr.) Dies. have total volumes exceeding $1\,mm^3 \cdot l^{-1}$ although their character is moderately nutrient-poor and dystrophic.

Fig. 2. Acidified regions in Sweden based on mapping of indicator mayflies White areas = pH > 5.4, ruled areas = pH 5.0—5.4, cross-ruled areas = pH < 5.0, (after LINGDELL & ENGBLOM, in BERNES 1985)

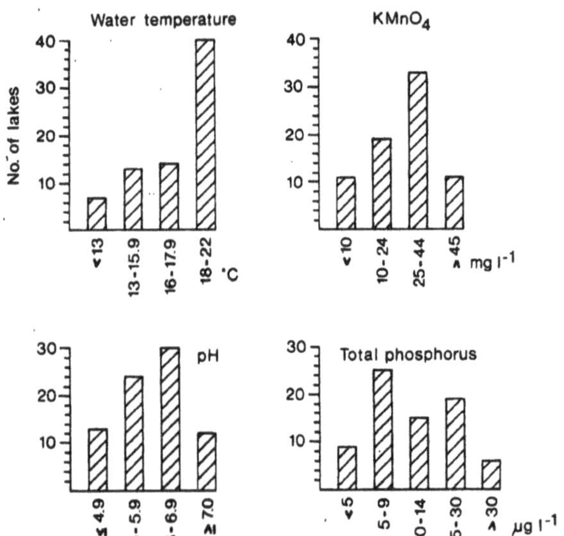

Fig. 3. Frequency of 73 Swedish oligotrophic lakes in 5 size-classes of characteristic environmental variables

The median total volume for all lakes is calculated to $0.27\,mm^3 \cdot l^{-1}$. The minimum volume, $0.007\,mm^3 \cdot l^{-1}$, is registered in an alpine lake in the high mountain region (72. Lake Påtjujaure) and the maximum, $7.1\,mm^3 \cdot l^{-1}$, in a south Swedish humic lake, situated in a fissure valley

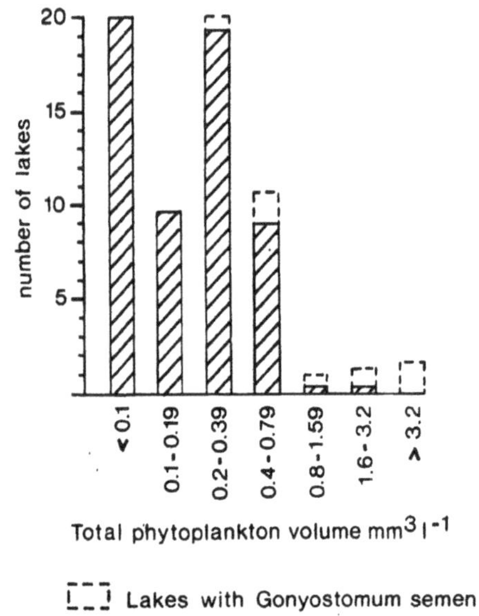

Fig. 4. Frequency of 73 Swedish oligotrophic lakes in 7 size-classes of total phytoplankton volumes

landscape (38. Lake St. Fjällsjön) where mass-developing *Gonyostomum* occupies > 90% of the total biomass (cf. Table 1).

The distribution of the most characteristic phytoplankton groups in these lakes is summarized in Fig. 5. Limes norrlandicus turns out to be an indicative borderline not only concerning terrestrial vegetation but also when dealing with phytoplankton biomasses and the distribution and/or frequency of certain algal groups. The total phytoplankton volume is generally larger (factor of ten) south of limes. Chrysophycean and cryptophycean flagellates are proportionally of larger importance in lakes north of limes where about half of the investigated lakes contains proportions as indicated in Fig. 5, but only 15—20% of the lakes south of limes. Diatoms, however, have no regional preference but occur in large turbulent lakes with a certain amount of nutrients all over the country (cf. EARLE et al. 1986). Concerning the large flagellate *Gonyostomum semen*, limes norrlandicus really seems to be its northern borderline. Lakes with a large proportion of dinoflagellates belong to the area especially exposed to acidification (cf. Fig. 2).

The classification of species and lakes using the TWINSPAN technique resulted in the formation of seven groups of lakes, where the species distribution, following a pH gradient, seems to have been of decisive importance (Fig. 6, Table 2). The lakes are divided in two large classes, where groups 1 to 4 comprise 52 lakes ranging from a circumneutral to a slightly acid state. The other large class includes 21 lakes, all very acid. Group no. 1 contains mainly diatom-dominated lakes with a high pH (cf. Fig. 5). Group no. 2 includes lakes of a circumneutral, oligohumic and oligo-mesotrophic character with a good light climate. The green alga *Botryococcus braunii* KÜTZ. and small unicellular flagellates are typical plankton organisms. Most of these lakes are situated south of limes norrlandicus. Those in group 2 situated north of limes are, actually, classified as borderline lakes to group no. 3 which includes deep oligotrophic coldwater lakes, among others those in the high mountain region. Common species in this group are small dinoflagellates, chrysophycean flagellates and the colony-forming green alga *Oocystis submarina* LAGERH. v. *variabilis* SKUJA. Group no. 4 con-

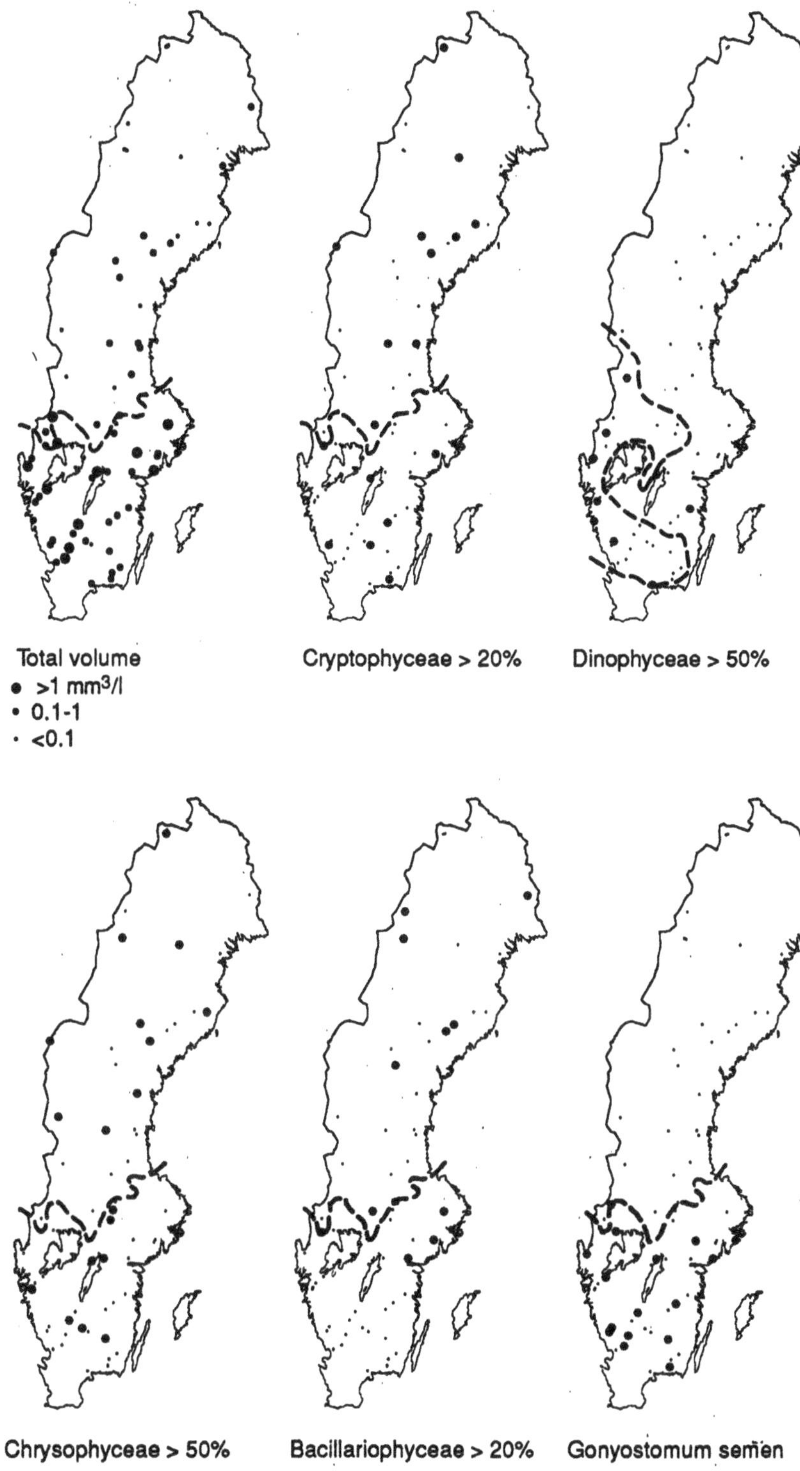

Fig. 5. Total phytoplankton volumes in 3 size-classes, regional distribution of important algal classes and the occurrence of *Gonyostomum* in 73 Swedish lakes investigated in August 1983 and 1986.
The position of limes norrlandicus is indicated and the region most vulnerable to acidification (Dinophyceae)

sists mainly of humic south Swedish lakes with both *Gonyostomum* and *Botryococcus* (Table 2). This group turns out to be the weakest in the discriminant analysis (only 50% agreement between the chemical and planktic classification). Many of the lakes (23%) fit just as well into group no. 2 because of the occurrence of *Botryococcus*, and another 15% could be classed into group no. 6 which includes *Gonyostomum*-dominated lakes. Groups no. 5 and 6 comprise very acid lakes, the former with clear waters and the latter with humic lakes. Both groups have *Peridinium inconspicuum* LEMM. and *P. pusillum* (PENN.) LEMM. as important flora components, but among the humic lakes in group 6 also *Gonyostomum* is very frequent. In group no. 7 an outlier is classified; a eutrophic lake with green algae, euglenoids but also *Gonyostomum* (46. Lake Ämten, Table 1). According to the discriminant analysis, the strongest validity (> 80%) is given to groups no. 5 and 6 which contain

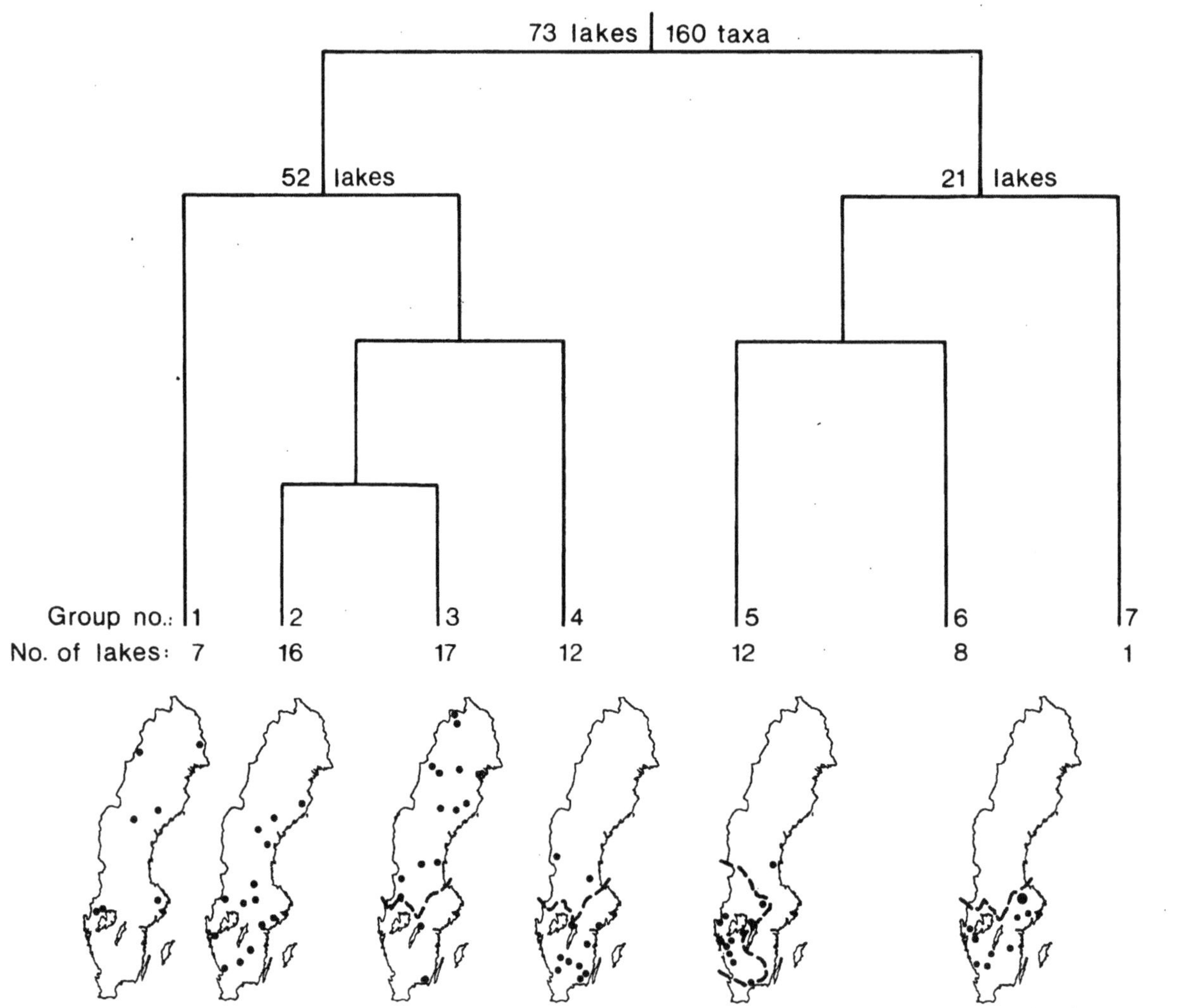

Fig. 6. Results of a TWINSPAN classification of 160 phytoplankton taxa in 73 Swedish lakes investigated in August 1983 and 1986. Areas marked as in Fig. 5

Table 2. The seven TWINSPAN groups with characteristic phytoplankton species from 73 Swedish oligo- and mesotrophic lakes sampled in August 1983 and 1986. Mean values and standard deviation of important environmental variables in each group.

TWINSPAN-groups:	1	2	3	4	5	6	7
Most decisive taxa in the TWINSPAN-grouping	*Aster. form.*, *Tab.flocc.* v. *ast.*, *Uroglena am.*	*Chrys. monads*, *Botr. br.*, *Rhod. lac.*	*Pseudoped.* sp., *Ooc. subm.* v. *var.*, *Per inconsp.* f. *arm.*	*Botr. br.*, *Ooc. subm.* v. *Ooc. subm. V. var.*	*Per. inconsp.*, *Per.pus.* — *Gymn. uber.*, *Din. cren.*	*Gon. sem.*	*Chlorella*, *Sph. schroet.*, *Mel. dist.*, *Trachelom.* spp., *Gon. sem.*
Number of lakes	7	16	17	12	12	8	1
Light climate[1] (Z mix · turbidity)	0.3 (0.1)	0.25 (0.1)	0.43 (0.3)	0.31 (0.3)	0.14 (0.2)	0.46 (0.3)	0.85
Secchi depth (m)	4.0 (2.3)	4.0 (1.5)	4.1 (2.0)	2.9 (1.7)	7.6 (5.2)	1.9 (0.6)	0.5
$KMnO_4$ ($mg \cdot l^{-1}$)	27 (14)	27 (11)	27 (16)	37 (15)	14.2 (15)	47 (18)	94
Water temp. (°C)	15.4 (3.6)	17.4 (1.9)	14.8 (3.8)	18.3 (2.4)	19.2 (1.7)	18.7 (1.3)	19.7
pH	7.1 (0.5)	6.8 (0.4)	6.5 (0.8)	5.8 (0.5)	4.9 (0.6)	5.2 (0.4)	6.0
Tot-P ($\mu g \cdot l^{-1}$)	22 (18)	14 (8)	8 (3)	16 (7)	5 (5)	17 (5)	33
NH_4-N ($\mu g \cdot l^{-1}$)	15 (10)	12 (12)	9 (7)	13 (10)	60 (60)	18 (13)	9
NO_3-N ($\mu g \cdot l^{-1}$)	8 (6)	11 (11)	11 (12)	15 (20)	110 (110)	7 (2)	7
Conductivity ($mS \cdot m^{-1}$)	5.85 (4.6)	5.2 (2.1)	3.09 (1.6)	5.14 (2.4)	5.84 (2.4)	4.87 (1.3)	2.65

[1] High values indicate poor light climate

species prevailing in the two types of very acid lakes. No. 5 are the extremely acidified and transparent Swedish west-coast lakes and no. 6 are acid humic lakes also affected by acidifying substances which have reduced their alkalinity to 0 and further lowered the pH by 1—2 units since the 1930's (ANDERSSON 1980; MALMER 1961; THUNMARK 1937).

Validity problems connected with a strict classification of species and sites are avoided in an ordination. As a result of the CANOCO test, lakes and species are placed in gradual transitions along compositions of environmental variables (Fig. 7). This is more in accordance with the real circumstances, where each lake seems to be an individual of its own, not exactly similar to any of the others. The general character of lakes and species, indicated in this diagram, is interpreted in relation to the direction of the lines of the various environmental variables or their opposite position to these

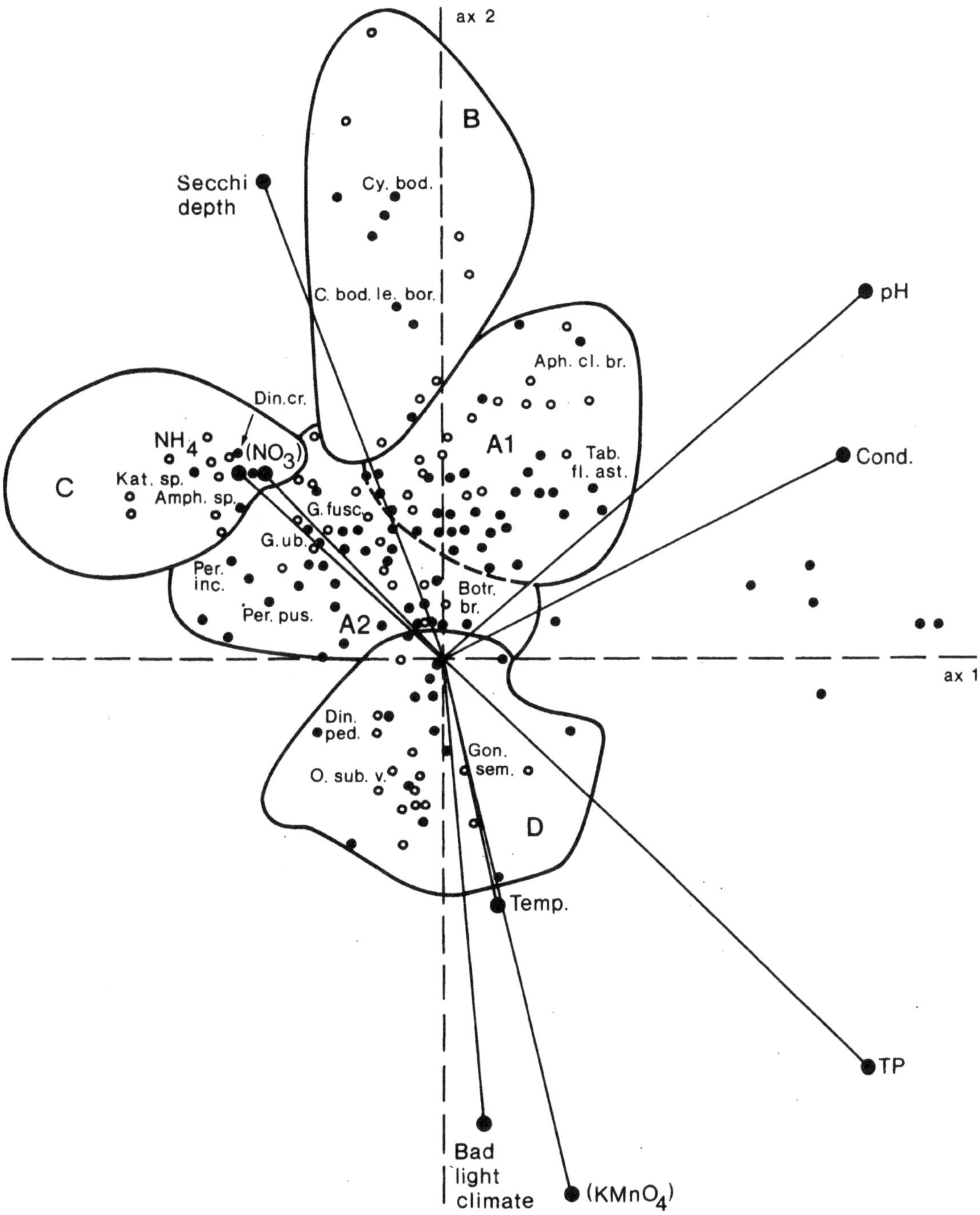

Fig. 7. Results of a CANOCO ordination of 160 phytoplankton taxa, 73 lakes and 9 environmental variables in Swedish lakes investigated in August 1983 and 1986. Filled circles refer to species, unfilled to lakes. A1—A2 = Various dystrophic lakes, B = high latitude mountain lakes, C = acidified clearwater lakes, D = dystrophic lakes with *Gonyostomum*. Axes are explained in the text. (*Amph.* sp. = *Amphidinium* sp., *Aph. c. br.* = *Aphanothece clathrata* v. *brevis*, *Botr. br.* = *Botryococcus braunii*, *Cy. bod.* = *Cyclotella bodanica*, *C. bod. le. bor.* = *Cyclotella bodanica* v. *lemanensis* f. *borealis*, *Din. cr.* = *Dinobryon crenulatum*, *Din. ped.* = *Dinobryon pediforme*, *Gon. sem.* = *Gonyostomum semen*, *G. fusc.* = *Gymnodinium fuscum*, *G. ub.* = *Gymnodinium uberrimum*, *Kat.* sp. = *Kathodinium* sp., *O. sub. v.* = *Oocystis submarina* v. *variabilis*, *Per. inc.* = *Peridinium inconspicuum*, *Per. pus.* = *Peridinium pusillum*, *Tab. fl. ast.* = *Tabellaria flocculosa* v. *asterionelloides*)

Table 3. Median values (25—75 percentiles) of tested environmental variables of lakes encircled in A—D in the CANOCO ordination of 73 Swedish oligo- and mesotrophic lakes, August 1983 and 1986. Plankton characteristics are given in Fig. 7 and in the text.

CANOCO-groups	A1	A2	B	C	D
No. of lakes	22	19	8	10	14
Light climate (Z mix · turbidity)	0.22 (0.2—0.3)	0.37 (0.3—0.5)	0.17 (0.1—0.4)	0.04 (0.01—0.15)	0.42 (0.3—0.6)
Secchi depth (m)	3.8 (3.1—5.1)	2.6 (2.3—3.5)	5.1 (3.8—8.2)	8.5 (2.8—11.4)	2.0 (1.6—3)
$KMnO_4$ ($mg \cdot l^{-1}$)	27 (20—34)	30 (26—36)	11 (8—21)	5 (4—11)	43 (36—55)
Water temp. (°C)	18.5 (15.6—19)	17.1 (15.8—18.4)	10.9 (10.1—13.1)	19.2 (17—20)	18.8 (17.8—19.7)
pH	6.8 (6.6—7)	5.8 (5.3—6.2)	7.3 (7—7.3)	4.5 (4.4—5)	5.5 (5.4—6.1)
Tot-P ($\mu g \cdot l^{-1}$)	10 (8—17)	8 (7—16)	7 (5—10)	4 (3—5)	18 (13—22)
NH_4-N ($\mu g \cdot l^{-1}$)	8 (5—13)	6 (4—19)	12 (6—15)	34 (24—109)	9 (8—19)
NO_3-N ($\mu g \cdot l^{-1}$)	6 (5—8)	7 (5—14)	7 (6—16)	84 (11—180)	8 (6—10)
Conductivity ($mS \cdot m^{-1}$)	5.2 (4.3—7.4)	2.6 (2—3.5)	2.9 (2.6—3.3)	5.9 (5.2—8)	4.2 (3.9—5.6)

lines. Axis 1 is mainly determined by variables whose lines coincide with the direction of that axis, i.e. all lines except bad light climate and temperature. The last two determine axis 2. In this material pH, TP and NH_4-N are of decisive importance for both axes. (Environmental variables within paranthesis were not significant in an approximate t-test of their regression coefficients in a multiple regression of each axis against all nine environmental variables). To facilitate the interpretation of Fig. 7 some lakes and species are encircled and the dashed line indicates special obstacles to make a strict separation. The large number of lakes and species within A reflects the difficulties to set bounds to a certain group of lakes which in reality presents a variety of transitional types, from those placed to the right in A1 with a high pH, comparatively high conductivity and an oligohumic character, to those to the very left in A2, being very acidified brown waters with a low conductivity (Table 3). The large lakes containing colony-living or filamentous diatoms are enclosed in A1, where also the most nutrient-rich lakes (without *Gonyostomum*) in this investigation are placed (Table 1). Many of these lakes, situated in central and northern Sweden, have a good light climate where the small colonial bluegreen alga *Aphanothece clathrata* WEST & WEST v. *brevis* BACHM. is a very characteristic species. This alga is also frequently found in the high mountain lakes encircled in B (Fig. 7). Several dinoflagellates indicated in A2 are found in almost all dystrophic Swedish lakes from south to north, but those indicated close to circle C (*Peridinium inconspicuum*, *P. pusillum*, *Gymnodinium uberrimum* ALLM.; KOF. & SWEZY) also belong to the planktic flora of the very acidified lakes whether clear or humic. The position of *Botryococcus braunii* centrally in Fig. 7 reflects its general occurrence especially in lakes within A and D, which is in accordance with its regular appearance in Swedish lakes of a dystrophic, low-alkaline character (cf. ROSÉN 1981). The same is true for *Stichogloea doederleinii* (SCHMIDLE) WILLE, *Sphaerocystis schroeteri* CHOD. and a variety of naked cryptophycean, chrysophycean and dinophycean flagellates which also have central positions in Fig. 7.

Lakes encircled in B are all situated in the northernmost part of Sweden. All except two are considered as high mountain lakes. The two exceptions are located in the coniferous woodland. They are more humic and somewhat more enriched (Table 1, no. 60, 67 Lakes Lilltjärnen and Laxtjärnen, Reivo) which allows a diverse flora including elements occurring in alpine lakes as well as species with a general occurrence in circumneutral lakes with a slightly humic character. Chrysophytes provided with a lorica, small centric diatoms of the genus *Cyclotella* and small flagellates are common features in these two types of lakes.

Ultraoligotrophic acidified lakes are encircled in C. Their water is very transparent due to the small particle content and the flora is extremely impoverished. In addition to the *Peridinium* and *Gymnodinium* species in common with the acidified, humic lakes recorded in A2, they have a diverse flagellate flora and small loricated chrysophytes. *Dinobryon crenulatum* WEST & WEST and *Dinobryon sociale* EHR. v. *americanum* (BRUNNT.) BACHM., for example, occur in such an environment.

The last assemblage of lakes, encircled in D, are those containing *Gonyostomum semen*. They are easily separated because they are humic, comparatively nutrient-rich, warmwater lakes with a poor light climate. The median value of pH is 5.5 with a total variation range from 4.6 to 6.7. Common subdominants in these lakes in addition to other flagellates, are *Oocystis submarina* v. *variabilis* and *Dinobryon pediforme* (LEMM.) STEINECKE — species which also may occur in a variety of humic lakes where *Gonyostomum* is missing. The same also applies to the other possible subdominant *Botryococcus*. Table 3 gives median values and ranges (25—75 percentiles) of tested environmental variables for each of the CANOCO groupings.

Discussion

When comparing the two multivariate techniques used in this investigation the difference in the results is obvious. The TWINSPAN classification does not, for example, discriminate the high mountain lakes which is clearly the case in the CANOCO ordination (as B). These lakes are, instead, pooled in TWINSPAN class 3 together with a mixture of oligohumic lakes encircled

A1 in the ordination diagram, Fig. 7. On the other hand, large turbulent, diatom-containing lakes are classified in a special group (no. 1) with the TWINSPAN technique, but in the ordination programme they fit into A1 as the furthest assemblage of lakes, closely connected with variables such as high pH, high conductivity and a certain amount of total phosphorus. The ordination of *Gonyostomum* lakes linked to variables characterizing dystrophy, nutrient enrichment and high water temperature, indicates a larger unity than revealed by the species classification, where they are separated in three different groups (Table 2, Fig. 7). The most weakly discriminated TWINSPAN groups 2, 4 and 6, as revealed by their environmental characters, follow a gradient along axis 2 in the CANOCO ordination and constitute very common types of nutrient-poor Swedish lakes (Table 3, lakes in A1—A2 closely connected with axis 2 and lakes in D). In spite of the divergences between these two multivariate methods, they elucidate the pattern of lakes and species in a complementary way and strengthen the interpretation of the tested material. The superiority of the CANOCO ordination, in this context, is connected with the direct reflection of species and lakes in relation to water quality.

It was usually not possible to distinguish typical lakes connected with certain ecoregions based on the phytoplankton classifications from the 73 lakes. They represented too low a number, spread over a large area. Some representatives were, however, recognized. The most evident separation zone coincides with limes norrlandicus. This was revealed by a difference in water temperature, which is the only climatic and latitudinal variable used in this investigation. In northern Sweden, a larger portion of small flagellated species was expected and was in fact found. These algae do not attain biomasses of the size of large species. Small phytoplankton organisms have the following advantages in a harsh environment: a more rapid regeneration time than large organisms, a larger surface to volume ratio giving a better capacity of nutrient uptake, and generally lower K_s-values for nutrient uptake and growth (MUNK & RILEY 1952; SMAYDA 1980; SOURNIA 1982). Besides, the mobile state, exhibited by flagellates, facilitates active movements to patches in the water where light and nutrients are most favourable. Algae in the ultraoligotrophic high latitude mountain lakes, where limiting concentrations of both nitrogen and phosphorus occur (cf. HOLMGREN 1984), adapt to a cold water by reaching optimum growth at lower temperatures and light conditions during the increased day length in summer (BONIN et al. 1982). In general, though, a ten degree lowering of the temperature reduces the growth rate by 50% (EPPLEY 1972). Such a temperature decrease is measured in the high mountain lakes compared to the many inland lakes.

The high mountain region is an easily separated ecoregion concerning phytoplankton, and would have been still more significant if more lakes had been included in the test. The large group of different dystrophic lakes, however, does not fit within a certain ecoregion in Sweden, but forms a variety of transitional types found both in northern and southern Sweden. But the very acidified clearwater lakes, belonging to the coastal area most exposed to westerly winds, represent an easily distinguished ecoregion.

In the inland the contribution of acid brown water lakes increases and some of them have been further acidified (cf. Fig. 2). *Gonyostomum* will successively occupy a larger number of humic lakes if they contain enough nutrients. As the number of acidified lakes increases, this alga obtains better possibilities for spreading, which is observed both in Sweden and Norway (CRONBERG et al. 1988; HONGVE et al. 1988).

The pattern of phytoplankton life-forms, to some extent discussed above, gives a complementary reflection of the growth conditions in the lakes. This is illustrated in Fig. 8, where life-forms in five of the chosen monitoring lakes are visualized. Good reproductive possibilities for a large number of phytoplankton life-forms are, in this investigation, attained in circumneutral lakes which are not too nutrient impoverished (e.g. 6. Lake St. Skärsjön in southwest Sweden). As soon as the environment becomes more biased, in one way or another, the number of life-forms decreases and an extreme situation is reached as in the very acidified clearwaters (21. Lake Härsvatten). It are mainly slow-growing dinoflagellates that survive in such an extremely nutrient-deficient water, where high total concentrations of aluminum also inhibit growth (HÖRNSTRÖM et al. 1984; PETTERSSON et al. 1985). When a certain humic content is available, the nutrient concentration rises and the number of life-forms increases, even if the water is acid (45. Lake Övre Skärsjön).

Considering results of both the chemical and biological investigations in the studied lakes, a total of 25 were chosen as representatives of 13 different ecoregions out of the 16 distinguished by classification of water-chemical variables (Table 1). Lakes representative of 3 regions

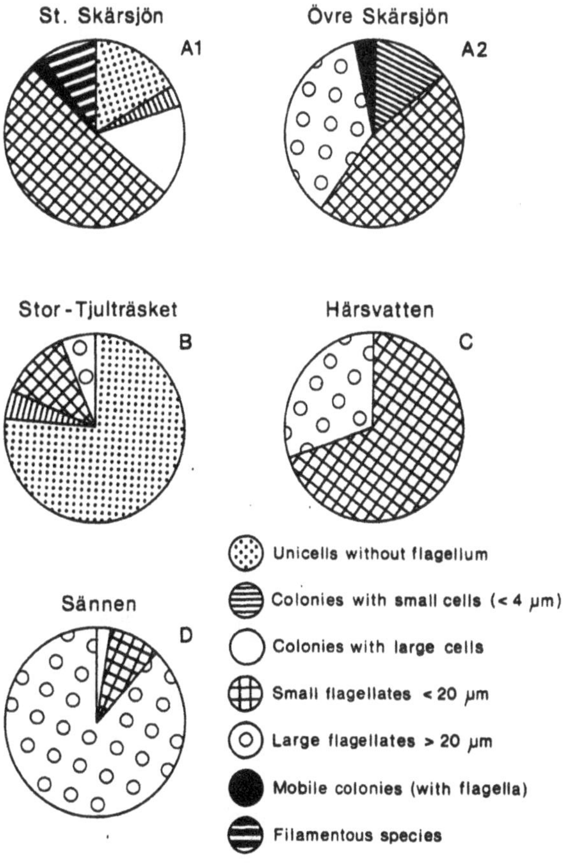

Fig. 8. Pattern of phytoplankton life-forms from some lakes selected for long-term monitoring, representing the groups A—D in the CANOCO ordination (cf. Fig. 7)

are thus missing. Very heterogeneous regions, and those especially exposed to acidification, are represented by a larger number of lakes than other areas. Since 1988, phytoplankton has been sampled in these 25 lakes once a month during the ice-free period, together with water-chemical variables, zooplankton and bottom fauna. Planktic species are analysed qualitatively as well as quantitatively. A more dense sampling during the spring will soon be an urgent need, in spite of considerable practical problems, because it is a very dynamic season when the annual peak in biomass and production is reached in many oligotrophic Swedish lakes (HOLMGREN 1984; OLSÉN & WILLÉN 1980; RAMBERG 1979).

This water quality monitoring of reference lakes is the only one which covers the whole country and will include comprehensive biological studies dealing with phytoplankton, zooplankton, bottom fauna, attached algae, macrophytes and fish (AHL 1985). Experiences from long-term limnological studies are however, based on more than twenty years of investigation of the four largest lakes in Sweden (PERSSON et al. 1989; WILLÉN 1984, 1987). Phytoplankton has, in this context, turned out to be a very valuable biological variable due to its rapid reflection of a changing water quality and of divergencies in the food web.

References

AHL, T. (ed. 1982): Hydrochemical analysis. Nordic IHD Report 3. Oslo.

— (1985): Water quality monitoring in Sweden. Water Qual. Bull. **10**, 82—90.

ANDERSSON, G. (1980): Long-term changes in basic water chemistry in some Swedish lake regions. Doct. Diss. Institute of Limnology, University of Lund, Sweden (in Swedish with English summary).

BERNES, C. (ed. 1985): Monitor. The national Swedish environmental monitoring programme — PMK. Stockholm.

BONIN, D., MAESTRINI, S., & LEFTLEY, J. (1981): Some processes and physical factors that affect the ability of individual species of algae to compete for nutrient partition. In: PLATT, T. (ed.), Physiological bases of phytoplankton ecology. Can. Bull. **210**, 292—309.

CRAMER, W., & HYTTEBORN, H. (1987): The separation of fluctuation and long-term change in vegetation dynamics of a rising seashore. Vegetatio **69**, 157—167.

CRONBERG, G., LINDMARK, G., & BJÖRK, S. (1988): Mass development of the flagellate *Gonyostomum semen* (Raphidophyta) in Swedish forest lakes — an effect of acidification? Hydrobiologia **161**, 217—236.

EARLE, J., DUTHIE, H., & SCRUTON, D. (1986): Analysis of the phytoplankton composition of 95 Labrador lakes, with special reference to natural and anthropogenic acidification. Can. J. Fish. Aquat. Sci. **43**, 1804—1811.

EPPLEY, R. (1972): Temperature and phytoplankton growth in the sea. Fishery Bull. **70**, 1063—1085.

FORSBERG, C., & RYDING, S-O. (1980): Eutrophication parameters and trophic state indices in 30 Swedish waste-receiving lakes. Arch. Hydrobiol. **89**, 189—207.

FRANSSON, S. (1965): The Borderland. Acta Phytogeogr. Suec. **50**, 167—175.

FÄNGSTRÖM, I., & WILLÉN, E. (1987): Clustering and canonical correspondence analysis of phytoplankton and environmental variables in Swedish lakes. Vegetatio **71**, 87—95.

GAUCH, H. (1982): Multivariate analysis in community ecology. Cambridge.

HILL, M. O. (1973): Reciprocal averaging: an eigenvector method of ordination. J. Ecol. **61**, 237—241.

— (1979): TWINSPAN, a FORTRAN program for arranging multivariate data in an ordered two-way table by classification of individuals and attributes. Cornell University, Ithaka, New York.

HOBRO, R., & WILLÉN, E. (1977): Phytoplankton countings. Intercalibration results and recommendations for routine work. Int. Rev. ges. Hydrobiol. **62**, 805—811.

HOGEWEG, P. (1976): Iterative character weighing in numerical taxonomy. Comput. Biol. Med. **6**, 199—211.

HOLMGREN, S. (1984): Experimental lake fertilization in the Kuokkel area, northern Sweden. Phytoplankton biomass and algal composition in natural and fertilized subarctic lakes. Int. Rev. ges. Hydrobiol. **69**, 781—817.

HONGVE, D., LØVSTAD, Ø., & BJØRNDALEN, K. (1988): *Gonyostomum semen* — a nuisance to bathers in Norwegian lakes. Verh. Int. Verein. Limnol. **23**, 430—434.

HÖRNSTRÖM, E., EKSTRÖM, C., & DURAINI, O. (1984): Effects of pH and different levels of aluminum on lake plankton in the Swedish west coast area. Rep. Inst. Freshw. Res., Drottningholm **61**, 115—127.

JOHNSON, R. (1989): Classification of profundal chironomid communities in oligotrophic humic lakes of Sweden using environmental data. Acta Biol. Debr. Oecol. Hung. **3**, 167—175.

— & WIEDERHOLM, T. (1989): Classification and ordination of profundal macroinvertebrate communities in nutrient poor, oligo-mesohumic lakes in relation to environmental data. Freshw. Biol. **21**, 375—386.

JONGMAN, R., TER BRAAK, C., & VAN TONGEREN, O. (1987): Data analysis in community and landscape ecology. Wageningen.

MALMER, N. (1961): Ecologic studies on the water chemistry of lakes in south Sweden. Bot. Not. **114**, 121—144.

MUNK, W., & RILEY, G. (1952): Absorption of nutrients by aquatic plants. J. Mar. Res. **9**, 215—240.

NORDISKA MINISTERRÅDET (1984): Naturgeografisk regionindelning av Norden (Physical-geographical regions of the Nordic countries). Arlöv (in Swedish).

OECD (1982): Eutrophication of waters. Monitoring, assessment and control. Paris.

OLSÉN, P., & WILLÉN, E. (1980): Phytoplankton response to sewage reduction in Vättern, a large oligotrophic lake in central Sweden. Arch. Hydrobiol. **89**, 171—188.

PERSSON, G., OLSSON, H., WIEDERHOLM, T., & WILLÉN, E. (1989): Lake Vättern, Sweden — A 20 year perspective. Ambio **18**, 208—215.

PETTERSSON, A., HÄLLBOM, L., & BERGMAN, B. (1985): Physiological and structural responses of the cyanobacterium *Anabaena cylindrica* to aluminum. Physiol. Plant. **63**, 153—158.

RAMBERG, L. (1979): Relations between phytoplankton and light climate in two Swedish forest lakes. Int. Rev. ges. Hydrobiol. **64**, 749—782.

ROSÉN, G. (1981): Phytoplankton indicators and their relations to certain chemical and physical factors. *Limnologica* (Berlin) **13**, 263—290.

ROSÉN, E., & SJÖGREN, E. (1988): Plant cover in alvar junipers on Öland. Distribution features correlated to shrub size and shape. Acta Phytogeogr. Suec. **76**, 101—112.

SJÖRS, H. (1965): Features of land and climate. Acta Phytogeogr. Suec. **50**, 1—12.

SOURNIA, A. (1982): Form and function in marine phytoplankton. Biol. Rev. **57**, 347—394.

STEVENSON, A., BIRKS, H., FLOWER, R., & BATTARBEE, R. (1989): Diatom-based reconstruction of lake acidification using canonical correspondence analysis. Ambio **18**, 228—233.

TER BRAAK, C. (1986): Canonical correspondence analysis: a new eigenvector technique for multivariate direct gradient analysis. Ecology **67**, 1167—1179.

THRONDSEN, J. (1978): Preservation and storage. In: SOURNIA, A. (ed.), Phytoplankton manual. Monographs on oceanographic methodology. UNESCO, pp. 69—74. Paris.

THUNMARK, S. (1937): Über die regionale Limnologie von Südschweden. Sveriges geol. undersökning. Ser. C. 410.

UTERMÖHL, H. (1958): Zur Vervollkommnung der quantitativen Phytoplanktonmethodik. Mitt. Int. Verein. Limnol. **9**, 1—38.

WILLÉN, E. (1976): A simplified method of phytoplankton counting. Br. Phycol. J. **11**, 265—278.

— (1984): The large lakes of Sweden: Vänern, Vättern, Mälaren and Hjälmaren. In: TAUB, F. (ed.), Lakes and reservoirs. Ecosystems of the world **23**, 107—134.

— (1987): Phytoplankton and reversed eutrophication in Lake Mälaren, Central Sweden, 1965—1983. Br. Phycol. J. **22**, 193—208.

Address of the authors:

Drs. EVA WILLÉN
SUSANNA HAJDU YLVA PEJLER
National Environmental Protection Board
Environmental Quality Laboratory
Box 7050
S — 750 07 Uppsala
Sweden

Limnologica (Berlin) **20** (2) 229—241, Juni 1990

Maj and Tor Nessling Foundation, Helsinki, Finland; Department of Biology, University of Joensuu, Joensuu, Finland

Characterisation of 151 Small Lakes in Eastern Finland Based on their Water Chemistry, Land-use on Catchment and Phytoplankton Communities

By V. Ilmavirta & P. Huttunen

With 12 Figures

Key words: Catchment; Eastern Finland; Land-use; Phytoplankton; PCA analysis; Water chemistry.

Abstract

In eastern Finland 151 lakes have been investigated during the high production period of the summer and the autumnal over-turn at the deepest area of the lakes. The aim of the study was to describe the relationships between lake water quality, phytoplankton communities and the land-use parameters on their catchments.

The lakes studied were mainly small, not very dark, and shallow. Major nutrient concentrations were low and pH circumneutral to acidic. However, the range of most parameters was broad and allowed for statistical treatment of the data.

Because the data set is large, there were many significant correlations between variables (mainly within water quality parameters), but particularly few of them explained well the variations between lakes. For example, forests, mainly coniferous, covered most of the lake catchments, but still had only weak explanatory value to water chemistry or to the structure of the phytoplankton communities in individual lakes. Of the variation e.g. in conductivity, only 49% was explained by shores in immediate connection to agricultured areas (mainly fields), peatlands (—) and maximum depth (—), and of the variation in cell numbers of Cryptophyceae, 42% was explained by total nitrogen, Ca and total phosphorus.

Lakes were further classified into 11 groups with principal component analysis (PCA, with varimax rotation) and the mean characteristics of these groups were used in stepwise multiple regression analyses. Variation of many average chemical characteristics in the PCA groups were significantly explained with catchment characteristics, i.e. P_{tot} by percentage of forest on the catchments (—), pH by peatlands (—), NO_3 by fshore (cf. p. 230) and catchment area, and conductivity by fields. The total biomass of phytoplankton was explained well by N_{tot}, pH and alkalinity (—), the cell numbers of Cryptophyceae by N_{tot}, pH and Secchi disc transparency, and Diatomophyceae by conductivity and alkalinity (—). P_{tot} or PO_4 concentrations had much lower explanatory value on biological parameters than N_{tot}.

Chlorophycean species and *Peridinium inconspicuum* dominated in acid lakes, but the biomass and the proportion of the total biomass of chrysophycean and cryptophycean flagellates increased with higher pH and nutrient concentrations. Diatoms were insignificant in most lakes and blue-greens bloomed only in a few lakes surrounded by extensive agriculture.

1. Introduction

In a country with more than 100 000 lakes the choice of proper methods for monitoring the quality of surface waters is of profound importance for correct environmental evaluations. Finnish lakes cover an extremely broad variety of lake types from oligotrophic to highly eutrophic, from clear waters to extremely brown waters, from ancient acid-bedrock basins to clayish basins and from southern lakes to northern ones. There are thus several environmental gradients which in association with the shallowness and the small water volume of Finnish lakes induce uncertainty in environmental evaluations.

The present, regular water-quality monitoring programs are yielding a great number of data; however, it is difficult or impossible to make reasonable evaluations, mainly because of the large time-scatter of samplings among and within the lakes, and because of low reliability of the data due to deviating sampling procedures. It is difficult to be convinced whether a measured variable has enough value for environmental evaluation, or whether the sampling procedures really allow for significant evaluations.

One way to figure out these problems is to investigate in a relatively restricted geographical area as many lakes as possible within a short time interval, and try to find the variables which best describe the most significant characteristics of the lakes and the obvious relationships between the environmental variables and the biological communities. Those variables and their principal influence on the biota can later be applied with particular emphasis for monitoring purposes. Most important is, however, to do this with special reference to the characteristics of the lake catchment areas.

In this respect 151 lakes in eastern Finland were investigated. The program was started in 1978. The main goals of the "151 lake project" were to, in a comprehensive way, approach the relationships between water chemistry, land-use, and biological assemblages in lakes, and to evaluate the community ecology of water organisms, mainly phytoplankton, as a tool in the character analysis of lakes. It also attempted to find principles for the classification of lakes on the basis of some simply measurable chemical and biological parameters and describe the linkages between the structure of the living phytoplankton communities in their environments and the paleolimnological data stored in the lake sediments.

The research strategy is more closely presented in Meriläinen et al. (1981). Relations of some taxonomical groups to lake water properties were earlier presented on surficial diatom assemblages (Meriläinen & Huttunen 1984; Huttunen & Meriläinen 1986 a, b, c), surficial Cladoceran assemblages (Cotten 1985; Huttunen et al. 1988), surficial Chrysophyta scales (Christie et al. 1988) and phytoplankton (K. Ilmavirta et al. 1984; Ilmavirta 1984, 1988).

The present paper attempts to analyse the large data set of the "151 lake project" to describe the mean characteristics of the waters and the catchments of the lakes studied and relate those parameters to the structure of the phytoplankton communities in the pelagial. This paper is a basic framework for further treatments of the data with focus on the community structure of phytoplankton at the species level, the autecology of species, indices of natural acidity in Finnish lakes, the characteristics of humic lake ecosystems, and to the propositions for further development of the monitoring practices on humic lakes, being performed later. However, some concluding remarks on biological studies are presented now.

2. Study area

The present paper is based on data from 151 lakes in eastern Finland located partly on granite complex, partly on Karelidic schists with some on apatite bedrock (Fig. 1). More than 60 physico-chemical parameters were measured and analysed from the catchments, lake waters and bottom sediments. However, in this paper the sediment results are not treated.

The lakes in the study were selected semi-randomly by avoiding lakes with heavy point-source loading or with greater settlements on their catchments. Large water-course lakes were also excluded. Small or medium sized headwater lakes were preferred to characterize more or less "natural" lakes, which mainly are affected by run-off from their own sub-catchment, and not so much by run-off from lakes higher in the system.

The area of the lakes (Fig. 2A) ranged from 0.036 to 15.13 km^2 (mean 1.30 km^2 ± SD 2.39 km^2). Most of lakes (121) were smaller than 1.50 km^2, which is seen as low median 0.46 km^2; 22 lakes were 1.50—5.00 km^2, but only 8 lakes were greater than 5.00 km^2. Maximum depth of the lakes (Fig. 2B) was 13.3 ± 8.0 m (median 11.5, min. 3.1, max. 45 m); 62% of the lakes were shallower than 15 m, 32% were 15—30 m deep, and only 3 lakes were deeper than 40 m. Depth of the epilimnion (Fig. 2C) varied from 1 to 9 m (mean 3.1 ± 1.5 m, median 3.0 m), 79% of lakes having an epilimnion deeper than 5.5 m. The epilimnion in 48 lakes was between 1.5 and 2.5 metres. Most lakes were stratified in summer (dimictic).

The catchment area (excluding the lake area, Fig. 2D) varied from 0.2 to 179.0 km^2 (mean 13.5 ± 28.5 km^2, median 4.13 km^2). Most of the catchments were small: 108 were smaller than 8.2 km^2 and only four were greater than 100.0 km^2. Lake/catchment-ratio was mainly low (Fig. 2E). Fields dominated the land-use on the catchments of four lakes, peatlands of 15 lakes and other land-use forms, mainly forests, of 125 lakes (Fig. 2 G—I). Of the catchments, 42 were without fields and nine without peatlands. All had more than 20% forest coverage. Median, minimum, average and maximum percentages were for fields 7.1%, 0%, 11.9% and 71%, respectively, for peatlands, 0.1%, 0%, 18.1% and 80%, and for forests, 74%, 19%, 70% and 100%.

The percentage of shoreline bound by fields (fshore, Fig. 2F) was also measured. Few lakes were entirely surrounded by fields (134 lakes with < 50%). The average fshore was 20.1 ± 26.5%, median 5%, minimum 0% (42 lakes) and maximum 95% (two lakes).

Frequency histograms in Fig. 2 show that the material is skewed on most parameters and the superimposed best-fitting normal distribution does not appear to fit the data very well. This was, however, expected, because the lakes in the study were not selected.

3. Methods

Water samples were collected from the middle of each lake both during the summer stagnation period in 1980 (13. 7. 1980—12. 8. 1980, all 151 lakes) and during the autumnal overturn in 1978 (50 lakes) and in 1979 (remaining 101 lakes). Each water sample per lake consisted of 4 to 10 independent subsample-series (volume

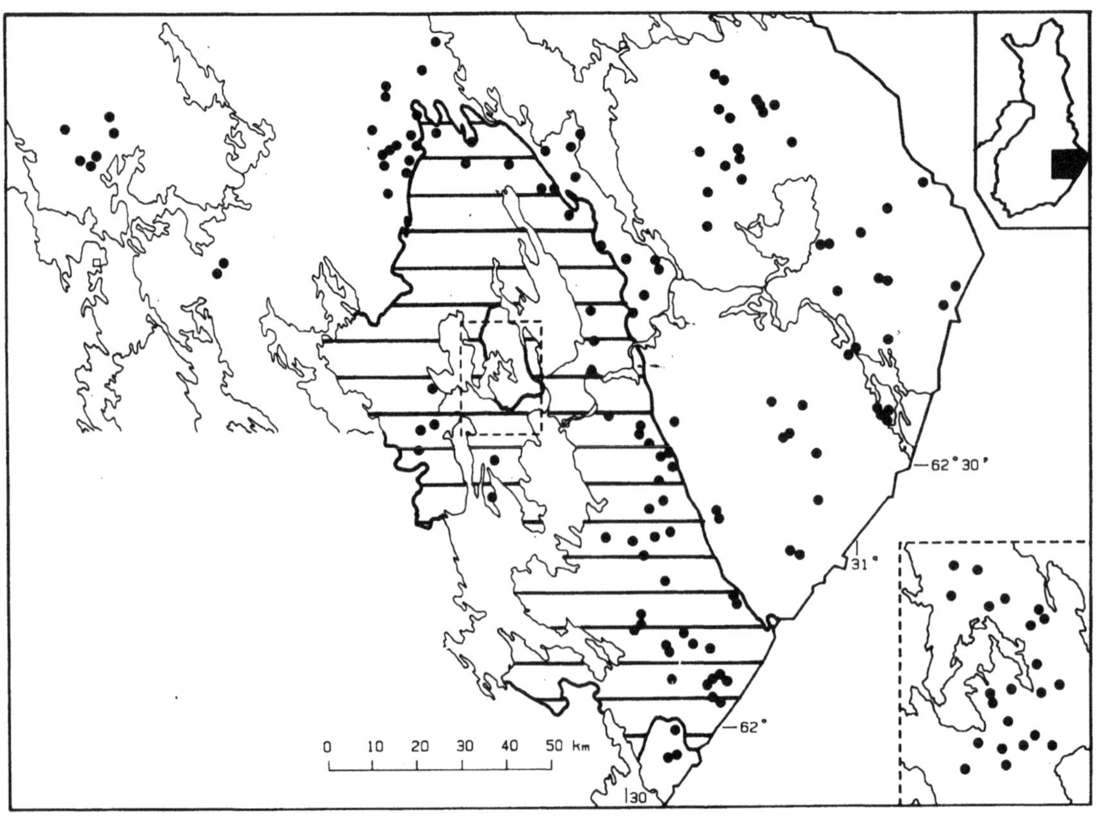

Fig. 1. The location of the 151 lakes studied in eastern Finland. The region of Karelidic schists is marked by the hatched area

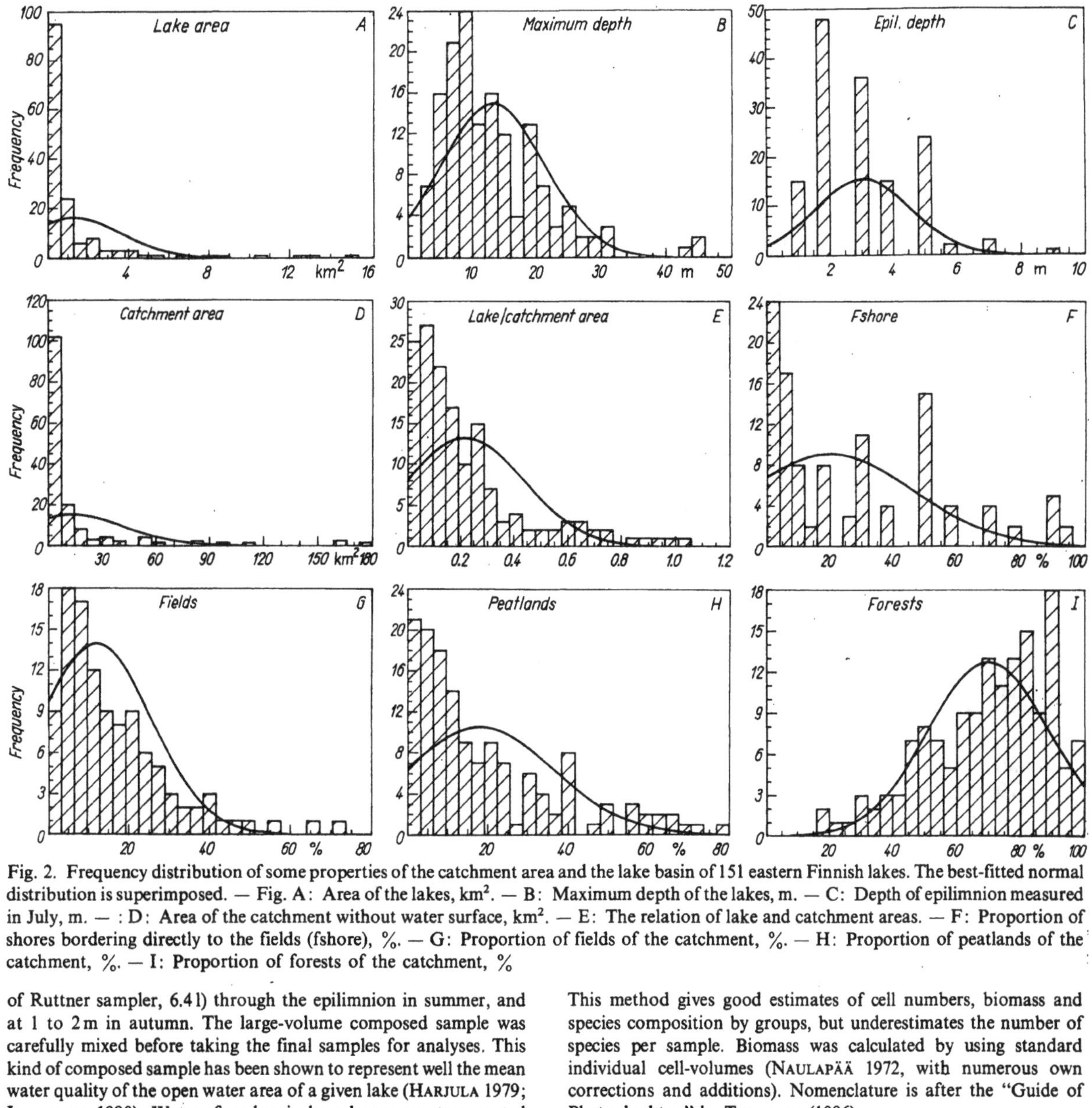

Fig. 2. Frequency distribution of some properties of the catchment area and the lake basin of 151 eastern Finnish lakes. The best-fitted normal distribution is superimposed. — Fig. A: Area of the lakes, km^2. — B: Maximum depth of the lakes, m. — C: Depth of epilimnion measured in July, m. — : D: Area of the catchment without water surface, km^2. — E: The relation of lake and catchment areas. — F: Proportion of shores bordering directly to the fields (fshore), %. — G: Proportion of fields of the catchment, %. — H: Proportion of peatlands of the catchment, %. — I: Proportion of forests of the catchment, %

of Ruttner sampler, 6.4 l) through the epilimnion in summer, and at 1 to 2 m in autumn. The large-volume composed sample was carefully mixed before taking the final samples for analyses. This kind of composed sample has been shown to represent well the mean water quality of the open water area of a given lake (HARJULA 1979; ILMAVIRTA 1980). Waters for chemical analyses were transported to the laboratory in polyethene bottles in dark cold-boxes. Phytoplankton samples were fixed immediately with acetic Lugol solution and preserved the next day with buffered formaldehyde solution.

Water analyses were done by standard methods of the National Board of Waters in Finland (ERKAMAA et al. 1977).

Phytoplankton was counted in sedimented chambers using an inverted microscope (UTERMÖHL 1958). More than 500 units (cells, colonies, 100 μm filaments) were counted from random visual fields. This method gives good estimates of cell numbers, biomass and species composition by groups, but underestimates the number of species per sample. Biomass was calculated by using standard individual cell-volumes (NAULAPÄÄ 1972, with numerous own corrections and additions). Nomenclature is after the "Guide of Phytoplankton" by TIKKANEN (1986).

4. Results

4.1. Water quality in summer and autumn

Some selected water properties are given in Table 1. Variations in concentrations of total phosphorus were

Table 1. Selected properties of water in the studied lakes in Eastern Finland. $n = 150$.

		Summer					Autumn				
		Median	Mean ± SD		Max	Min	Median	Mean ± SD		Max	Min
Colour	($mg\ Pt \cdot l^{-1}$)	40.0	55.4	45.5	200	5	—	—	—	—	—
Secchi	(m)	2.6	3.5	2.1	11.0	0.9	3.0	3.5	2.1	11	0.9
P_{tot}	($\mu g \cdot l^{-1}$)	13.0	16.5	11.2	60	0	10.0	14.7	11.5	61	2.0
PO_4	($\mu g \cdot l^{-1}$)	1.0	2.0	2.4	18	0	—	—	—	—	—
N_{tot}	($\mu g \cdot l^{-1}$)	515.0	552.0	214.1	1410	190	340	361.8	117.4	1440	90
NO_3	($\mu g \cdot l^{-1}$)	2.0	8.8	26.2	170	0	—	—	—	—	—
Cond.	($mS \cdot m^{-1}$, 25 °C)	4.6	5.1	2.9	17.4	1.2	4.4	5.0	2.9	17.2	1.1
Alk.	($meq \cdot l^{-1}$)	0.160	0.168	0.123	0.52	0	0.170	0.174	0.131	0.6	0
pH		7.13	6.96	0.74	9.1	4.52	6.84	6.71	0.61	7.52	4.49
Ca	($mg \cdot l^{-1}$)	4.50	5.01	3.09	19.4	0.8	3.40	3.99	2.98	19.8	0.2

large, from zero to 61 $\mu g \cdot l^{-1}$. PO_4 concentrations in summer were low, the maximum values being only 18 $\mu g \cdot l^{-1}$. Summer and autumn values of total P were similar, but total nitrogen concentrations rose significantly during summer with the accumulation of autochthonous biomass in the water bodies from an average of 360 to 550 $\mu g \cdot l^{-1}$. Maximum concentrations detected were similar (over 1400 $\mu g \cdot l^{-1}$), and significant amounts of NO_3 were present in most lakes.

Of the lakes, 101 had total phosphorus concentrations lower than 16.3 $\mu g \cdot l^{-1}$ and only 16 lakes over 30 $\mu g \cdot l^{-1}$. The distribution of data on nitrogen is also skewed towards lower values.

The average summer values of pH were 0.25 units higher than autumn values due to primary production and the three most eutrophic lakes dropped significantly out of the 95% confidence limits of the regression between autumn and summer pH:

$$pH_{summer} = 1.09\,pH_{autumn} - 0.41 \quad (1)$$

Variation explained was 82%. Alkalinity of waters was constant through the growing season, the average being 0.167 meq $\cdot l^{-1}$:

$$ALK_{autumn} = 0.99\,ALK_{summer} + 0.0049 \quad (2)$$

Variation explained was 92.6%. Conductivity remained constant, but Ca concentration was higher in summer than in autumn.

Water colour was measured only in summer. Few of the lakes were very dark since 58% of the lakes had colouration lower than 45 mg $Pt \cdot l^{-1}$. The maximum value, 200 mg $Pt \cdot l^{-1}$ was measured from two lakes, and ten lakes were between 145 and 165 mg $Pt \cdot l^{-1}$. The Secchi reading, which largely was determined by water colour (Fig. 3), was slightly lower in summer:

$$Secchi_{autumn} = 0.89\,Secchi_{summer} + 0.54 \quad (3)$$

Variation explained was 83%, and four of the lakes dropped out of the 95% confidence limits because of the increase of turbidity by algae.

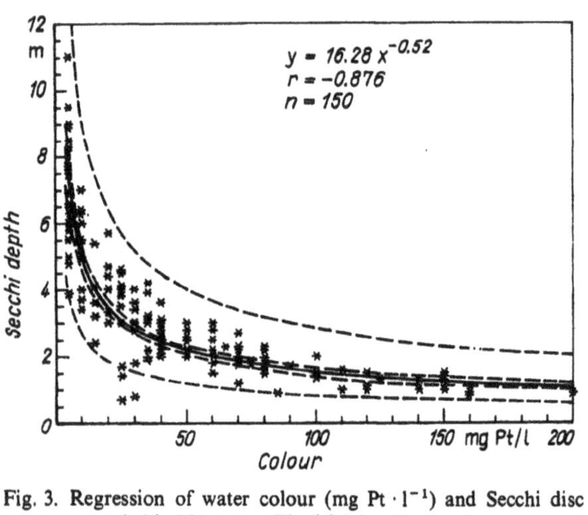

Fig. 3. Regression of water colour (mg Pt $\cdot l^{-1}$) and Secchi disc transparency (m) in 151 eastern Finnish lakes. Broken lines represent the 95% confidence and prediction limits ($P \leq .0000$)

4.1.1. The effects of land-use on water quality

Because of the large number of lakes sampled, there were in the simple linear regression analyses many significant correlations within water quality parameters, but few between water quality and catchment characteristics. Few of those explained more than 60% of the observed variation ($r = > 0.800$). However, there was a clear trend towards lower pH values in summer with increasing peatland percentage, and towards smaller deviation of pH between lakes with increase in field and fshore percentages (Fig. 4 A, B). The percentage of forests, which is high on most catchments, had, however, low explanatory value (Fig. 4 C).

The low explanation of most observed correlations was mainly caused by the highly variable nature of the lakes and partly also by the possible small error caused by sampling and analyses without replicates. It is also known that many relationships, such as depth of epilimnion/Secchi disc transparency and some biological parameters are nonlinear. It is also obvious that any single variable which describes the land-use on catchment can hardly alone maintain relationships since few of the lakes had a clear predominance of other land-use type than forest.

The stepwise selection of the land-use and catchment area variables in multiple regression analyses gave better explanations of water chemistry. Some results of calculations with summer data from 151 lakes are given in Table 2. Also in stepwise selection of variables, the percentage of forests on catchments had a reasonable influence only on Secchi depth, P_{tot} and N_{tot}.

Of the variation in alkalinity, Ca explained only 56%, which indicates that instead of Ca, the main buffering system in these lakes is the carbon system, and that dissolved organic matter also acts as a buffer (see KÄMÄRI 1984). Ca was mainly washed into the lakes directly from fields, but forests on catchments influenced alkalinity by reducing the Ca buffering capacity of the water.

Factors influencing the depth of the epilimnion in the lakes are interesting. Of the single variables, Secchi best described the epilimnion depth with 50% of the variation explained (linear model), and water colour 40% (multiplicative model). In stepwise variable selection, Secchi

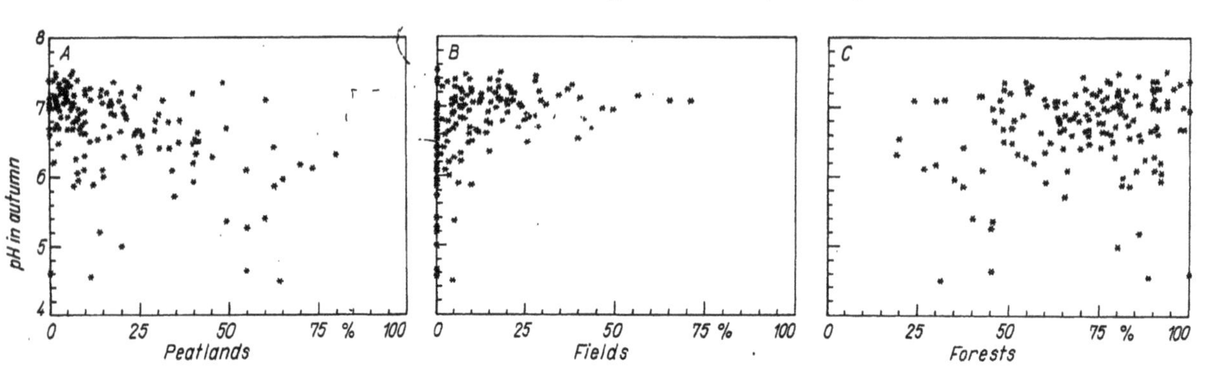

Fig. 4. The autumnal pH plotted against the percentage of (A) peatlands, (B) fields and (C) forests of the catchment area

Table 2. Stepwise multiple regression analysis of some water quality parameters (n = 151) to catchment area characteristics. Cumulative r^2 is given after independent variable and (—) when negative. Catchment = catchment area; depth = maximum depth of the lake; fshore = percentage of the shore line bordered directly to fields and others are percentages of the catchment area.

Dependent	Independent	Var. expl. (%)
Conductivity	fshore (.44), peatl. (—.46), depth (—.48)	48
Colour	peatl. (.37), fshore (.41), catchment (.43), area (—.48)	48
pH	peatl. (—.29), field (.37), area (.38)	38
Secchi	forest (.22), depth (.26), catchment (—.28), area (.32), fshore (—.36)	36
N_{tot}	fshore (.28), forest (—.34), depth (.36)	36
P_{tot}	forest (—.23), depth (—.30), fshore (.35)	35
Alkalinity	fshore (.23), peatl. (.34)	34
Ca	fshore (.25), peatl. (.27)	27
SO_4	fshore (.25)	25
PO_4	peatl. (.16), catchment (.19)	19
NO_3	catchment (.09), fshore (.15)	15

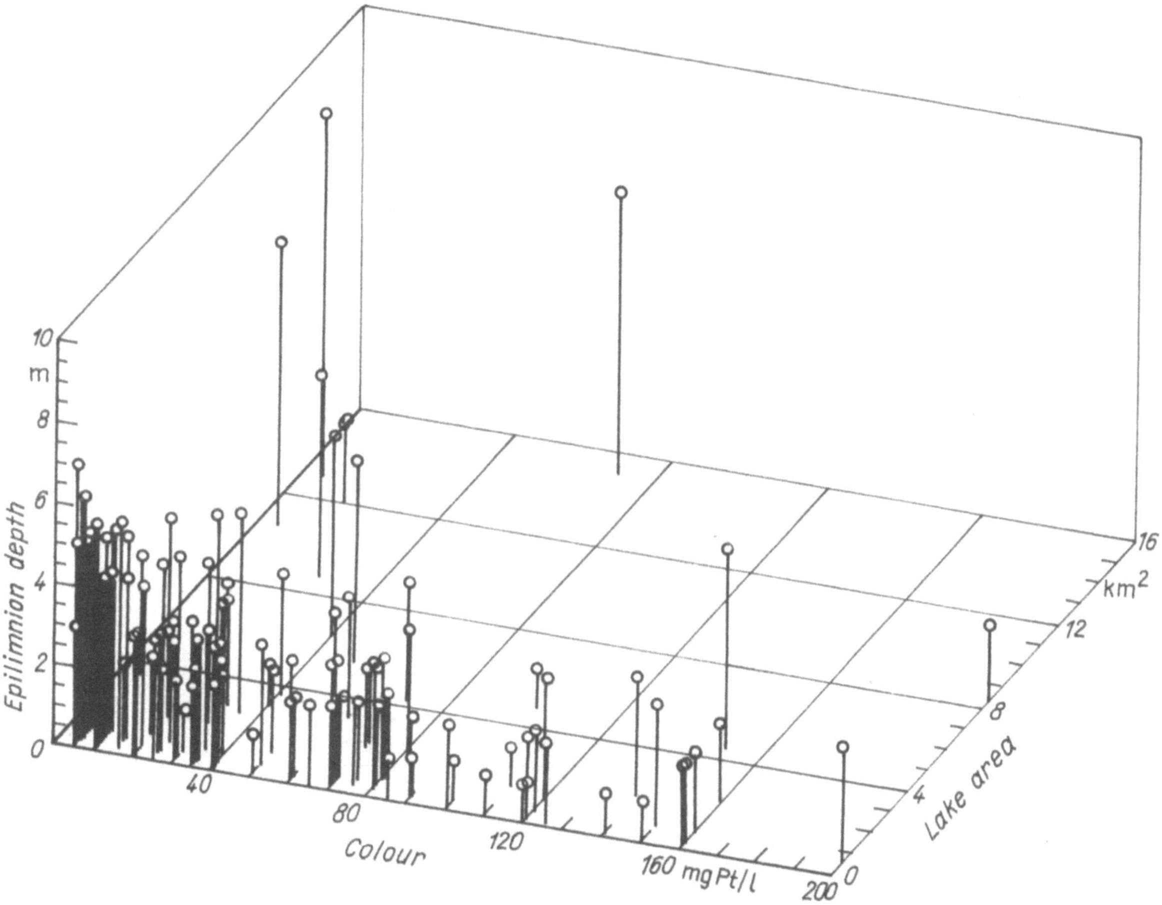

Fig. 5. The three-dimensional plot of epilimnion depth (m) to water colour (mg Pt · l^{-1}) and lake area (km²)

and lake area explained 61% of variation (colour and maximum depth of lakes remained excluded). When Secchi, which is a descriptive variable, was leaved out, colour (linear) and lake area explained 39% of variation (Fig. 5). The additional explanation of maximum depth as third variable was only 3%.

4.2. Phytoplankton of the lakes in summer

Of the summer samples counted, 262 taxa or counting units (cells or cysts, filaments, colonies etc.) were identified. The highest number of taxa per lake was 55, the lowest three, and the average was 31.8 ± 9.5 taxa and median 32. Of the species, only one (*Katablepharis ovalis* SKUJA) had a frequency of over 95% (Table 3). Only *Uroglena americana* CALK., *Monoraphidium minutum* (NÄG.) KOM.-LEGN., a small *Chromulina* CIENK., and *Cryptomonas* EHR. had a frequency higher than 75%. Thirteen additional taxa occurred in 50% of the lakes, 64 taxa in only one lake, and 27 taxa in two lakes.

The total number of cells or units per sample varied between 1.9 × 10^4 and 2.7 × 10^6 cells per 100 ml. Average cell number was 3.9 × 10^5 and median 2.7 × 10^5. Total cell number didn't follow biomass closely (Fig. 6A) due to the pronounced differences in the species composition of samples. There was, however, a significant correlation (r^2 = +0.43), but it was strongly determined by one extreme data point in the upper right corner (without it r^2 = +0.29). Also chlorophyll a concentration had similarly only a low true correlation with total biomass (Fig. 6B).

Table 3. The occurrence of the most frequent phytoplankton taxa in the 151 eastern Finnish lakes during summer stagnation period plotted against total nitrogen concentration in the lake water. Ranking is according to the number of lakes where the species were detected. Horizontal bars represent the variation range of total nitrogen concentrations in lakes where the species was detected; asterisks = the weighted mean of those values. Some colourless flagellates are included in the list.

Taxon	No. of lakes in which taxa was found
Katablepharis ovalis	144
Uroglena americana (cell 1.)	128
Monoraphidium minutum	112
Chromulina sp.	107
Cryptomonas sp. (small)	105
Cryptomonas sp. (big)	98
Elakatothrix gelatinosa	93
Oocystis sp. (small)	89
Dinobryon acuminatum	86
Chroomonas sp (small)	85
Chrysomonadineae (small)	84
Cyclotella sp. (small)	80
Rhizosolenia longiseta	78
Chlamydomonas sp. (small)	78
Chroomonas sp (big)	78
Bitrichia chodatii	75
Sphaerocystis schroeteri	74
Monoraphidium dybowskii	74
Chlamydomonas sp. (extra small)	69
Crucigenia tetrapedia	67
Merismopedia warmingiana	63
Aphanothece clathrata	62
Mallomonas akrokomos	60
Chrysamoeba sp.	60
Mallomonas tonsurata	59
Peridineae (small)	57
Cryptomonas sp. (medium)	56
Crucigenia quadrata	55
Asterionella formosa	53
Microcystis reinboldii	50
Dinobryon divergens	48
Oocystis rhomboidea	48
Synura uvella (cell)	48
Dinobryon bavaricum	47
Chrysomonadineae (big)	46
Melosira distans	46
Mallomonas acaroides	46
Pseudokephyrion entzii	46
Spiniferomonas sp.	45
Mallomonas sp.	41
Chlamydomonas sp. (medium)	39
Chrysomonadineae (medium)	38
Gloeotila fennica	38
Oocystis sp.	37
Tetraedron minimum	36
Salpingoeca sp.	36
Kephyrion littorale	36
Cyclotella sp. (big)	35
Gloeocapsa limnetica	35
Quadrigula pfitzeri	34
Dinobryon suecicum	33
Gyromitus cordiformis	33
Anabaena flos-aquae (100um)	33
Chlamydocapsa planctonica	32
Elakatothrix viridis	32
Peridinium inconspicum	31
Salpingoeca frequentissima	30
Dinobryon borgei	29
Dinobryon suecicum v. longispinum	29
Scenedesmus bicellularis	28
Chlamydomonas sp. (big)	28
Stichogloea doederleinii	27
Epipyxis polymorphum	27
Gomphosphaeria lacustris	26
Monoraphidium setiforme	25
Schroederia judayi	25

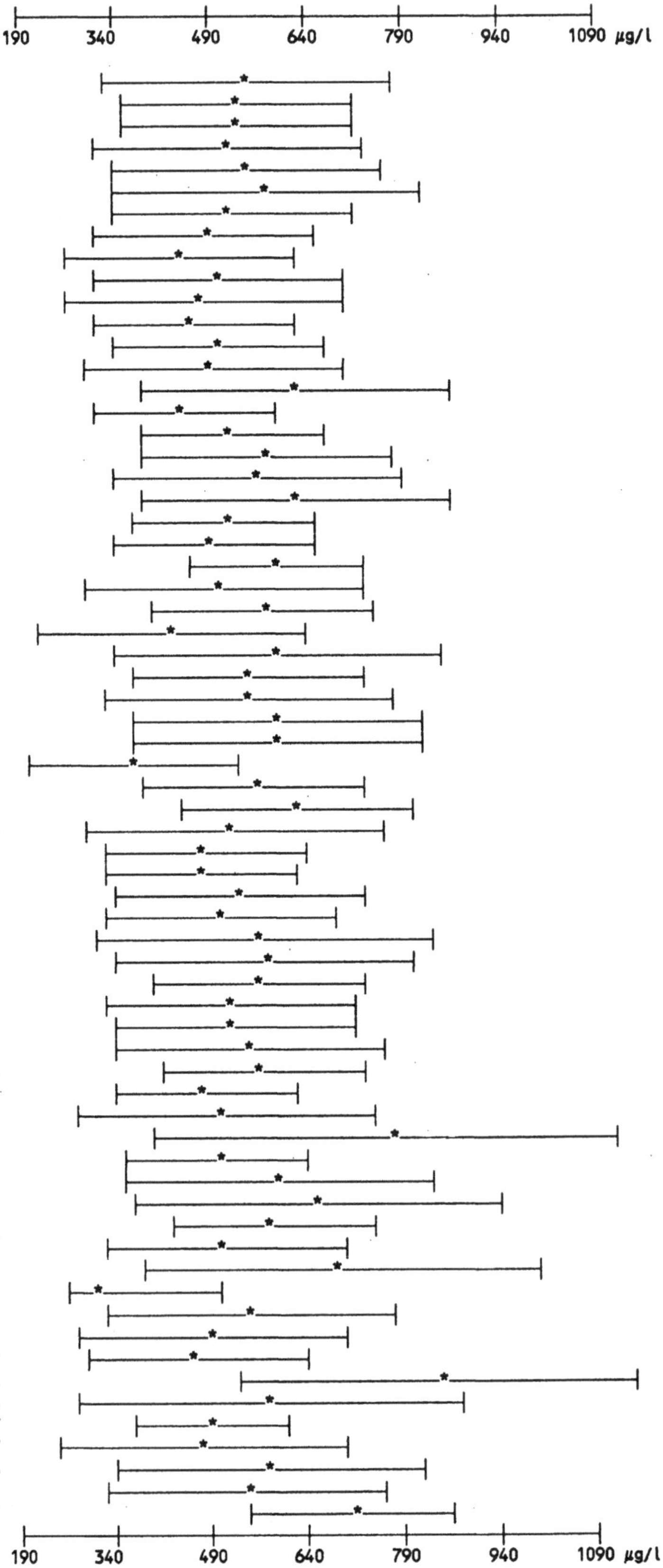

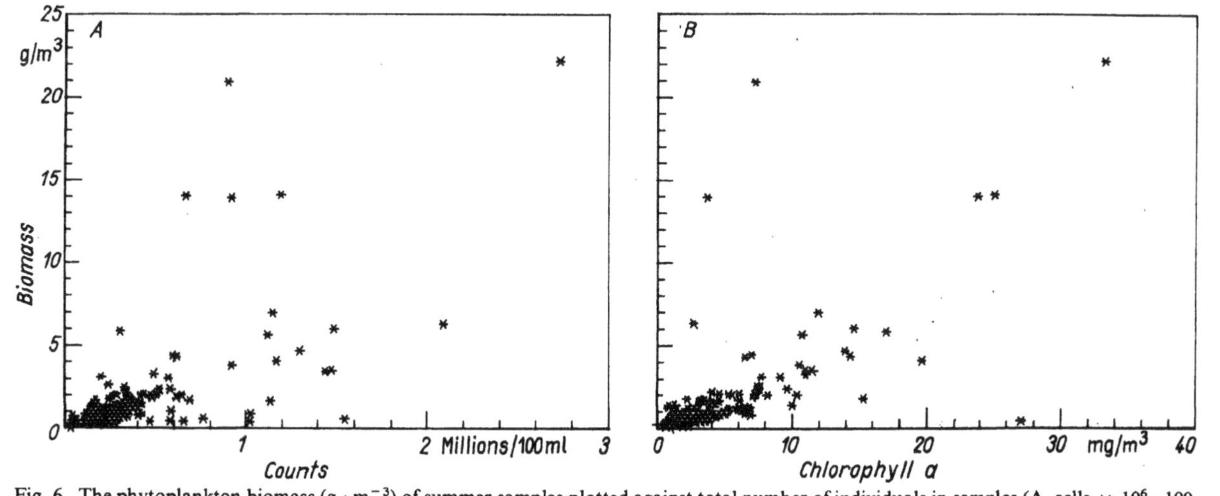

Fig. 6. The phytoplankton biomass ($g \cdot m^{-3}$) of summer samples plotted against total number of individuals in samples (A, cells $\times 10^6 \cdot 100\ ml^{-1}$) and chlorophyll **a** concentrations (B, $mg \cdot m^{-3}$)

The highest biomass recorded was $22.2\,g \cdot m^{-3}$ f. w. and the lowest was only 0.025 g (mean 1.85 ± 3.2 g, median 0.95 g). Of the 151 lakes, 86 had a biomass lower than 1.0 g and 39 lakes, 1.0—2.2 g. Two lakes exceeded 20 g and three, 14 g. The material is thus strongly weighted on oligotrophic lakes, which was also seen in the chemistry of the lakes.

Chlorophyll **a** concentrations varied from values under the detection limit to $33.2\,\mu g \cdot l^{-1}$. Average chlorophyll concentration was $4.8 \pm 5.2\,\mu g$ and median 3.1 µg. Of the lakes, 122 had chlorophyll concentration between 1.8 and 7.3 µg and only four lakes exceeded 20 µg.

Among phytoplankton groups the Chrysophyceae had the highest average cell number, 1.53×10^5 cells per 100 ml (median 1.0×10^5), which is twice the cell number of Cryptophyceae, 6.03×10^4 (median 3.3×10^4). Both groups are flagellated and the cells of the species are mainly small, cryptophytes living singly, and many chrysophytes occurring in colonies and with the individual cell covered by theca. The Chlorophyceae also had a relatively high average cell number, 7.61×10^4 (median 3.99×10^4) and the highest maximum cell number, 2.08×10^6. In some lakes the average proportion of the blue-greens, divided into Chroococcales and Nostocales, were also relatively high, 5.58×10^4 colonies per 100 ml (median 0.67×10^4) and 1.03×10^4 filaments (100 µm) per 100 ml (median 0.0), respectively. Data on blue-greens were extremely strongly scewed towards higher values as was expected. Low cell numbers of diatoms were found in most samples (average, 1.96×10^4 cells, median 0.76×10^4), but species of Zygnematales and Euglenophyceae occurred by chance.

4.2.1. Phytoplankton vs. water quality

In the linear regression analyses using the basic data of single variables, water chemistry parameters explained a minor proportion of the variation in biological variables. This is seen clearly in the plots of total phosphorus and total nitrogen on ascending biomass of phytoplankton (Fig. 7). High concentrations were detected in all biomass classes without any visual order. The figure was similar for other chemical parameters.

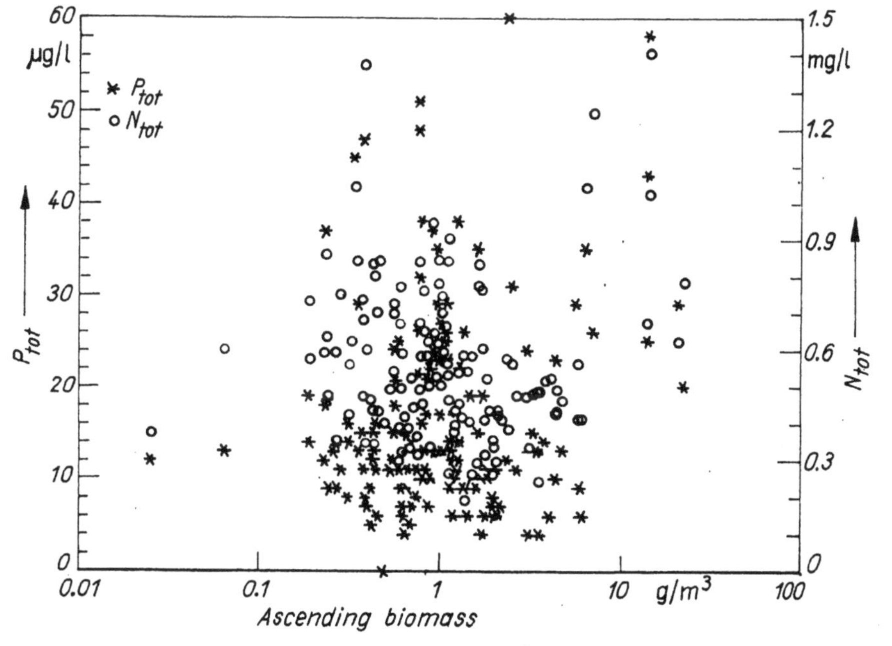

Fig. 7. Total phosphorus (*, $\mu g \cdot l^{-1}$) and total nitrogen (o, $mg \cdot l^{-1}$) plotted against ascending biomass of phytoplankton ($g \cdot m^{-3}$) in 151 lakes

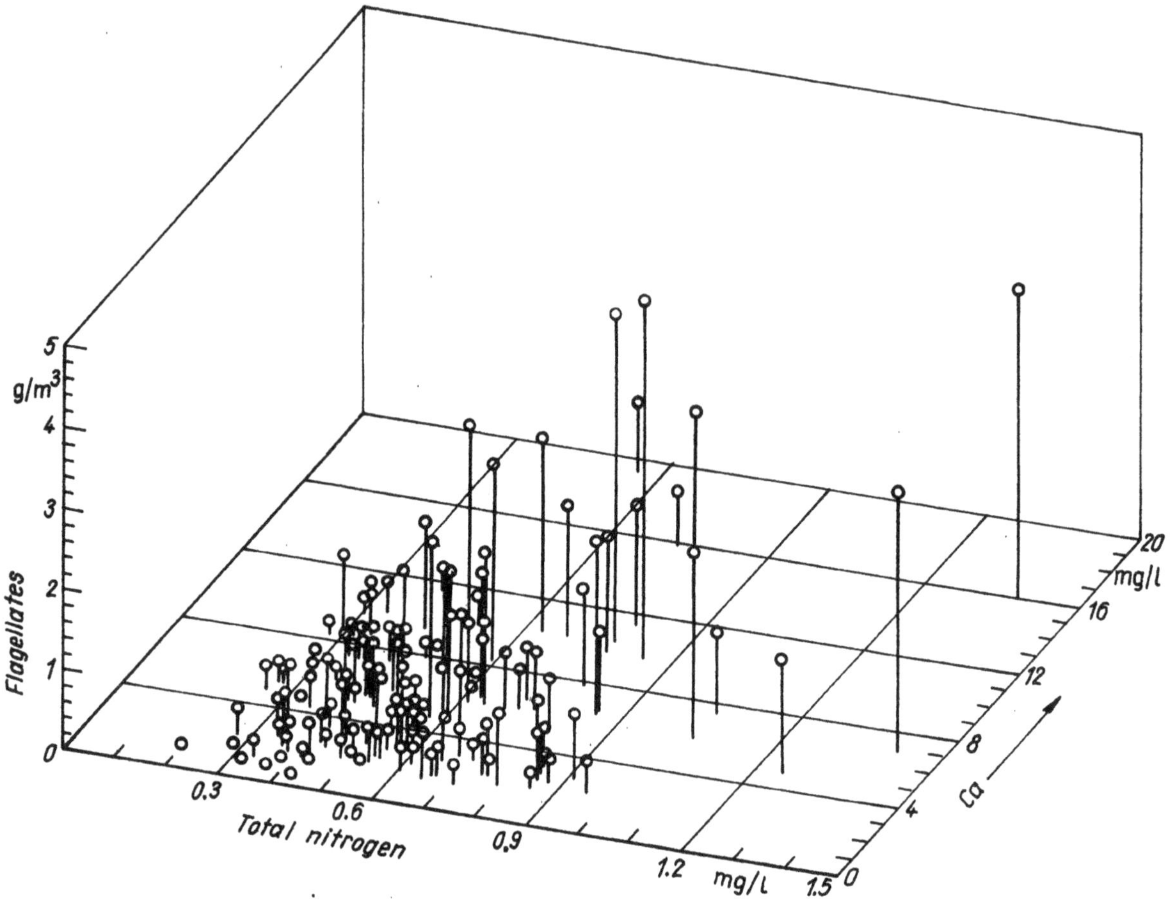

Fig. 8. The three-dimensional plot of biomass of flagellates in Chrysophyceae and Cryptophyceae ($g \cdot m^{-3}$) to the total nitrogen and Ca concentrations ($mg \cdot l^{-1}$) in summer samples

The stepwise selection of variables in multiple regression analyses also gave low explanations of variation in biological variables. The best, but still low explanation, 48%, was detected between the sum biomass of cryptophycean and chrysophycean flagellates and total nitrogen, Ca and water colour, which described, however, a clear growing tendency of biomass with higher total nitrogen and Ca concentrations (Fig. 8). Higher water colouration tended to increase flagellate biomass slightly and more clearly the percentage of flagellates in total biomass or cell numbers.

4.2.2. Phytoplankton vs. catchment characteristics

In the multiple regression analyses, there was no strong correlation of biological variables with characteristics of the catchment area (Table 4). Chrysophyceae and Cryptophyceae, which were the main contributors of total cell number ($r = +0.613$ and $+0.667$, respectively), but have particularly different patterns of cell numbers in the lakes ($r = +0.518$) both still had fshore as the most weighting variable. Results of these analyses did not show any other trends, not even weak ones.

4.3. PCA groups of the lakes

The high number of lakes in the study and their relatively high variability, both in water chemistry and land-use parameters, and in phytoplankton community structure, made it difficult to find good explanations between variables in individual lakes. To get similar lakes to-

Table 4. Stepwise multiple regression analysis of some phytoplankton measures to characteristics of catchment area. Cumulative r^2 is given after every independent variable and (—) when negative.

Dependent	Independent	Var. expl. (%)
Flagellate sum[1]	fshore (.33), peatl. (.35)	35
Total biomass	fshore (.30), depth (—.33)	33
Chlorophyll **a**	fshore (.27), peatl. (.33)	33
Chrysophyceae cells[2]	fshore (.24), field (—.26)	26
Cryptophyceae cells	fshore (.24)	24
Total cell number[3]	fshore (.18), field (—.21), area (.23)	23
Number of taxa[4]	field (.09), catchment (.12)	12

[1] Sum biomass of chrysophycean and cryptophycean flagellates.
[2] Partly colonies like *Uroglena*.
[3] Cells, colonies, filaments.
[4] Includes also counts of some chrysophycean cysts and undetermined taxa.

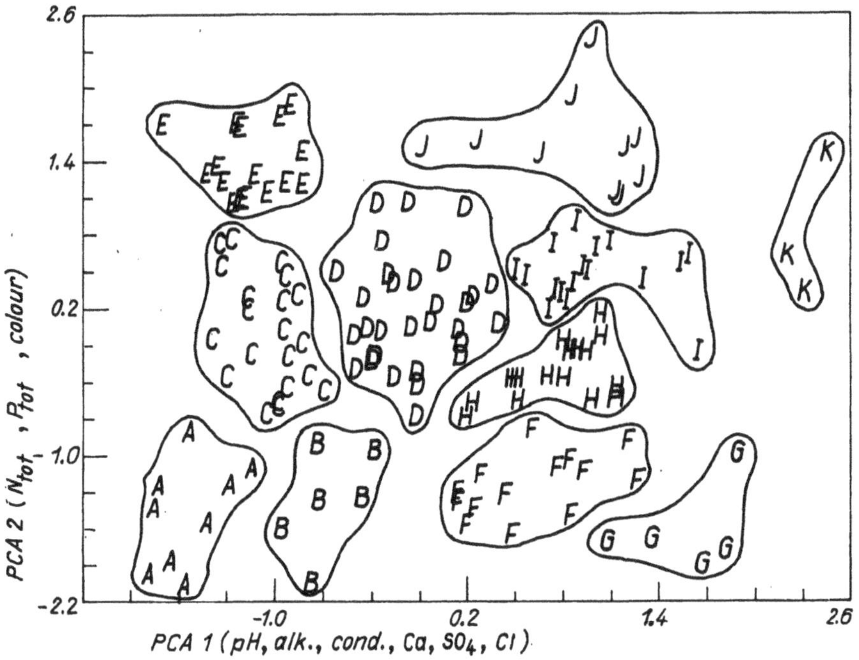

Fig. 9. The PCA groups A—K of the 151 lakes in eastern Finland. Grouping is based on all chemical parameters studied in autumn samples. The most weighting variables are given

gether the data were further treated by grouping the lakes according to the autumnal chemical analyses by the principal component analysis (PCA, with varimax rotation). The 11 PCA groups, from A to K (Fig. 9), were composed on the basis of the chemical data from autumn, but not from summer. This was done to describe the more or less stable average chemical balance of the lakes in the cool turbulent water just after the production period rather than the rapidly changing situations in the lakes during the most active part of the growing season in July. Catchment variables were not included in the analysis.

The PCA grouping method was selected since the treatment of data with, e. g. moving average or smoothing, will dissipate a too significant part of the individual variation, which partly remains in the averages of PCA groups.

The x-axis or PCA factor 1 (Fig. 9) is mainly determined by autumnal pH, alkalinity, conductivity, Ca, SO_4 and Cl and could thus be called the "Acidity axis". The Y-axis or factor 2 is the "nutrient and humus axis" since the most heavy loads are from autumnal total nitrogen, total phosphorus, and water colour. Axes 1 and 2 together explained 75.6% of the total variation in the data set. Also the summer values of the variables fitted the PCA model excellently since 95% of the variation in factor 1 was explained with summer conductivity, pH, N_{tot}, alkalinity and P_{tot}, and 88% of the variation in factor 2, by summer Secchi, P_{tot}, colour and N_{tot}.

The mean characteristics and the number of lakes of the groups are given in Table 5. Group A is an acidic clear-water group with low nutrient concentrations, E are more or less acidic nutrient-rich humic waters, F are neutral, high alkalinity and conductivity lakes with low nutrient concentrations, J are coloured nutrient-rich lakes with reasonable alkalinity and pH, and K are productive, indifferent lakes with high alkalinity, pH, conductivity and nutrient concentrations. Some of the groups, like B and C, are more or less similar in spite of different positions in the PCA field. This is mainly related to the greater Secchi depth of group B (7.2 m) compared to group C (2.6 m). Group G is determined by a higher pH (7.34), alkalinity ($0.46\,meq \cdot l^{-1}$) and conductivity ($8.34\,mS \cdot m^{-1}$ at 25 °C) compared to group H (7.16, 0.26 and 6.51, respectively).

Many of the PCA groups still overlap, but some groups differ significantly from each other as is seen in the notched Box-and-Whisker plots of total nitrogen in summer and autumn (Fig. 10) where the notches of the median values do not overlap. There is still significant

Table 5. Characterization of the PCA groups A—K based on their water quality. Number of lakes in different groups is given.

A. Acidic oligotrophic clear water lakes	(9)
B. Acid-sensitive oligotrophic clear water lakes	(6)
C. Sensitive oligotrophic clear water lakes, turbid	(21)
D. Sensitive oligotrophic mesohumic lakes	(31)
E. Acidic mesotrophic polyhumic lakes	(16)
F. Neutral oligotrophic clear water lakes	(15)
G. Buffered oligotrophic clear water lakes	(5)
H. Buffered oligo-mesotrophic clear water lakes	(17)
I. Neutral mesotrophic mesohumic lakes	(17)
J. Neutral eutrophic polyhumic lakes	(10)
K. Buffered eutrophic mesohumic lakes	(3)

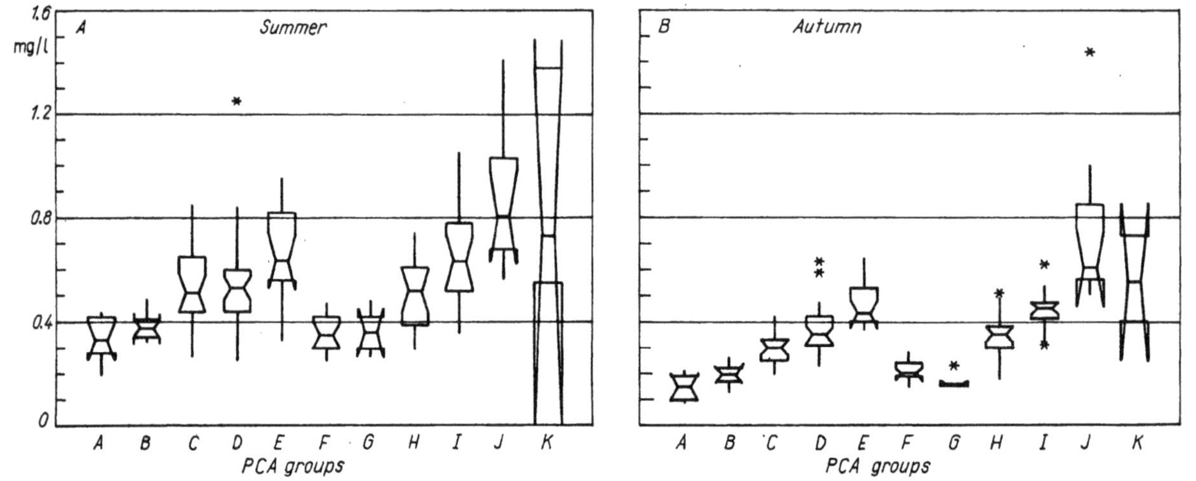

Fig. 10. The notched Box-and-Whisker plot of total nitrogen concentrations (mg · l^{-1}) in PCA groups in summer (A) and autumn (B). A notch corresponds to the width of a confidence interval for the median (horizontal line), while the width of the box is proportional to the square root of the number of observations in the data set. Whiskers extend to those maximum and minimum points which are within 1.5 times the interquartile range. Extreme values are indicated with points. Two boxes with notches that do not overlap indicate significantly different median values at 95% confidence level

Table 6. Stepwise multiple regression analysis of some average water quality measures of PCA groups to characteristics of catchment area. Cumulative r^2 is given after every independent variable and (—) when negative. For explanations see Table 2.

Dependent	Independent	Var. expl. (%)
N_{tot}	forest (—.83), fshore (.96)	96
Colour	forest (—.79), peatl (.91)	91
P_{tot}	forest (—.88)	88
Secchi depth	forest (.82)	82
pH	peatl. (—.80)	80
NO_3	fshore (.55), catchment (.75)	75
PO_4	peatl. (.71)	71
Conductivity	field (.65)	65
Alkalinity	field (.44)	44
Ca	none	0

variation left in some groups shown by the long whiskers and large boxes. Summer N_{tot} values of some groups are significantly higher than autumn values.

The stepwise selection of variables in multiple regression analyses with average catchment characteristics in lake groups in summer explained highly variations in many chemical parameters (Table 6). Of the catchment variables, lake area and, in most cases, catchment area did not play any significant role. Peatlands significantly explained only the pH of water.

4.4. Phytoplankton in PCA groups

Highest median biomasses and cell numbers (Fig. 11) were recorded in groups J and I, which were the nutrient-

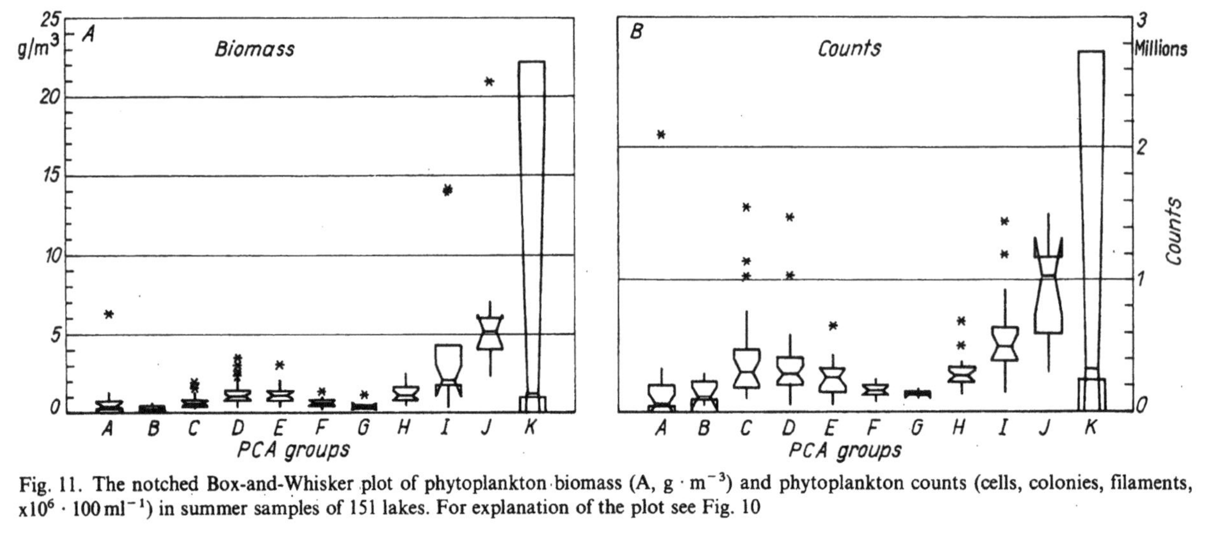

Fig. 11. The notched Box-and-Whisker plot of phytoplankton biomass (A, g · m^{-3}) and phytoplankton counts (cells, colonies, filaments, x10^6 · 100 ml^{-1}) in summer samples of 151 lakes. For explanation of the plot see Fig. 10

rich clear-water groups. Lowest biomasses were in nutrient-poor acid group B, and lowest cell numbers (Fig. 11 B) in group G with high pH. Since biomass is partly an estimated variable having severe possible errors in individual volume measurements, the cell count data are used in the further treatment of data.

Chrysophytes had significant cell numbers in all lake groups excluding A, where Chlorophyceae species clearly dominated and C, where Chroococcales dominated (Fig. 12). The second dominants in groups E—K were mainly Chlorophyceae or Cryptophyceae, but in group A and C Chrysophyceae, and in B and D Chroococcales bluegreens. Third dominant were mostly Cryptophyceae or Chlorophyceae and in group A Chroococcales. Diatomophyceae, Peridiniales or Nostocales did not dominate in any group, but were, however, dominant in a few single lakes. The cell numbers of Euglenophyceae were mainly low and they were completely lacking from most groups (A, C, F, G, H and K), which partly described the more or less oligotrophic nature of the present data set and that there are no lakes where sewage loading has polluted the lake water.

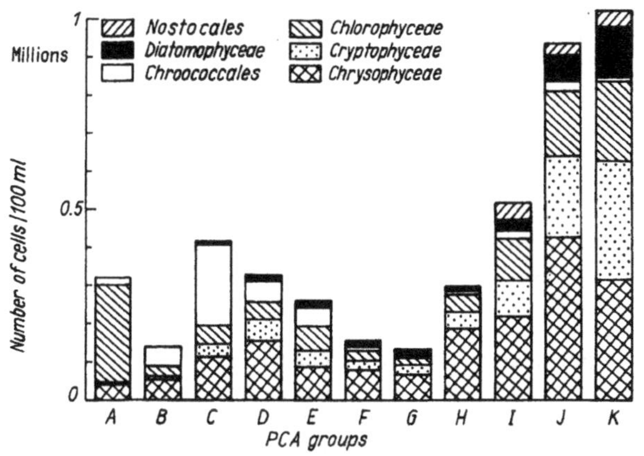

Fig. 12. Number of cells (or colonies and filaments) of different phytoplankton groups in PCA groups A—K in summer samples. Phytoplankton groups with minor cell numbers (Euglenophyceae, Dinophyceae, Zugnematales) have been left out.

4.4.1. Phytoplankton in PCA groups vs. water chemistry

Stepwise selection of variables in multiple regressions gave fair explanations to total biomass, cell numbers and chlorophyll concentration and to many phytoplankton groups (Table 7). Total nitrogen seems to be the most important variable in describing the cell number distributions in these lakes. There is obviously some autocorrelation between biomass and total nitrogen since nitrogen is largely accumulating into the biomass during the growing season and does not recycle as rapidly as phosphorus. However, this autocorrelation can be accepted since we attempted to find some chemical variables, among other factors which describe the changes in the structure and functioning of the phytoplankton communities in small lakes, and these variables should be commonly used and easy and accurate to measure.

pH is an important variable for many phytoplankton groups composed of small-celled species, but variables describing the mean nutrient balances (conductivity) or other production resources (alkalinity) are more significant for phytoplankton species bearing hard cell-covers. It is also worthy to note that the occurrence of Chlorophyceae species was impossible to explain even with average PCA variables. This could be partly due to their exceptionally high dominance in group A. Also Si had no correlation with the cell number of diatoms ($r = -0.053$).

Euglenophyceae occurred so sporadically that statistical analyses were not made. Zygnematales were also omitted because of their low cell numbers in samples and the resulting inaccuracy of counts associated with them.

4.4.2. Phytoplankton in PCA groups vs. catchment characteristics

Table 8 shows, that fshore is the most important independent factor explaining the variations in biological factors. Forest percentage is out of all regressions. Maximum depth of the lake is significant for Peridiniales and Nostocales and lake area for total number of cells and number of taxa.

5. Discussion

The 151 lakes in this study are mainly small and shallow and they are headwater lakes. Thus, their water quality could reflect the quality of the soil and bedrock on their catchment and the land-use therein. There is very little recently published material on large scale surveys of Finnish lakes which can be compared to the 151 lake results. The data collected by the National Board of

Table 7. Stepwise multiple regression analysis of some average phytoplankton measures of PCA groups to water quality. Cumulative r^2 is given after every independent variable and (—) when negative. For explanations see Table 4.

Dependent	Independent	Var. expl. (%)
Chlorophyll a	N_{tot} (.94), pH (.98), cond. (—.99)	99
Total biomass	N_{tot} (.77), pH (.96), alk. (—.98)	98
Total cells	N_{tot} (.79), colour (—.91), Ca (.97)	97
Flagellate sum	N_{tot} (.78)	78
Number of taxa	Alk. (.46), pH (—.66)	66
Cell numbers of phytoplankton groups:		
Cryptophyceae	N_{tot} (.78), pH (.94), Secchi (.97)	97
Chrysophyceae	N_{tot} (.77), pH (.87), cond. (—.93), alk. (.97)	97
Diatomophyceae	Cond. (.83), alk. (—.94)	94
Peridiniales	Cond. (.36), alk. (—.68), pH (.84)	84
Nostocales	N_{tot} (.66), pH (.79)	79
Chroococcales	pH (—.39), Ca (.74)	74
Chlorophyceae	none	0

Table 8. Stepwise multiple regression analysis of some average phytoplankton measures of PCA groups to characteristics of catchment area. Cumulative r^2 is given after every independent variable and (—) when negative. For explanations see Table 4.

Dependent	Independent	Var. expl. (%)
Chlorophyll a	fshore (.87), peatl. (.98)	98
Flagellate sum	fshore (.87), field (—.95)	95
Total biomass	fshore (.91), depth (—.95)	95
Number of taxa	field (.43), area (.68)	68
Cell numbers of phytoplankton groups:		
Chrysophyceae	fshore (.92)	92
Total cell number	fshore (.82), area (—.89)	89
Nostocales	fshore (.78), depth (—.86)	86
Cryptophyceae	fshore (.84)	84
Diatomophyceae	fshore (.73)	73
Peridiniales	depth (—.39), catchment (—.65)	65
Chlorophyceae	area (—.42)	42

Waters and Environment as a part of the environmental monitoring program of 826 lakes in all parts of Finland (HEINONEN 1980) contained mainly water-course lakes which were excluded from the present study. In that material, about 160 lakes were investigated from the same water area as our 151 lakes. The species composition of our lake groups showed similarly to HEINONEN's results the overall dominance of Chrysophyta (incl. diatoms) and low biomasses of blue-greens and euglenophytes. The mean phytoplanktonic biomass of the 160 lakes was also of the same magnitude, 1.94 mg $\cdot$ l^{-1}, but the list of the 50 most frequent phytoplankton species was largely different. There were more diatoms than in our material and many flagellated taxa, like *Katablepharis ovalis*, *Uroglena americana*, *Chromulina* sp. and *Chlamydomonas* EHR. species, were completely lacking. This is obviously caused by the preservation of the monitoring material with formaldehyde solution (HEINONEN 1980) without preceding Lugol solution fixation. This method has shown to effectively break thin-walled flagellate cells, and if formaldehyde is not buffered, diatom frustules will dissolve within a few years (HÄLLFORS et al. 1979). Those lacking flagella could have been destroyed before counting and, therefore, there is perhaps no true discrepancy between the presented results.

The survey of 35 small lakes of the Kangasala area in southern Finland (ILMAVIRTA 1980, 1983, 1984, 1988; ILMAVIRTA & TOIVONEN 1986) contained data from some slightly more eutrophic lakes and several brown-water lakes. Dominant species in those lakes were also flagellates, like *Cryptomonas* species, small chrysophycean flagellates, *Chroomonas acuta*, *Katablepharis ovalis*, some *Chlamydomonas* species and *Uroglena americana*.

In the 50 humic lakes of the Evo area in southern Finland, (ARVOLA 1986), flagellates, such as *Chlamydomonas* species, *Chroomonas acuta*, small *Cryptomonas* species, *Peridinium inconspicuum* LEMM., and *P. willei* HUITF.-KAAS dominated in early spring samples.

The number of taxa per sample was lower in ARVOLA's lakes (mean, 18) but higher in HEINONEN's lakes (61) than in our 151 lakes (32).

An investigation of 61 lakes in National Parks in southern and central Finland (ELORANTA 1988) provides a good comparative material. Lakes dominated with chrysophytes were slightly acidic and not very dark. Cryptophytes preferred more acidic and dark lakes with peatland on their catchment than chrysophytes. Diatoms had higher biomasses in larger lakes with clear water. All these results fit well with the 151 lake material.

In the study of 58 lakes in central Finland (ELORANTA 1986) the biomass values and species compositions were similar as in our 151 lakes. The relation of total biomass or different phytoplankton groups to water quality parameters was diverse, but many correlations were reported. However, all of those explained, like in our material, very poorly variations between lakes.

It seems that even single or only few samples can give a quite reliable overview of the phytoplankton species in humic waters. It is more difficult to show valid relationships between species composition and chemical parameters and especially catchment characteristics. ROSÉN (1981) made a very extensive investigation of phytoplankton in 1000 Swedish lakes throughout the country. His approach was mainly autecological concerning the frequency plots of species in the lakes along several chemical gradients. Also some attention was paid to the relationships of total biovolume of phytoplankton to water chemistry, but no large-scale multivariate analyses were done. ROSÉN's paper is an infinite source of detailed knowledge for comparisons.

The recent work of FÄNGSTRÖM & WILLÉN (1987) nicely attempted to analyse phytoplankton communities of 23 Swedish lakes with correspondence analyses. They were able to classify the lakes on the basis of the similarity in phytoplankton communities and to find the load and direction of the most weighting chemical parameters which influenced phytoplankton species distribution. Some similarities were described on the dominance of species compared to the 151 lakes, e. g. *Rhodomonas lacustris* PASCH. & RUTTN. (collectively *Cryptomonas* sp. in our data) was present in almost all lakes, and *Peridinium inconspicuum* prefered highly acidic lakes.

Analyses of phytoplankton composition in 95 lakes in Labrador (EARLE et al. 1986) and of 97 headwater lakes in insular Newfoundland (EARLE et al. 1987) present a complete-linkage cluster analysis of 80 selected phytoplankton species and SPEARMAN correlation coefficients and associated probabilities between seven derived factors and the abundance of selected phytoplankton species. In these papers, several flagellates favored dystrophic lakes, which was discussed in relation to the heterotrophy of these species in such environments where high concentrations of organic matter may serve as supplementary nutrition under light-limited conditions. This discussion is quite true since high heterotrophic activity has been measured in small flagellate cells in dystrophic lakes in the Evo area, Southern Finland (K. SALONEN, unpublished data). In the present 151 lakes, flagellates were similarly dominating in the dark lakes (see K. ILMAVIRTA et al. 1983; ILMAVIRTA 1983), where they also can profit from the accumulated phosphorus in the hypolimnion below a shallow thermocline by swimming there and returning to illuminated water-layer (SALONEN et al. 1984). Diatoms in Canadian lakes preferred shallow, unstratified lakes to maintain themselves in the euphotic zone. In our data a direct indication of this pattern cannot be seen; more careful treatment of the data is needed.

6. Acknowledgements

We are indebted to Dr. JOUKO MERILÄINEN for cooperation during the work, to Mrs. KAIJA ILMAVIRTA for counting the phytoplankton samples and many colleagues, home and abroad, who have given valuable comments on the manuscript. Warm thanks are due to all those several persons, who have performed the huge sampling operations on 151 lakes and the laboratories of The Water District of North Karelien and The Karelian Research Institute, Section of Ecology in University of Joensuu.

This study was financially supported by The Academy of Finland and The Maj and Tor Nessling Foundation, which is sincerely acknowledged.

References

ARVOLA, L. (1986): Spring phytoplankton of 54 small lakes in southern Finland. Hydrobiologia (The Hague) 137, 125—134.

CHRISTIE, C. E., SMOL, J. P., HUTTUNEN, P., & MERILÄINEN, J. (1988): Chrysophyte scates recorded in lake sediments from eastern Finland. Ibid. **161**, 237—243.

COTTEN, C. (1985): Cladoceran assemblages related to lake conditions in eastern Finland. Indiana Univ. (U.S.A.), Dept. Biol., Doctoral thesis, 97 pp.

EARLE, J. C., DUTHIE, H. C., & SCRUTON, D. A. (1986): Analysis of the phytoplankton composition of 95 Labrador lakes, with special reference to natural and anthropogenic acidification. Can. J. Fish. Aquat. Sci. **43**, 1804—1811.

— (1987): Factors influencing the distribution of phytoplankton in 97 headwater lakes in insular Newfoundland. Ibid. **44**, 639—649.

ELORANTA, P. (1986): Phytoplankton structure in different lake types in central Finland. Holarct. Ecol. **9**, 214—224.

— (1988): Etelä- ja Keski-Suomen kansallispuistojen kasviplanktonista heinäkuussa 1987 (Summary: Phytoplankton in national park lakes of South and Central Finland in 1987). Jyväskylän yliopiston biologian laitoksen tiedonantoja **51**, 1—39.

ERKAMAA, K., MÄKINEN, J., & SANDMAN, O. (1977): Methods of water analysis by authorized and water authority laboratories. Nat. Bd Waters Finland, Rep. **121**, 1—54 (in Finnish).

FÄNGSTRÖM, I., & WILLÉN, E. (1987): Clustering and canonical correspondence analysis of phytoplankton and environmental variables in Swedish lakes. Vegetatio **71**, 87—95.

HARJULA, H. (1979): Analysis of errors in estimating phytoplankton primary productivity and chlorophyll a with special reference to Lake Päijänne. Ann. Bot. Fennici **16**, 307—337.

HEINONEN, P. (1980): Quantity and composition of phytoplankton in Finnish inland waters. Publ. Wat. Res. Inst. Finland **37**, 1—91.

HUTTUNEN, P., & MERILÄINEN, J. (1986a): Applications of multivariate techniques to infer limnological conditions from diatom assemblages. Developm. Hydrobiol. **29**, 201—211.

—— (1986b): A comparison of different pH indices derived from diatom assemblages. Univ. Joensuu, Publ. Karelian Inst. **79**, 41—46.

—— (1986c): Diatom response to pH and humic matter of the water. Ibid. **79**, 47—54.

——, COTTEN, C., & RÖNKKÖ, J. (1988): Attempts to reconstruct lake water pH and colour from sedimentary diatoms and Cladocera. Verh. Int. Verein. Limnol. **23**, 870—873.

HÄLLFORS, G., MELVASALO, T., NIEMI, Å., & VILJAMAA, H. (1979): Effect of different fixatives and preservatives on phytoplankton counts. Publ. Wat. Res. Inst. **34**, 25—34.

ILMAVIRTA, K., HUTTUNEN, P., & MERILÄINEN, J. (1984): Phytoplankton in 151 eastern Finnish lakes: species composition and its relations to water chemistry. Verh. Int. Verein. Limnol. **22**, 822—828.

ILMAVIRTA, V. (1980): Phytoplankton in 35 Finnish brown-water lakes of different trophic status. Developm. Hydrobiol. **3**, 112—120.

— (1983): The role of flagellated phytoplankton in chains of small brown-water lakes in southern Finland. Ann. Bot. Fennici **20**, 187—195.

— (1984): The ecology of phytoplankton in brown-water lakes. Verh. Int. Verein. Limnol. **22**, 817—821.

— (1988): Phytoflagellates and their ecology in Finnish brown-water lakes. Hydrobiologia (The Hague) **161**, 255—270.

—, & TOIVONEN, H. (1986): Comparative studies on macrophytes and phytoplankton in ten small brown-water lakes of different trophic status. Aqua Fennica **16**, 125—142.

KÄMÄRI, J. (1984): Suomen karujen pienvesistöjen happamoitumisherkkyys. (Sensitivity to acidification of small oligotrophic lakes of Finland.). Nat. Bd Wat. Finland, Rep. **239**, 1—112.

MERILÄINEN, J., & HUTTUNEN, P. (1984): Ecological interpretation of diatom assemblages by means of two-way indicator species analysis (TWINSPAN). In: MANN, D. G. (ed.), Proceedings of the Seventh Diatom Symposium 385—391.

— HUTTUNEN, P., & PIRTTIALA, K. (1981): Research strategy for the ecology of diatoms: the diatom study of 151 lakes in eastern Finland. In: ROSS, R. (ed.), Proceedings of the Sixth Symposium on Recent and Fossil Diatoms, pp. 413—423 Koenigstein.

NAULAPÄÄ, A. (1972): Eräiden Suomessa esiintyvien planktereiden tilavuuksia. (Summary: Mean volumes of some plankton organisms found in Finland.). Nat. Bd Wat. Finland, Rep. **40**, 1—47.

ROSÉN, G. (1981): Tusen sjöar. Växtplanktons miljökrav. Stat. naturvårdverk. Rapp., 1—110.

SALONEN, K., JONES, R. I., & ARVOLA, L. (1984): Hypolimnetic phosphorus retrieval by diel vertical migrations of lake phytoplankton. Freshwat. Biol. **14**, 431—438.

TIKKANEN, T. (1986): Kasviplanktonopas (Guide of Phytoplankton). Suomen Luonnonsuojelun Tuki Oy. Forssa.

UTERMÖHL, H. (1958): Zur Vervollkommnung der quantitativen Phytoplanktonmethodik. Mitt. Int. Verein. Limnol. **9**, 1—38.

Address of the authors:

V. ILMAVIRTA
Maj and Tor Nessling Foundation
P. Hesperiankatu 3 A 4
SF — 00260 Helsinki
Finland

P. HUTTUNEN
Department of Biology
University of Joensuu
P. O. Box 111
SF — 80101 Joensuu
Finland

Limnologica (Berlin) **20** (2), 243—251, Juni 1990

Lammi Biological Station, University of Helsinki, Lammi, Finland

Chemical Budgets for a Small Dystrophic Lake in Southern Finland

By Lauri Arvola, Kalevi Salonen & Martti Rask

With 5 Figures

Summary

Budgets of carbon, nitrogen, phosphorus, sulphur and cations (Ca, Na, Mg, K, Fe and Al) in a small forest lake with two inflows were monitored during three consecutive summer seasons. The chemical load from the catchment correlated strongly with rainfall, and the chemical inputs to the lake thus varied with a large amplitude between the summers. Around 25—50% of the phosphorus load and 10—30% of the nitrogen load was retained in the lake. About 50% of the input of iron was also retained, whereas for other cations concentrations in the input, the lake and the outflow were similar and their fluctuation between years was low. Of the organic carbon input to the lake, 85—97% was allochthonous. During the study period there was a pronounced increase in the ammonium concentration in inflowing waters possibly due to clearcutting and prescribed burning in the catchment. In contrast, the concentrations of organic carbon and phosphate coincided better with water runoff, having higher concentrations during the wet summer than during the dry ones.

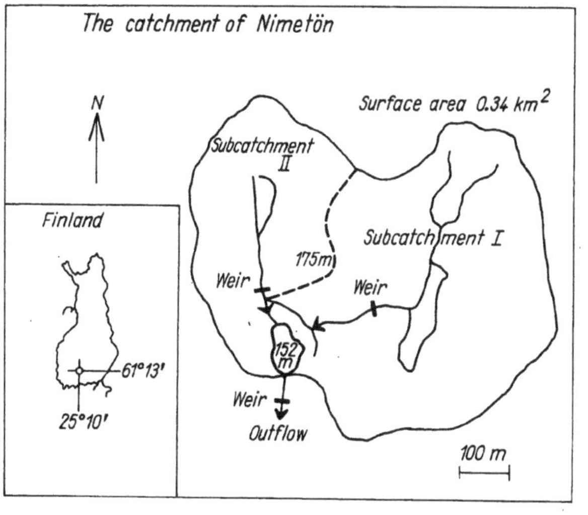

Fig. 1. Lake Nimetön and its catchment

Introduction

Humic lakes of the boreal coniferous zone typically have high allochthonous inputs of organic matter from their catchments, which may completely regulate the physical and chemical characteristics of a lake (Wetzel 1983) and especially its carbon budget (Sarvala et al. 1981) and trophic status (Carpenter et al. 1983). Humic substances are known to have an important role in the phosphorus and iron cycles (Francko & Heath 1979; Jackson & Hecky 1980) as well as in the cycle of aluminium (Driscoll 1980; Rosén 1982).

This paper presents chemical budgets for a small, extremely humic lake. Besides the nutrients attention was also paid to organic carbon because of its known importance in determining the bioavailability of phosphorus and some cations. Further, because the forest surrounding the lake was clearcut and burned during the study period, the results give information about the effects of forestry on the runoff chemistry of a small headwater lake.

Study Area

The study area (Fig. 1) is situated in the Evo State Forest in Lammi, southern Finland, ca. 150 to 160 m above sea level. Forestry is practically the only human activity in the catchment, which has morainic soil overlying granodiorite and gneiss bedrock (Arvola 1983). Two small inflows to the lake drain two subcatchments (Fig. 1). The spruce and Scots pine forest surrounding the lake was clearcut, except just around the lake, after the first study summer in winter 1981—1982, and ca. 46% of the catchment area was then burned in May 1983. The catchment had been fertilized a decade earlier (1972) with 5000 kg of nitrogen (50% NO_3^{2-} and 50% NH_4^+), 800 kg of calcium and 400 kg of magnesium. Additionally, subcatchment II had been fertilized with 270 kg of potassium and 1400 kg of phosphorus (as K_2PO_4).

Lake Nimetön (Table 1) is a small and relatively deep headwater lake lying in a depression between low hills (Fig. 1), which shelter the lake against wind. The shoreline is formed by a floating *Sphagnum* peat mat and in some places also *Fontinalis antipyretica* grows on the bottom. There is one outlet from the lake. The lake area (0.004 km²) is 1.2% of the area of its catchment (0.34 km²).

Table 1. Lake Nimetön. The physico-chemical data represent average for the uppermost 1 m water layer in May—September (1981—1983).

	Mean	Range
Surface area (km²)	0.004	
Volume (m³)	3.6×10^3	
Maximum depth (m)	11	
Mean depth (m)	8	
pH	5.9	5.6—6.0
Conductivity (mS · m⁻¹ 20 °C)	3.5	3.3—3.7
Alkalinity (meq · l⁻¹)	0.08	0.07—0.09
Colour (mg Pt · l⁻¹)	255	225—300

Material and Methods

Most chemical analyses were made weekly or every fortnight (cations) from the inlets, the outlet and the lake. However, in 1981 cations were determined only 3 times during the study season. Sampling started after ice-break in May and continued until September. Both inlets have weirs for sampling and waterflow measurements. In addition, there was a permanent fluviometer in the outflow allowing continuous monitoring of the outflow.

Samples for chemical analysis were taken from the inlets and outlet directly to sample bottles. In 1981 lake water samples were taken from the surface with a 1 m long (diameter 45 mm) acrylic tube. In 1982—1983 separate samples were taken with a 0.5 m long Ruttner sampler from depths of 0—0.1 m (sampler used horizontally), 0.25—0.75 m and 0.75—1.25 m. Phosphate and nitrate + nitrite were determined from filtered (Whatman GF/C) samples, using the method of MURPHY & RILEY (1962) for phosphate and that of WOOD et al. (1967) for nitrate + nitrite. Ammonium nitrogen was determined in 1981 from unfiltered samples with an ammonium electrode and in 1982—83 by the method of SOLORZANO (1969). Total phosphorus and total nitrogen were determined after KOROLEFF (1979) as PO_4—P and NO_3—N in an AKEA analyzer following alkaline persulphate digestion in an autoclave for 30 minutes. Cations were determined in an atomic absorption spectrophotometer and sulphate turbidimetrically after evaporation and combustion of the samples to oxidize organic compounds (cf. RASK et al. 1986). Water colour was determined from absorbance at 420 nm and compared with platinum-cobalt standard solutions. Organic carbon was calculated by using water colour measurements and the conversion factor based on the correlation (r^2 = 0.84) between water colour and organic carbon concentration in the lake (1979). Organic carbon was analysed by combustion of unfiltered samples according to the method of SALONEN (1979). Water pH was measured by using an Orion (Model 701) pH-meter.

Rainfall data were collected at the Evo Station of Finnish Game and Fisheries Research Institute, Game Division, about 5 km SW from the study area. Water temperature and oxygen concentration were measured with a combined probe (Yellow Springs Instruments). Respiration and primary production of plankton were measured weekly, the former with the DIC-method (see SALONEN et al. 1983) and the latter with the acidification and bubbling ^{14}C method (SCHINDLER et al. 1972). In 1981 the measurements were made from 0—1 m samples incubated at 0.5 m (ARVOLA et al. 1986) and in 1982—1983 at the depths of 0.1, 0.5 and 1 m.

Inputs to the lake from the inflows were calculated by taking into the consideration the average concentration of each element at the beginning and at the end of each period (one week or two weeks) and the accumulated water runoff at the same time in both inflows. Losses through the outflow were calculated in the same manner by using the runoff data from the permanent fluviometer.

Results

Hydrological conditions

Total and monthly rainfall varied greatly between summers (Table 2). In 1981 the rainfall during May—September was 1.4 times that in 1982 and 1983. Water outflow was closely related to rainfall (Fig. 2). According to the input and output measurements of water, about 65% of outflowing water originated from the main inflows in 1981 and 1982, and 90% in 1983. If evaporation rates had been measured, the percentages would have been still larger.

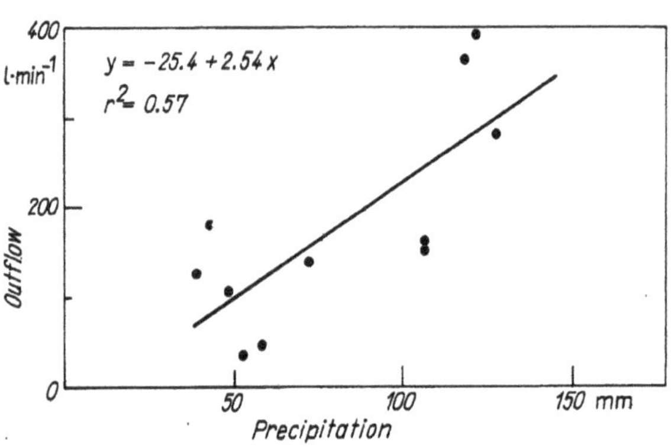

Fig. 2. Relationship between average monthly rainfall and water outflow from the lake

Table 2. Monthly rainfall and water runoff.

	Rainfall (mm)						Outflow ($10^3 \cdot m^3$)
Year	May	Jun	Jul	Aug	Sep	May—Sep	May—Sep
1981	7	118	128	122	53	428	84.7
1982	32	72	39	107	49	298	30.0
1983	64	43	58	53	107	324	25.3

Because of steep thermal stratification and a shallow (about 0.5 m) outlet, only the uppermost epilimnion of the lake was flushed by throughflow. Thus the residence time was different for the epilimnion and hypolimnion. During maximum runoff it was only 2 to 3 days for the epilimnion, while during dry periods it was several weeks. For the hypolimnion the residence time must be months or even years (see SALONEN et al. 1984). The theoretical residence time for the whole water mass of the lake varied between 2 months (1981) and half a year (1983).

Organic carbon

In 1981—1983 the flushing of organic carbon from the catchment was 1400—5000 kg $C \cdot km^{-2}$ (Table 3). This high load of organic carbon from the catchment was reflected in the water colour. In 1981 the mean colour in the inlet II was 435 mg $Pt \cdot l^{-1}$ and in 1983 310 mg $Pt \cdot l^{-1}$. The respective values for the lake were 300 mg $Pt \cdot l^{-1}$ (1981) and 225—240 mg $Pt \cdot l^{-1}$ (1982—1983). In 1981—1983 the load of organic carbon varied from 480 to 1740 kg $C \cdot lake^{-1}$ (Table 4), which represents 0.8 to 3.0 g $C \cdot m^{-2} \cdot d^{-1}$. Autochthonous primary production was 3 to 15% of the total load of organic carbon entering the lake (Table 5).

Outflow was the main (85—90%) output of carbon from the lake (Table 4). Respiration of plankton in the epilimnion comprised only 10—18% of the total organic carbon loss. Because the carbon budgets were negative in 1981 and 1982 carbon inputs to the lake were probably underestimated.

Compared to phosphorus the load of organic carbon decreased from 1981 to 1982—83, but compared to nitrogen it was slightly higher in 1983 than in 1981 (Table 3). Relatively, the carbon load was lowest in 1982.

Table 3. Losses ($kg \cdot km^{-2}$) of organic carbon, total nitrogen and total phosphorus and phosphate from the catchment of Nimetön. The elements entering the lake with groundwater (calculated to be 35% in 1981—1982 and 10% in 1983 from total water runoff) were assumed to have the same concentrations as in the inflow.

Year	Org. C	N	P	PO_4-P	C:N:P-ratio
1981	4976	192	19.1	17.1	261:10.5:1
1982	1400	69	7.4	4.2	189: 9.4:1
1983	1482	54	6.6	4.2	225: 8.4:1

Table 4. Budgets (kg · $lake^{-1}$) for organic carbon, total nitrogen (N), total phosphorus (P) and phosphate-phosphorus (PO_4-P); IN = Inflow, GW = Calculated groundwater, Pp = Primary production, R = Respiration of plankton, B = Biomass loss of phytoplankton by outflow, OUT = Outflow. In groundwater the concentration of organic carbon, total phosphorus and total nitrogen were assumed to be the same as in the inflow.

Organic carbon

Year	Input				Output				Input-Output
	IN	GW	Pp	Total	R	B	OUT	Total	
1981	1282	461	51	1794	143	5	1728	1876	−133
1982	350	126	70	546	92	3	642	737	−191
1983	461	43	86	590	95	5	458	558	32

N

Year	Input			Output	Input-Output
	IN	GW	Total		
1981	30.6	34.7	65.3	51.6	13.7
1982	14.0	9.4	23.4	16.4	7.0
1983	16.8	1.7	18.5	16.5	2.0

P

Year	IN	GW	Total	Output	Input-Output
1981	3.9	2.6	6.5	4.9	1.6
1982	1.6	1.0	2.5	1.4	1.2
1983	2.0	0.3	2.2	1.2	1.1

PO_4-P

Year	IN	GW	Total	Output	Input-Output
1981	3.5	2.4	5.8	3.0	2.8
1982	0.9	0.6	1.4	0.7	0.7
1983	1.3	0.1	1.4	0.5	1.0

Table 5. The proportion of photosynthetic carbon (Pp) in the total load of organic carbon to the lake.

Year	Pp (kg C · $lake^{-1}$)	Tot. load/lake (kg C · $lake^{-1}$)	Pp/load (%)
1981	51	1794	2.9
1982	70	546	12.8
1983	86	590	14.6

Nitrogen

In 1981 nitrogen input and output was 3 to 3.5 times higher than in 1982—1983 (Table 4). The daily input fluctuated in 1981 between 12.5 and 138 mg N · m^{-2}, while during the two other years (1982—1983) it was below 20 mg N · m^{-2} (Fig. 3) after June. In 1981 the concentration of total nitrogen at the surface of the lake varied from 460 to 850 µg N · l^{-1} (mean 610), in 1982 from 470 to 650 µg N · m^{-2} (mean 560) and in 1983 from 520 to 820 µg N · l^{-1} (mean 620). In the inflows the concentration of total nitrogen showed more pronounced variation: in 1981 it varied in inflow I between 240—780 µg N · l^{-1} and in inflow II between 490—980 µg N · l^{-1}; in 1982 between 380—1500 and 380—1080 µg N · l^{-1}; and in 1983 between 460—1010 and 480—970 µg N · l^{-1}. In comparison to phosphorus the load of nitrogen decreased by 20% from 1981 to 1983 (Table 3). Retention of nitrogen varied between 10 and 30% of the input.

The proportion of NH_4-N in total nitrogen was low in both inflows (Table 6) and in the lake, where it varied

Table 6. Proportions (%) of ammonium, nitrate and phosphate in total nitrogen and total phosphorus.

Year	Inflow I			Inflow II			Outflow		
	NH_4-N	NO_3-N	PO_4-P	NH_4-N	NO_3-N	PO_4-P	NH_4-N	NO_3-N	PO_4-P
1981	<4.3	<3.0	67	<2.8	<2.0	94	—	—	—
1982	5.2	2.9	33	3.5	<1.0	69	4.4	<1.3	51
1983	8.0	6.9	44	5.0	<1.0	71	5.3	<1.2	33

between 3.2—3.6%. In 1981 NH_4-N was often undetectable (<15 µg · l^{-1}) with the NH_4 electrode. NO_3-N was only occasionally detectable (>7 µg · l^{-1}) in inflow II and in the lake, so that its contribution to total nitrogen was usually below 1%. The contribution of nitrate showed a slight increase during the study in inflow I (Table 6), while ammonium showed a more pronounced increase in both inflows. In summer 1983 the average concentrations of ammonium and nitrate in inflow I were 49 µg · l^{-1} and 42 µg · l^{-1}, respectively. These values were 2.5 to 3 times higher than in 1981.

Phosphorus

The relative retention for phosphorus in the lake was much higher than for nitrogen and carbon. In 1981 phosphorus retention was 25% and in 1982—1983 about 50%. However, no notable difference between the phosphorus net inputs was detected.

Molybdate reactive phosphorus (MRP probably represents some organic P as well as PO_4-P) represented 60 to 90% of the total phosphorus load, while in the lake it was only 40 to 60% and in the outflow 30 to 50% (Table 6). In inflows I and II, the concentrations of phosphate varied greatly; in 1981, when the proportion of phosphate in total phosphorus was highest, it varied between 80—210 µg PO_4-P · l^{-1} (mean 140 µg l^{-1}) in inflow II, but in inflow I it fluctuated only from 3 to 30 µg PO_4-P · · l^{-1} (mean 14 µg · l^{-1}).

The phosphorus load showed a similar decreasing trend to nitrogen during the summers 1982—1983 (Fig. 3). The daily phosphorus load varied between 0.2—15 mg P · m^{-2} · d^{-1}. In the rainy summer of 1981, the average phosphate concentration at the surface of the lake and in inflow II was twice as high as in 1982—1983, while in inflow I there was no difference in the concentration of phosphate between these years.

Cations

Concentrations of Ca, Na, Mg, K, Fe, and Al differed very little between input and output or between study years (Table 7). Seasonal variation was also low except for Fe and Al (Fig. 4).

Calcium was the most abundant cation. Its load was 2.0—2.4 times higher than that of sodium, the next commonest cation (Table 7). The retention of most cations was rather low. The only exception was iron, the retention of which varied in 1982 and 1983 between 49 to 58% of the input. In 1981 the budget was negative for iron, as for most of the elements (Table 7).

There was a distinct difference between the two inflows only in the concentration of aluminium and potassium (Fig. 4).

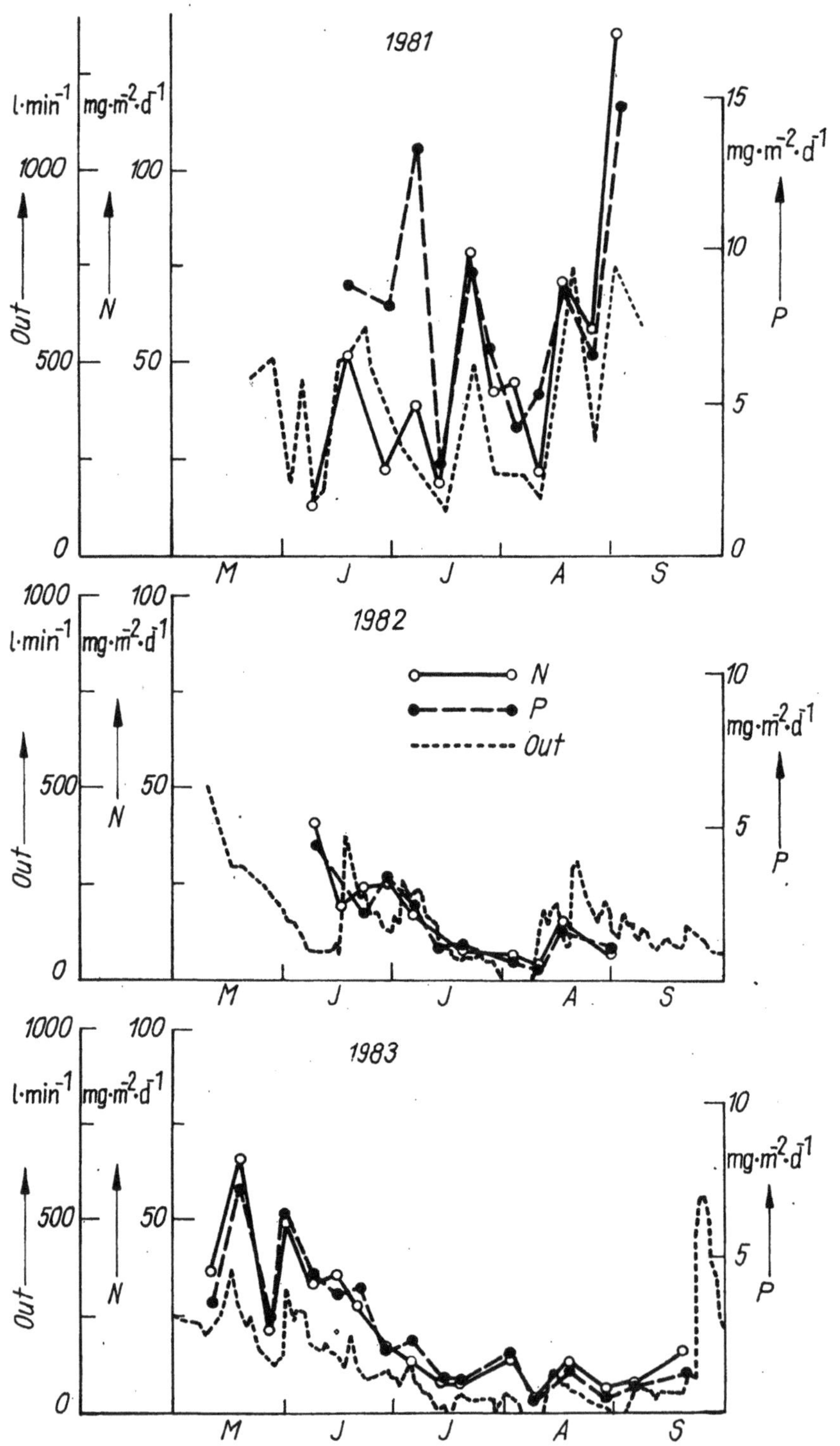

Fig. 3. The daily inputs of total phosphorus (P) and total nitrogen (N) to the lake, and water outflow (Out) from it

pH and sulphate

In 1981, pH in the inflows fluctuated between 5.2 and 6.1 (mean 5.7) and at the surface (0—1 m) of the lake between 5.4 and 5.8 (mean 5.6). In 1982—1983, the average pH values were 5.9—6.0 for the lake and slightly higher for the inflows.

Annual sulphate concentrations (Table 9) were at the same level from year to year both in the inflows and in the lake, although there was a substantial variation in sulphate load during the seasons (Tables 7 and 9). Compared to calcium, the sulphate concentration was much higher in inflow I than in inflow II (cf. Table 9, Fig. 5). A similar trend was also seen in the ratio between (Ca + Mg)/SO_4 (Table 9). The relationship between the ratio of Ca/SO_4 and pH also differed between the inflows (Fig. 5).

Table 7. The budgets of some cations (kg · lake^{-1}) for Nimetön. IF = through inflow, GW = calculated ground water.

Year	Input			Output	Input-Output
	IF	GW	Total		
	Ca^{2+}				
1981*	198.0	110.0	308.0	353.1	−45.1
1982	65.4	36.9	102.3	97.3	5.0
1983	88.2	9.0	97.2	86.2	11.0
	Mg^{2+}				
1981*	57.5	32.0	89.5	101.6	−12.1
1982	19.5	11.0	30.5	30.0	0.5
1983	25.6	2.63	28.3	27.3	1.0
	$Fe^{2+,3+}$				
1981*	24.5	15.1	39.7	44.9	− 5.7
1982	15.0	8.0	23.0	11.7	11.3
1983	16.5	1.7	18.1	7.6	10.6
	Al_{tot}				
1981*	29.8	16.9	46.7	44.5	2.2
1982	3.6	2.0	5.5	4.9	0.7
1983	5.9	0.6	6.5	5.4	1.1
	Na^{+}				
1981*	83.6	46.8	130.4	124.5	5.9
1982	32.3	18.1	50.4	43.8	6.5
1983	39.1	4.0	43.1	38.7	4.4
	K^{+}				
1981*	71.3	42.7	114.0	116.0	−2.0
1982	27.7	16.1	43.8	41.4	2.4
1983	33.5	3.3	36.8	37.2	0.3
	SO_4^{-}				
1981*	224.1	124.6	348.6	698.5	−349.9
1982	118.3	63.4	181.7	216.2	− 34.5
1983	141.3	12.3	153.6	186.7	− 23.1

* In 1981 only three measurements were made.

Table 8. Molar cationic compositions (%) in the input and output waters of Nimetön.

Year	Ca^{2+}	Na^{+}	Mg^{2+}	K^{+}	$Fe^{2+,3+}$	Al_{tot}
	Input					
1981	35.2	24.8	16.6	12.4	3.3	7.7
1982	33.5	28.5	16.4	13.6	5.4	2.6
1983	34.8	26.6	17.0	13.5	4.6	3.4
	Output					
1981	36.8	22.7	17.8	12.5	3.4	6.9
1982	34.6	27.0	17.8	15.1	3.0	2.6
1983	34.5	26.8	18.0	15.3	2.2	3.2

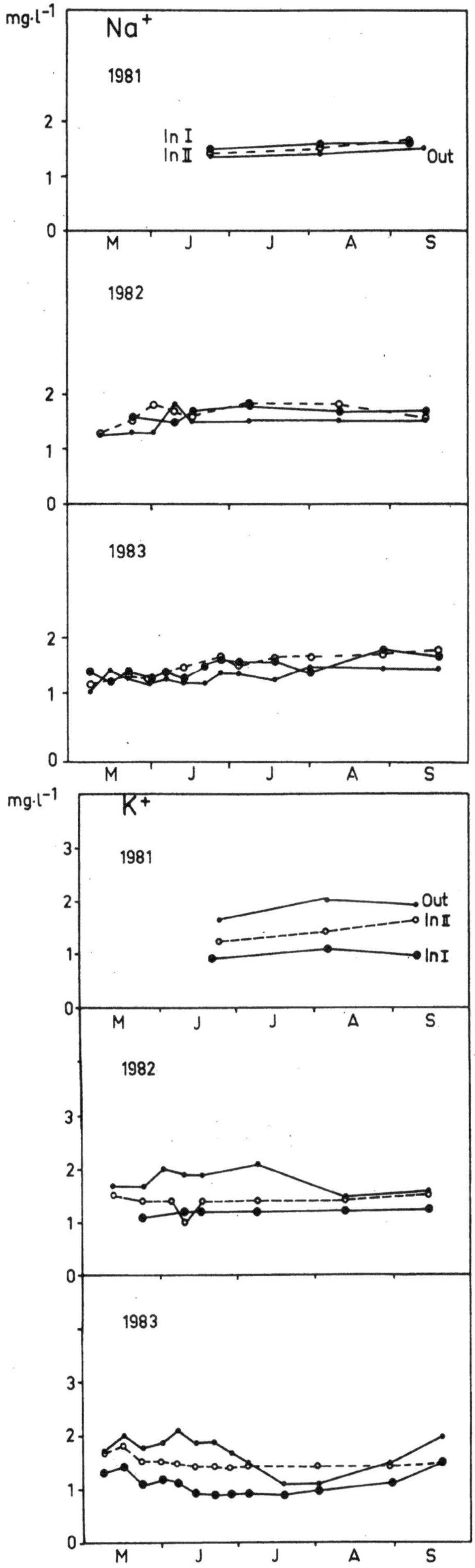

Fig. 4. Temporal and seasonal variation in the concentrations of major cations in the inflows and in the lake

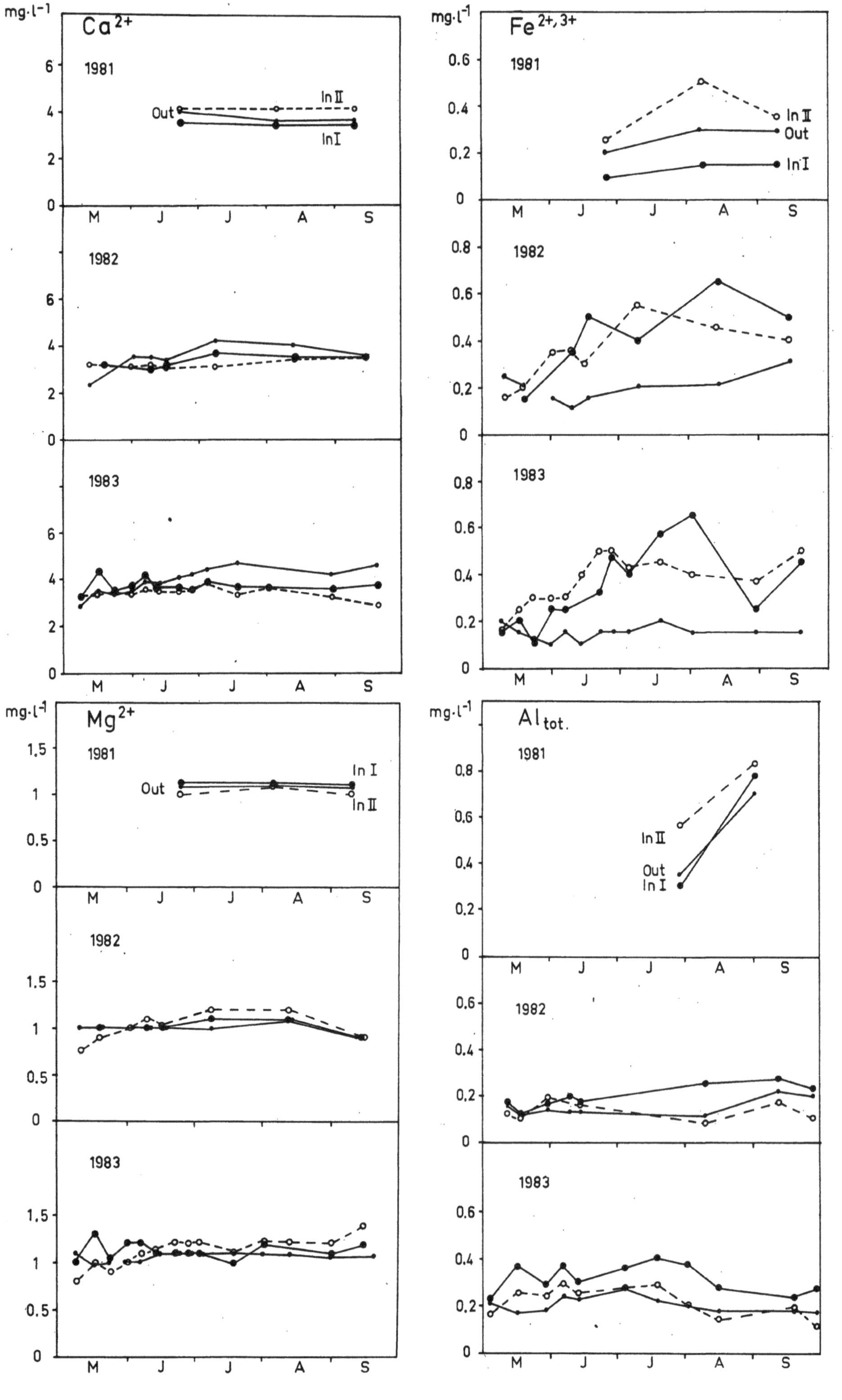

mg·l⁻¹
Ca2+
1981
1982
1983
In II
Out
In I
M J J A S
Fe2+,3+
Mg2+
Al tot.

Table 9. Average pH, concentrations (µmol · l^{-1}) of calcium, calcium + magnesium, sulphate and ratios of Ca/SO_4 and $Ca + Mg/SO_4$.

Year/Catchment	pH	Ca	Ca + Mg	SO_4	Ca/SO_4	$Ca + Mg/SO_4$
1981						
Inflow I	5.66	95	141	85	1.12	1.66
Inflow II	5.74	88	129	47	1.87	2.74
Lake	5.58	105	155	86	1.22	1.80
1982						
Inflow I	5.93	85	128	77	1.10	1.66
Inflow II	6.02	86	128	46	1.87	2.78
Lake	5.93	82	142	74	1.11	1.92
1983						
Inflow I	5.92	94	141	81	1.16	1.74
Inflow II	6.02	98	144	53	1.85	2.72
Lake	6.01	87	132	78	1.13	1.69

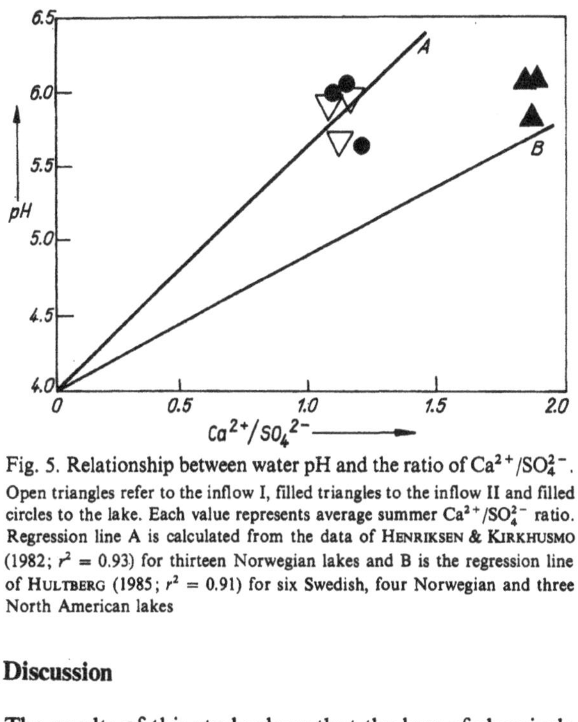

Fig. 5. Relationship between water pH and the ratio of Ca^{2+}/SO_4^{2-}. Open triangles refer to the inflow I, filled triangles to the inflow II and filled circles to the lake. Each value represents average summer Ca^{2+}/SO_4^{2-} ratio. Regression line A is calculated from the data of HENRIKSEN & KIRKHUSMO (1982; $r^2 = 0.93$) for thirteen Norwegian lakes and B is the regression line of HULTBERG (1985; $r^2 = 0.91$) for six Swedish, four Norwegian and three North American lakes

Discussion

The results of this study show that the loss of chemicals from the catchment of Lake Nimetön was mainly determined by rainfall. This agrees well with the observations of HOBBIE & LIKENS (1973) and TATE & MEYER (1983), who found, for example, that organic carbon losses from small catchments are strongly influenced by variations in hydrological conditions. TATE & MEYER (1983) suggested that the change in carbon export from a catchment due to differences in annual runoff appears to be at least a 2-fold change, while the change due to vegetation succession and different forest treatment histories (clear-cutting etc.) is only in the order of 10—20%.

Ammonium was the only chemical whose loss markedly increased after the first study year. This might be due to the forestry activities in the catchment of Lake Nimetön. RUUHIJÄRVI et al. (1986) observed at another prescribed burning site in the same area that ammonium concentration increased 10-fold in the humus layer and 3.5-fold in the upper layer of soil (podsol) after burning. In contrast to ammonium, the loss of phosphate decreased after the first year in this study, which is contradictory to the results of RUUHIJÄRVI et al. (1986), who found a 2-fold phosphate increase in the humus layer after burning.

In Lake Gårdsjön, Sweden, HULTBERG (1985) observed a similar high (73—74%) retention of iron and PERSSON & BROBERG (1985) a 20—25% retention of phosphorus. High phosphorus and iron retention could be connected with humic substances (cf. WRIGHT & HENRIKSEN 1980), which form humic-iron-phosphorus complexes (FRANCKO & HEATH 1979). Dissolved organic matter and/or bacteria and algae may encourage their precipitation into the hypolimnion (see TIPPING 1985). However, aluminium had a rather low retention in Nimetön, although it too should be associated with organic matter (DRISCOLL 1980; ROSÉN 1982).

Compared to atmospheric precipitation at Lammi (JÄRVINEN 1986) the losses of Ca, Mg and K from the catchment seemed to be higher than their input to it. In contrast, the deposition of nitrogen, phosphorus (except in 1981) and sodium was higher than their output from the catchment, which is in accordance with the results of LEPISTÖ (1984).

The loss of major cations from the catchment of Nimetön was very similar to that from the catchment of Hubbard Brook (Table 11). Compared to some other catchments in Scandinavia and North America (Table 12), it seemed that the concentration of calcium (Ca) and potassium (K) were rather high. This and a quite high phosphorus loss from the catchment (inflow II, cf. Table 9) of Nimetön, indicate possible residual effects from fertilization with K_2PO_4 about a decade before our study.

Data reviews by HULTBERG (1985) for the northern hemisphere showed a good correlation between the molar concentrations of Ca and SO_4 (Ca/SO_4) and runoff pH. The results for inflow II agree well with the results of HULTBERG while the results for inflow I and the lake agree better with the results reported by HENRIKSEN & KIRKHUSMO (1982) and may indicate that the main factors regulating pH in the inflows were not necessarily the same.

The strongly negative sulphate budget in 1981 (Table 7) might indicate that the concentration of sulphate in groundwater was possibly higher than estimated or that the *Sphagnum* mat along the shore increased the sulphate concentration when water flowed through it. Another explanation could be that the greater flushing of the hypolimnion in this rainy summer increased epilimnetic SO_4 concentrations (mixing of H_2S subsequently oxidized to SO_4) as well as the concentrations of most cations.

The allochthonous organic carbon load (85 to 97% of the total) is so high in Lake Nimetön that it may regulate the whole metabolism of the lake. One indication of

Table 10. Organic carbon, total nitrogen and total phosphorus losses (kg · km^{-2} · yr^{-1}) from various catchments.

Area	org. C	N	P	Ref.
Nimetön* (S Finland)	1400—5100	54—192	6.6—19.1	This study
23 small drainage basins (Finland)	—	79—740	4—44	KAUPPI (1978)
Teerensuonoja (S Finland)	—	130—370	2.5—6.7	LEPISTÖ (1984)
Huhtisuonoja (SE Finland)	—	89—300	5.2—25	LEPISTÖ (1984)
Vähä-Askanjoki (N Finland)	—	75—230	2.5—26	LEPISTÖ (1984)
Forest rivers (Sweden)	—	90—160	8	AHL & WIEDERHOLM (1977)
Gårdsjön (SW Sweden)	6744	207	3.1	BROBERG & PERSSON (1984)
Storsjön (SW Sweden)	—	235	2	LIKENS et al. (1977)
Kuokkel (N Sweden)	—	97	4.1	JANSSON (1979)
Dellen (N Sweden)	—	80	4.3	BRINK (1979)
Langtjern (SW Norway)	8740	155	3.3	WRIGHT & HENRIKSEN (1980)
Mirror Lake (USA)	1860**	106***	13.8***	LIKENS et al. (1985)

* May-September; ** Runoff input; *** Fluvial input

Table 11. Comparison of cationic losses (May—September) from the catchments of Nimetön and Mirror Lake (LIKENS et al. 1985).

	Ca	Mg	Na	K
	(kg · km^{-2})			
Nimetön (1981—1983)	497	146	220	191
Mirror Lake (1970—1975)	698	148	361	126

this is the high plankton respiration, which often exceeded the primary production (SALONEN et al. 1983). The ratio of autochthonous to allochthonous inputs of carbon into the lake resembles that of Wood's Pond, a small dystrophic lake in Maritime Canada (CARPENTER et al. 1983), although those authors suggested that the biological processes of Wood's Pond could be driven almost entirely by autotrophic production.

In view of the temporal variability, a three year study is too short to quantify the effects of different catchment manipulations on runoff chemistry. Moreover, the sampling intervals during the study and the lack of sampling in early spring and late autumn make it difficult to compare the results of different years (cf. KOHONEN 1982).

In general, carbon, nitrogen and phosphorus losses from our study site were at about the same level as from some other forest catchments in Finland, Norway, Sweden and elsewhere. The likely indications of the forest treatments (fertilization, clear-cutting and burning) were the high potassium and phosphate concentrations in inflow II, and the pronounced increase in the ammonium loss from the catchment after clear-cutting and burning.

Acknowledgements

We thank L. RASK for carrying out the nutrient analyses, and R. I. JONES, T. KAIRESALO and R. RUUHIJÄRVI for their comments on the manuscript. R. I. JONES revised the English of the manuscript.

Table 12. Average concentrations and proportions of major constituents in runoff from some catchments in Scandinavia and North America.

Catchment	Ca^{2+}		Mg^{2+}		Na^{+}		K^{+}		Ref.
	(mg · l^{-1})	(%)	(mg · l^{-1})	(%)	(mg · l^{-1})	(%)	(mg · l^{-1})	(%)	
Nimetön (1981—83)	3.7	46	1.1	14	1.8	23	1.1	18	This study
Gårdsjön	1.5	15	1.4	14	6.5	65	6.0	6	HULTBERG (1985)
Birkenes	1.3	27	0.5	10	2.8	57	0.3	6	WRIGHT & JOHANNESEN (1980)
Hubbard Brook	1.7	49	0.4	10	0.9	32	0.2	9	LIKENS et al. (1977)
Rawson Lake	2.2	47	0.8	17	1.3	28	0.4	9	SCHINDLER et al. (1976)
Sudbury	5.7	59	1.3	14	2.1	22	0.5	5	JEFFRIES et al. (1982)

The table is modified from the data presented by HULTBERG (1985).

References

AHL, T., & WIEDERHOLM, T. (1977): Swedish water quality criteria: nutrients — PM 918. Swed. Nat. Environm. Prot. Bd., Stockholm, 123 pp. (in Swedish).

ARVOLA, L. (1983): Primary production and phytoplankton in two small, polyhumic forest lakes in southern Finland. Hydrobiologia 101, 105—110.

— SALONEN, K., BERGSTRÖM, I., HEINÄNEN, A., & OJALA, A. (1986): Effects of experimental acidification on phyto-, bacterio- and zooplankton in enclosures of a highly humic lake. Int. Rev. ges. Hydrobiol. **71**, 737—758.

BRINK, N. (1979): Växtnäringsförluster från skogsmark. Ekohydrologi 5. Swed. Univ. of Acricult. Sci., Div. of Water Management, Uppsala, 1—11 (in Swedish).

BROBERG, O., & PERSSON, G. (1984): External budgets for phosphorus, nitrogen and dissolved organic carbon for the acidified Lake Gårdsjön. Arch. Hydrobiol. **99**, 160—175.

CARPENTER, J. W., GREEN, R. H., & PATERSON, C. G. (1983): A preliminary organic carbon budget for a small dystrophic lake in Maritime Canada. Hydrobiologia **106**, 275—282.

DRISCOLL, C. T. (1980): Aqueous speciation of aluminium in the Adirondack region of New York State, USA. In: DRABLOS, D., & TOLLAN, A. (eds.), Ecological Impact of Acid Precipitation. SNSF-project, Oslo-Ås, pp. 214—215.

FRANCKO, D. A., & HEATH, R. T. 1979: Functionally distinct classes of complex phosphorus compounds in lake water. Limnol. Oceanogr. **24**, 463—473.

HENRIKSEN, A., & KIRKHUSMO, L. A. (1982): Acidification of groundwater in Norway. Nordic Hydrol. **13**, 183—192.

HOBBIE, J. E., & LIKENS, G. E. (1973): Output of phosphorus, dissolved organic carbon, and fine particulate carbon from Hubbard Brook watersheds. Limnol. Oceanogr. **18**, 734—742.

HULTBERG, H. (1985): Budgets of base cations, chloride, nitrogen and sulphur in the acid Lake Gårdsjön catchment, SW Sweden. Ecol. Bull. **37**, 133—157.

JACKSON, T. A., & HECKY, R. E. (1980): Depression of primary productivity by humic matter in lake and reservoir waters of the boreal forest zone. Can. J. Fish. Aquat. Sci. **37**, 2300—2317.

JANSSON, M. (1979): Nutrient budgets and the regulation of nutrient concentrations in a small subarctic lake in northern Sweden. Freshw. Biol. **9**, 213—231.

JEFFRIES, D. S., SCHNEIDER, W. A., & SNYDER, W. R. (1982): Geochemical Interactions of Watersheds with Precipitation near Sudbury, Ontario. In: NRIAGU, J. O. (ed.), Environmental Impact of Smelters. New York.

JÄRVINEN, O. (1986): Laskeuman laatu Suomessa 1971—1982. Vesihallituksen monistesarja **408**, 1—142 (in Finnish).

KAUPPI, L. (1978): Effect of drainage basin characteristics on the diffuse load of phosphorus and nitrogen. Publ. of the Water Res. Inst. **30**, 21—41.

KOHONEN, T. (1982): Influence of sampling frequence on the estimates of runoff water quality. Publ. of the Water Res. Inst. **27**, 5—30.

KOROLEFF, F. (1979): Methods for the chemical analysis of seawater. Meri **7**, 1—60 (in Finnish).

LEPISTÖ, A. (1984): Ainetaseet pienillä valuma-alueilla. Vesihallituksen monistesarja **264**, 1—105 (in Finnish).

LIKENS, G. E., BORMANN, F. H., PIERCE, R. S., EATON, J. S., & JOHNSON, N. M. (1977): Biogeochemistry of a Forested Ecosystem. New York.

— EATON, J. S., JOHNSON, N. M., & PIERCE, R. S. (1985): Flux and balance of water and chemicals. In: LIKENS, G. E. (ed.), An Ecosystem Approach to Aquatic Ecology, Mirror Lake and its Environment. New York.

MURPHY, J., & RILEY, J. P. (1962): A modified singlesolution method for the determination of phosphate in natural waters. Analyt. Chim. Acta **27**, 31—36.

PERSSON, G., & BROBERG, O. (1985): Nutrient concentrations in the acidified Lake Gårdsjön: The role of transport and retention of phosphorus, nitrogen and DOC in watershed and lake. Ecol. Bull. **37**, 158—175.

RASK, M., HEINÄNEN, A., SALONEN, K., ARVOLA, L., BERGSTRÖM, I., LIUKKONEN, M., & OJALA, A. 1986: The limnology of a small, naturally acidic, highly humic forest lake. Arch. Hydrobiol. **106**, 351—371.

ROSÉN, K. (1982): Supply, loss and distribution of nutrients in three coniferous forest watersheds in central Sweden. Reports in Forest Ecology and Forest Soils **41**. Dept. of Forest Soils, Swed. Univ. of Agricult. Sci., Uppsala.

RUUHIJÄRVI, R., LINDHOLM, T., & VASANDER, H. (1986): Some consequences of using prescribed burning in forestry. Lammi Notes **13**, 1—5.

SALONEN, K. (1979): A versatile method for the rapid and accurate determination of carbon by high temperature combustion. Limnol. Oceanogr. **24**, 177—183.

— KONONEN, K., & ARVOLA, L. (1983): Total respiration of plankton in two small polyhumic lakes. Hydrobiologia **101**, 65—70.

— ARVOLA, L., & RASK, M. (1984): Autumnal and vernal circulation of small forest lakes in Southern Finland. Verh. Int. Verein. Limnol. **22**, 103—107.

SARVALA, J., ILMAVIRTA, V., PAASIVIRTA, L., & SALONEN, K. (1981): The ecosystem of the oligotrophic lake Pääjärvi 3. Secondary production and an ecological energy budget of the lake. Verh. Int. Verein. Limnol. **21**, 454—459.

SCHINDLER, D. W., SCHMIDT, R. V., & REID, R. A. (1972): Acidification and bubbling as an alternative to filtration in determining phytoplankton production by the ^{14}C-method. J. Fish. Res. Bd. Canada **29**, 1627—1631.

— NEWBURY, R. W., BEATY, K. G., & CAMPBELL, P. (1976): Natural water and chemical budgets for a small Precambrian lake basin in central Canada. J. Fish. Res. Bd. Can. **33**, 2526—2543.

SOLORZANO, L. (1969): Determination of ammonia in natural waters by the phenolhypochlorite method. Limnol. Oceanogr. **14**, 799—801.

TATE, C. M., & MEYER, J. L. (1983): The influence of hydrologic conditions and successional state on dissolved organic carbon export from forested watersheds. Ecology **64**, 25—32.

TIPPING, E. (1985): Accumulation of humic substances in the hypolimnion of a seasonally anoxic lake following co-sedimentation with iron oxide. Org. Geochem. **8**, 123—124.

WETZEL, R. G. (1983): Limnology. Philadelphia.

WOOD, E. D., ARMSTRONG, F. A. J., & RICHARDS, F. A. (1967): Determination of nitrate in sea water by cadmium-copper reduction to nitrite. J. Mar. Biol. Ass. U. K. **47**, 23—31.

WRIGHT, R. F., & HENRIKSEN, A. (1980): Hydrologic and chemical studies at Langtjern. IR 71/80. SNSF-project, Oslo-Ås, 108 pp.

— & JOHANNESEN, M. (1980): Input-output budgets of major ions at gauged catchments in Norway. In: DRABLOS, D., & A. TOLLAN (eds.). Ecological Impact of Acid Precipitation. SNSF-project, Oslo-Ås. pp. 250—251.

Adress of the authors:
L. ARVOLA
K. SALONEN
M. RASK
Lammi Biological Station
University of Helsinki
SF-16900 Lammi
Finland

Limnologica (Berlin) **20** (2), 253—258, Juni 1990

Department of Biology, University of Turku, Finland

Diel Changes in Horizontal Microdistribution of Littoral Cladocera

By M. WALLS, M. RAJASILTA, J. SARVALA & J. SALO

With 2 Figures

Key words: Horizontal microdistribution; Zooplankton; Cladocera; Predation; Patchiness.

Abstract

We studied the diel distribution of cladoceran zooplankton in a shallow bay of Lake Pyhäjärvi, SW Finland. Zooplankton samples were collected during one day and night from 24 locations distributed along three horizontal zones 0.5, 2.5, and 5.0 m from the shoreline.

The total number of species was significantly higher in the night samples, mainly due to an increased occurrence of benthic and periphytic cladocerans. The number of individuals also increased in the samples taken at night, but the increase was not statistically significant. There were differences between species; increased night activity was indicated in *Ceriodaphnia*, *Diaphanosoma*, *Sida*, *Bosmina*, *Acroperus*, and *Eurycercus*. In *Bosmina longispina* and *Ceriodaphnia quadrangula* the individuals caught at night were also significantly larger. These changes were probably due to a nighttime release from predation, as the planktivorous young fish stayed away from the shore during the night.

Introduction

The littoral zone is the most productive area in many lakes. The ensuing intensity of all life processes, together with the inherent physical heterogeneity of the habitat, accentuate the complexity of the predator-prey and competitive interactions between species. Rapid changes in community structure may occur over short distances and time intervals. The understanding of the littoral system thus requires detailed investigations on both microspatial and short-term temporal scales. Such detailed studies on the horizontal distribution and diel movements of zooplankton are relatively few (DUMONT 1967; EINSLE 1972; KAIRESALO 1980; FAIRCHILD 1981; TESSIER 1983; DAVIES 1985; DIFONZO & CAMPBELL 1988; KAIRESALO et al. 1989). In contrast, the diel vertical migration of zooplankton has been the subject of numerous studies (e. g. STICH & LAMPERT 1981; GLIWICZ 1986; VUORINEN 1986; DINI & CARPENTER 1988).

Predation pressure by visually feeding fish seems to be one of the ultimate reasons behind zooplankton vertical movements. In the shallow littoral areas where there is no effective depth gradient that could be used to escape from fish predation by migrating vertically, a different behavioural strategy may have evolved. Littoral zooplankton may utilize swarms and active onshore-offshore migration or vegetated areas as refuges to avoid predation.

Diel horizontal migration between the littoral and pelagial areas has been reported in many fish species (e.g. HALL et al. 1979; VØLLESTAD 1983; NAUD & MAGNAN 1987). Such fish movements most probably contribute to a diurnally variable feeding pressure on the littoral zooplankton. Indeed, variation in the abundance of littoral cladocerans related to changing fish predation has been documented in a few studies (FAIRCHILD 1982; BOIKOVA 1986; JAKOBSEN & JOHNSEN 1987; CRYER & TOWNSEND 1988; WINFIELD & TOWNSEND 1988). In the present paper we show that diel changes in horizontal distribution, accompanied by changes in size distribution, occur in many species of freshwater littoral cladocerans. We also speculate that these changes reduce the predation pressure of young fish on the cladocerans.

Material and Methods

We studied the diel microdistribution of zooplankton on 26—27 July 1984 in a shallow bay of mesotrophic Lake Pyhäjärvi, SW Finland (60°54′—61°06′N and 22°09′—22°25′E). Additional information about the lake is given in SARVALA et al. (1984). Zooplankton samples were taken from a total of 24 sites (8 locations along the shore in three horizontal zones 0.5, 2.5 and 5.0 m offshore). The depth was approximately 0.5 m at every location. Samples were taken both during the day (3—5 pm) and at night (0—2 am). A conical plankton net (diameter 45 cm, mesh size 100 μm) was carefully lowered close to the bottom, moved sidewards about 0.5 m, opened, and hauled vertically to the surface. Each sample was preserved in 70% ethanol.

The number of species and number of individuals belonging to each species were counted from a 10 ml subsample of each original 150 ml sample. We also measured the carapace length of the three most abundant species (*Bosmina longispina*, *Ceriodaphnia quadrangula*, and *Polyphemus pediculus*). The measurements were taken as the longest distance along a straight line between the hind point and the anterior edge of the carapace.

Data on the abundance and food of juvenile fish in the study area were available from the summers of 1982 and 1984 (SARVALA & RAJASILTA pers. comm.). Young fish were captured with a bag seine in the daytime throughout the summer, and over a diel cycle on two occasions in 1984. In the laboratory, the fishes were identified and measured, and the stomach contents analysed from a sample of approximately 20 individuals.

Results

Species diversity was higher at night in all three zones. The total number of cladoceran species (Table 1) differed significantly between day and night samples in all zones (Mann-Whitney test: zone 0.5 m, $U' = 54.4$, $P < 0.05$; zone 2.5 m, $U' = 51.0$, $P = 0.05$; zone 5.0 m, $U' = 61.0$, $P < 0.01$). However, the number of species observed is always a function of the number of individuals studied. Indeed, the total number of individuals was higher at night in all three zones, but due to the large within-zone variance, the differences were not statistically significant. To account for the variable number of individuals in the samples, the diversity differences were tested by the rarefaction technique (SIMBERLOFF 1978). There was significant heterogeneity within the zones, especially

Table 1. Numbers of individuals of different cladoceran species and total number of species in the day and night samples from 8 sites located at equal distances along the shoreline.

Distance 0.5 m offshore																
	Day								Night							
Species	1	2	3	4	5	6	7	8	1	2	3	4	5	6	7	8
C. quadrangula	22	95	—	—	2	—	—	—	20	47	38	110	13	515	110	154
Daphnia sp.	—	2	—	1	—	—	—	—	—	—	—	—	—	3	—	—
D. brachyurum	2	6	—	1	5	—	—	—	8	12	7	13	19	55	36	39
S. crystallina	3	2	—	17	36	11	40	24	1	35	29	22	19	29	4	3
B. longirostris	5	2	—	3	10	5	—	4	7	—	29	9	—	8	7	20
B. longispina	14	11	2	32	190	20	2	31	10	29	1133	332	2509	468	59	76
A. harpae	2	—	—	—	1	6	—	—	10	4	10	7	—	8	15	3
A. affinis	—	—	—	—	—	—	—	—	—	—	—	—	—	—	4	—
A. quadrangularis	—	—	—	—	—	—	—	—	1	—	—	—	—	—	—	—
A. excisa	—	—	—	—	—	—	—	—	—	—	—	—	—	8	1	—
A. nana	5	2	1	4	—	—	—	—	—	—	—	—	—	—	2	—
E. lamellatus	—	—	—	—	1	3	—	4	—	6	10	11	16	18	9	8
Other Chydoridae	2	—	—	—	—	2	—	2	1	—	9	3	3	5	7	—
P. pediculus	—	—	—	3	442	858	1592	1562	2	132	17	561	42	1768	48	34
Total # of spec.	8	7	2	8	8	7	3	6	9	7	11	10	7	12	15	8
Total # of ind.	55	120	3	61	687	905	1634	1627	60	265	1282	1068	2621	2885	302	337
Distance 2.5 m offshore																
C. quadrangula	1	7	25	1	10	—	81	82	9	110	23	26	73	92	22	24
Daphnia sp.	—	—	—	—	—	—	—	—	—	3	11	6	—	—	—	—
D. brachyurum	1	1	8	4	—	—	4	2	1	17	13	10	8	31	6	5
S. crystallina	1	1	3	—	4	2	11	2	1	11	15	6	10	31	11	12
B. longirostris	—	4	3	4	14	7	9	2	2	3	3	13	3	4	11	55
B. longispina	1	15	13	15	125	45	150	190	1	6	22	912	25	255	230	238
A. harpae	3	—	10	—	1	1	20	29	4	9	6	3	10	20	67	2
A. affinis	—	—	3	—	—	—	—	—	—	—	1	3	1	—	8	—
A. quadrangularis	—	—	—	—	—	—	2	—	—	—	—	—	1	—	—	—
A. excisa	1	—	—	—	3	1	—	2	—	1	—	10	1	—	8	—
A. nana	1	—	5	3	—	—	7	5	—	1	—	—	2	—	6	—
E. lamellatus	—	—	5	—	1	1	11	17	—	6	4	—	18	8	20	2
Other Chydoridae	—	—	3	—	1	—	2	2	—	1	1	6	4	1	—	2
P. pediculus	—	—	—	—	11	1	4	—	2	4	1	3	—	31	3	10
Total # of spec.	7	5	10	5	9	7	11	10	7	12	11	12	14	9	11	9
Total # of ind.	9	28	78	27	170	58	301	333	20	172	100	998	156	473	392	350
Distance 5.0 m offshore																
C. quadrangula	—	21	12	5	5	13	4	2	1	44	5	21	18	234	48	41
Daphnia sp.	—	—	—	—	—	—	1	2	4	2	3	1	1	—	—	2
D. brachyurum	—	8	—	1	—	—	1	2	4	16	4	3	2	54	6	2
S. crystallina	—	—	1	—	—	—	4	—	—	7	11	3	10	60	3	14
B. longirostris	—	3	—	13	24	14	1	110	3	—	2	3	5	24	6	7
B. longispina	2	20	8	14	48	22	52	221	4	17	11	113	17	558	123	103
A. harpae	—	—	1	1	—	4	1	2	4	1	1	10	13	12	12	14
A. affinis	1	—	—	1	1	—	—	—	—	—	—	—	—	6	3	—
A. quadrangularis	—	—	1	—	—	2	—	2	—	1	—	—	2	—	—	—
A. excisa		-	—	—	1	1	—	—	1	—	—	1	1	6	3	—
A. nana	—	1	2	2	1	—	3	2	—	—	—	1	2	—	12	5
E. lamellatus	—	3	—	—	1	—	1	—	—	4	5	4	8	6	3	2
Other Chydoridae	1	—	1	1	—	—	—	—	—	1	1	1	4	—	6	5
P. pediculus	—	—	—	—	1	—	—	—	14	3	—	—	—	6	9	19
Total # of spec.	3	6	7	8	8	6	9	8	8	10	9	11	13	10	12	11
Total # of ind.	4	56	26	38	82	56	68	343	35	96	43	161	83	966	234	214

close to the shoreline. The observed number of species in a sample fell outside the 95% confidence limits for the expected number of species, based on the combined data of the respective zone and time of day, in 5, 3, and 1 (day; zones from 0.5 to 5.0 m) and in 4, 4, and 1 (night) out of the 8 possible cases. This heterogeneity prevented any meaningful comparisons of diversity at the level of individual samples. But when the total number of individuals at each time of day was drawn from the combined data for the zone, the number of species observed during the day was significantly lower than expected in the 0.5 m zone; in the 2.5 and 5.0 m zones no differences were significant.

Variance to mean ratio ($I = s^2/\bar{x}$; ELLIOTT 1971) also indicated significant aggregation in all zones both in the daytime and at night in almost all cases where the numbers were sufficient for statistical testing (F-test).

The following cladoceran species differed in abundance between day and night: *Ceriodaphnia quadrangula, Diaphanosoma brachyurum, Sida crystallina, Bosmina longispina, Acroperus harpae*, and *Eurycercus lamellatus* (Table 2). In all species the number of individuals was always higher at night.

The size distributions of *Bosmina longispina* and *Ceriodaphnia quadrangula* at the different horizontal zones were different between day and night (Table 3). In *B. longispina*, the individuals were largest close to the shoreline and were significantly larger at night in all zones (Fig. 1). In *C. quadrangula*, a similar trend was found, but only the mean carapace length of the individuals sampled from the 0.5 m zone was significantly larger at night. In *P. pediculus*, there were no statistically significant differences between the day and night samples; in the two offshore zones the number of individuals measured was too small for statistical analyses.

Table 2. Significance of the difference between day and night abundance in each species in the three zones: Mann-Whitney test; $n_1 = n_2 = 8$ in each case tested.

Species	Mann-Whitney test results		
	Zone 0.5 m	Zone 2.5 m	Zone 5.0 m
C. quadrangula	**	NS	*
Daphnia sp.	NS	NS	NS
D. brachyurum	***	*	*
S. crystallina	NS	*	*
B. longirostris	NS	NS	NS
B. longispina	*	NS	NS
A. harpae	*	NS	*
A. affinis	NS	NS	NS
A. quadrangularis	NS	NS	NS
A. excisa	NS	NS	NS
A. nana	NS	NS	NS
E. lamellatus	**	NS	*
Other Chydoridae	NS	NS	NS
P. pediculus	NS	NS	NS

NS: $P \geq 0.05$, *: $P < 0.05$, **: $P < 0.01$, ***: $P < 0.001$

Table 3. Significance of the difference in carapace length between day and night samples in *Bosmina longispina* and *Ceriodaphnia quadrangula* in the three zones 0.5 m, 2.5 m, and 5.0 m offshore (Mann-Whitney test).

Species	0.5 m	2.5 m	5.0 m
B. longispina	***	***	***
C. quadrangula	***	NS	NS

NS: $P \geq 0.05$; ***: $P < 0.001$

A comparison between the occurrence of *B. longispina* and *B. coregoni* in the plankton and in the stomachs of young perch on 16 and 30 July 1982, revealed that the perch selected for the largest size classes of both species of *Bosmina* (Fig. 2) (Mann-Whitney: *B. coregoni* 16 July, $U' = 626$, $P < 0.001$; *B. longispina* 16 July, $U' = 1194$, $P < 0.001$; *B. longispina* 30 July, $U' = 3411.5$, $P < 0.001$).

Discussion

The differences between the day and night samples in the six cladoceran species indicate increased night activity, which occurred especially close to the shore. Although *Bosmina* is easily transported with even weak water currents (KAIRESALO et al. 1990) we consider it unlikely that water currents or turbulence could have caused the observed diel patterns. In the central parts of the lake the mean values of wind-induced current speed may be $5-10\ \mathrm{cm \cdot s^{-1}}$ (SARKKULA 1984), but these values are certainly not representative of the sheltered bay areas, especially as the weather was calm throughout the whole sampling period. Nighttime cooling of water above a littoral slope creates convective circulation with a stable undercurrent away from the shore and an unstable return flow along the surface (HORSCH & STEFAN 1988). In our case, however, the uniform depth of the wide shallow shore area gives us reasons to believe that convectional currents should not result in a noticeable horizontal flow.

With the exception of *Bosmina* spp. and *Daphnia* spp., the other cladocerans encountered in the samples consist of littoral species that tend to dwell among the vegetation or stay on the bottom during the daylight hours. Species that have an appearance easily visible to predators (e.g. the thick carapaces of Chydoridae) are especially vulnerable to, for instance, juvenile fish that feed in schools. At night these species are probably less vulnerable to visually foraging predators and may therefore utilize a wider habitat. According to STENSON & OSCARSON (1985) both *Acroperus* and *Eurycercus*, typically living in bottom habitats, were found amongst plankton when fish were absent in Lake Gårdsjön.

In Pyhäjärvi, *Bosmina* species tend to form very dense, spherical swarms (diameter approximately 3—6 cm, pers. observation) during the day right at the shoreline. Larger swarms have been reported by DUMONT (1967) and KLEMETSEN (1970). Especially the dense swarms may increase the probability of escaping from predation by night they migrate some tens of meters offshore to feed. On the basis of stomach fullness analysis of young perch young fish. According to SARVALA & RAJASILTA (pers. comm.), both perch and roach juveniles feed heavily on littoral zooplankton in Lake Pyhäjärvi. These fish forage close to the shoreline especially during daylight hours. At sampled on 11 July 1984, there were no significant differences in the fullness indices (0—3: empty — full) between the four sampling hours (0600, 1200 am and pm) (Kruskal-Wallis, $H = 4.29$, $df = 3$, NS). This result indicates that the fish had been feeding both in the daytime and at night.

Especially *Bosmina longispina* is a very important source of food for littoral fish in Pyhäjärvi (SARVALA et al. 1989; SARVALA & RAJASILTA pers. comm.). The uneven distribution of *B. longispina* in the 0.5 m zone corres-

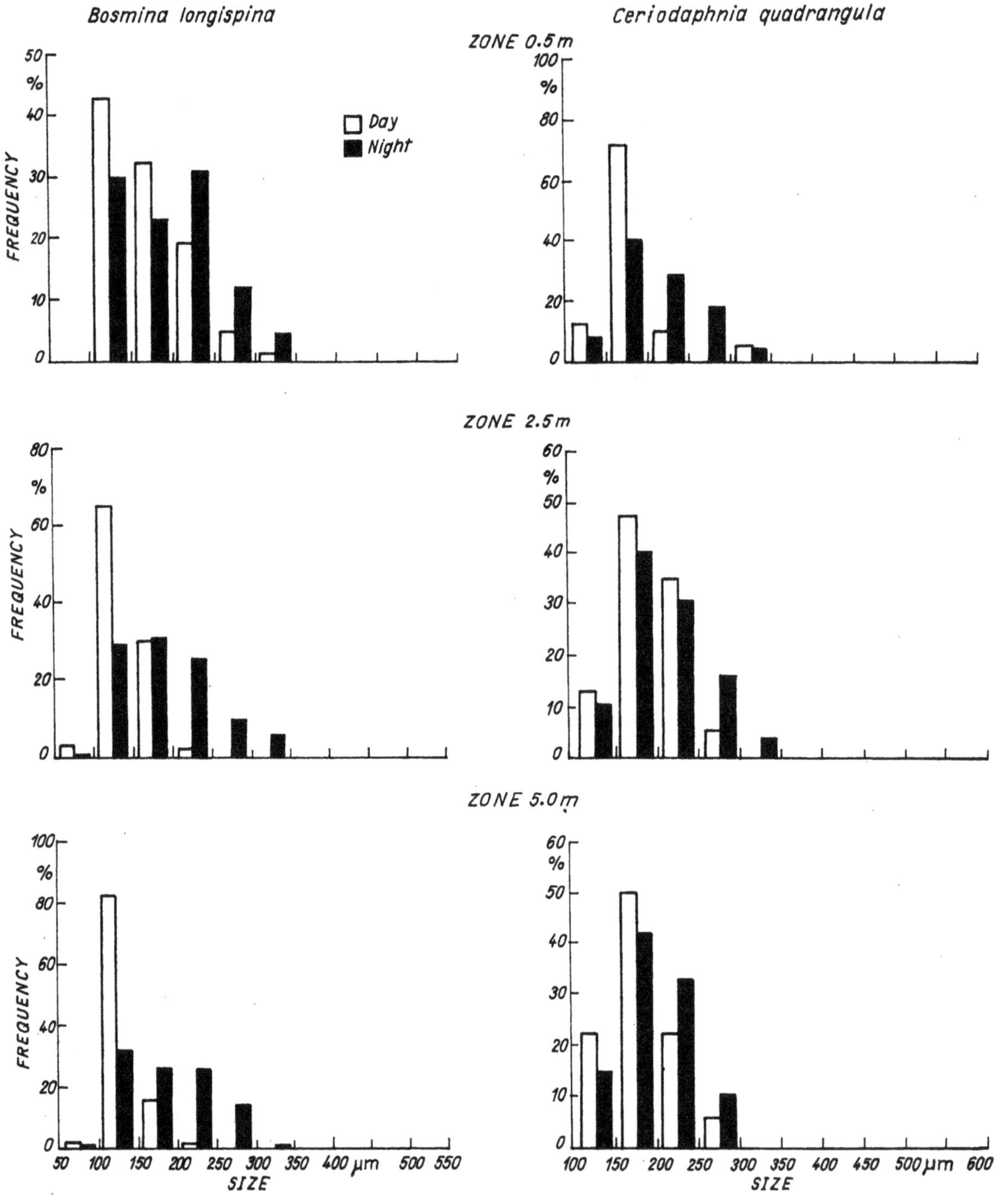

Fig. 1. Length distributions (% of total numbers) of *Bosmina longispina* (*n* varies between 84—250) and *Ceriodaphnia quadrangula* (*n* varies between 36—177) during day and night in the three zones

ponds to swarm formation during the day. Dense swarms are distributed more or less randomly right at the shoreline. Therefore, it may be unlikely that these swarms get sampled. The cladocerans might also stay on the bottom during the day and ascend to open water during the night, thus performing vertical movements on a microscale. According to JAKOBSEN & JOHNSEN (1988), swarming behaviour, that they found in *Bosmina longispina* to occur only in light, can be interpreted as an adaptive diurnal rhythm in the trade-off between reducing risk of predation and maximizing feeding rate.

Body size of cladoceran prey has been shown to be one of the most important criteria in the food selection of planktivorous fish (e.g. EGGERS 1982). Fish predation can result in an elimination of the largest individuals of prey species and in a decrease in the mean body size of the population. In Pyhäjärvi, we showed that young perch selected the largest individuals within the *Bosmina* species. In accordance with this, the average size of crustacean zooplankton was approximately similar in the pelagial and along the open shores, but significantly smaller on the vegetated shores, where the young fish were aggre-

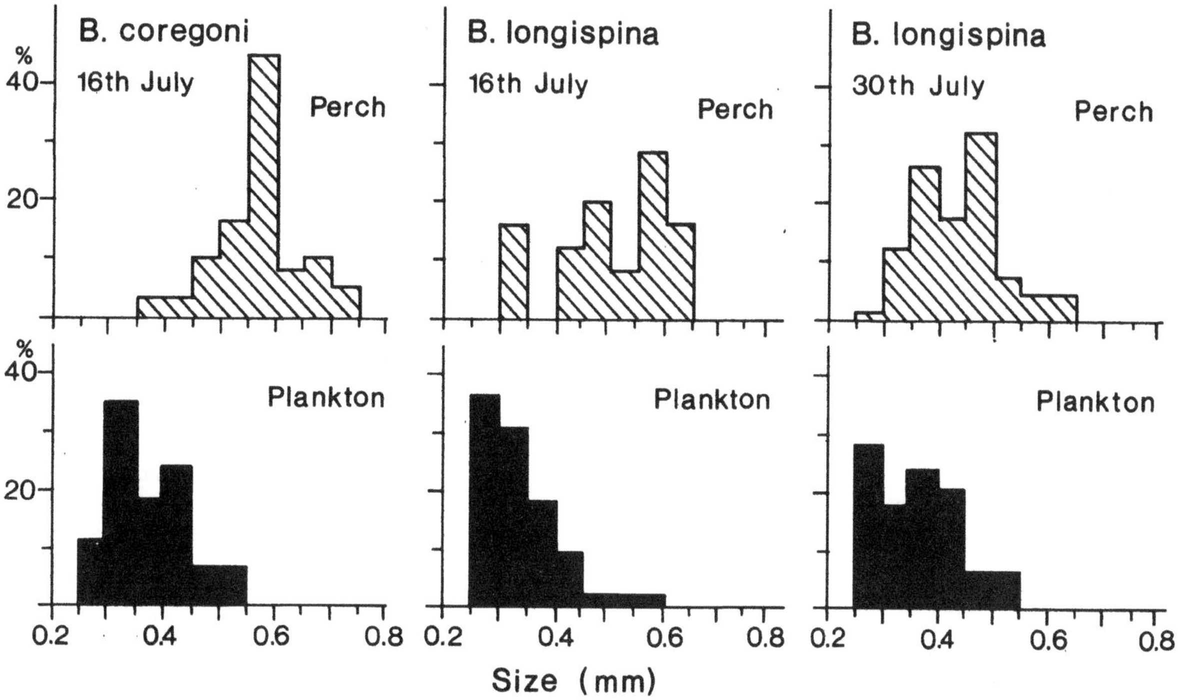

Fig. 2. Length distributions (% of total numbers) of *Bosmina coregoni* and *B. longispina* in the littoral plankton and in the perch stomachs in July 1982. In *B. coregoni*, $n = 17$ and $n = 38$ in the plankton and the perch stomachs, respectively. In *B. longispina*, $n = 55$ (plankton) and $n = 24$ (stomachs) on 16 July; $n = 54$ and $n = 85$ on 30 July, respectively

gated (SARVALA & RAJASILTA pers. comm.). Likewise, the abundance of littoral zooplankton in Pyhäjärvi was negatively correlated with macrophyte abundance. Moreover, in early summer cladocerans were more abundant in the littoral, but in the late summer the situation was reversed, most probably due to the heavy predation by young fish in the littoral. WERNER & HALL (1988) report a similar case in Lake Lawrence, where a decreasing density and size of *Daphnia galeata* from the pelagic edge of the littoral zone towards shore was probably due to predation by numerous young bluegill, confined to the littoral by predatory fish.

To conclude, on the basis of the diel changes in cladoceran horizontal microdistribution and size distribution, the perch stomach fullness and stomach content analyses, it seems evident that predation by juvenile fish is an important factor determining the structure of the littoral cladoceran community. Detailed studies focusing on each of the components are needed in order to reveal the regulatory mechanisms in the littoral community.

Acknowledgements

The study was supported financially by the Academy of Finland (research contract 09/109) and Maj and Tor Nessling Foundation. We also wish to thank TERTTU LAURIKAINEN for drawing the Fig. 2.

References

BOIKOVA, O. S. (1986): Horizontal distribution of crustaceans in Lake Glubokoe. Hydrobiologia **141**, 113—123.

CRYER, M., & TOWNSEND, C. R. (1988): Spatial distribution of zooplankton in a shallow eutrophic lake, with a discussion of its relation to fish predation. J. Plankton Res. **10**, 487—501.

DAVIES, J. (1985): Evidence for a diurnal horizontal migration in *Daphnia hyalina lacustris* SARS. Hydrobiologia **120**, 103—105.

DIFONZO, C. D., & CAMPBELL, J. M. (1988): Spatial partitioning of microhabitats in littoral cladoceran communities. J. Freshwat. Ecol. **4**, 303—313.

DINI, M. L., & CARPENTER, S. R. (1988): Variability in *Daphnia* behaviour following fish community manipulations. J. Plankton Res. **10**, 621—635.

DUMONT, H. J. (1967): A five day study of patchiness in *Bosmina coregoni* BAIRD in a shallow eutrophic lake. Mem. Ist. Ital. Idrobiol. **22**, 81—103.

EGGERS, D. M. (1982): Planktivore preference by prey size. Ecology **63**, 381—390.

EINSLE, U. (1972): Die räumliche Verteilung einiger pelagisch lebender Crustaceen im Litoral des Mindelsees. Schriften des Vereins für Geschichte des Bodensees und seiner Umgebung **90**, 259—269.

ELLIOTT, J. M. (1971): Some methods for the statistical analysis of samples of benthic invertebrates. Freshwat. Biol. Assoc. Sci. Publ. **25**, 1—148.

FAIRCHILD, G. W. (1981): Movement and microdistribution of *Sida crystallina* and other littoral microcrustacea. Ecology **62**, 1341—1352.

— (1982): Population responses of plant-associated invertebrates to foraging by largemouth bass fry (*Micropterus salmoides*). Hydrobiologia **96**, 169—176.

GLIWICZ, Z. M. (1986): Predation and the evolution of vertical migration in zooplankton. Nature **320**, 746—748.

HALL, D. J., WERNER, E. E., GILLIAM, J. F., MITTELBACH, G. G., HOWARD, D., DOER, C. G., DICKERMAN, J. A., & STEWART, A. J. (1979): Diel foraging behaviour and prey selection in the golden shiner (*Notemigonus crysoleucas*). J. Fish. Res. Board Can. **36**, 1029—1039.

HORSCH, G. M., & STEFAN, H. G. (1988): Convective circulation in littoral water due to surface cooling. Limnol. Oceanogr. **33**, 1068—1083.

JAKOBSEN, P. J., & JOHNSEN, G. H. (1987): The influence of predation on horizontal distribution of zooplankton species. Freshwat. Biol. **17**, 501—507.

— (1988): The influence of food limitation on swarming behaviour in the waterflea *Bosmina longispina*. Anim. Behav. **36**, 991—995.

KAIRESALO, T. (1980): Diurnal fluctuations within a littoral plankton community in oligotrophic Lake Pääjärvi, southern Finland. Freshwat. Biol. **10**, 533—537.

— & PENTTILÄ, S. (1990): Effect of light and waterflow on the spatial distribution of littoral *Bosmina* (Cladocera) Verh. Int. Verein. Limnol. **24**, (in press).

KLEMETSEN, A. (1970): Plankton swarms in Lake Gjökvatn, East Finnmark. Astarte **3**, 83—85.

NAUD, M., & MAGNAN, P. (1988): Diel onshore-offshore migrations in northern redbelly dace, *Phoxinus eos* (COPE), in relation to prey distribution in a small oligotrophic lake. Can. J. Zool. **66**, 1249—1253.

SARKKULA, J. (1984): Veden virtaukset Säkylän Pyhäjärvessä. (Water currents in Lake Pyhäjärvi.) Lounais-Suomen Vesiensuojeluyhdistys r.y. Julkaisu **60**, 6—8 [in Finnish].

SARVALA, J., AULIO, K., MÖLSÄ, H., RAJASILTA, M., SALO, J., & VUORINEN, I. (1984): Factors behind the exceptionally high fish yield in the Lake Pyhäjärvi, southwestern Finland — hypotheses on the biological regulation of fish production. Aqua Fennica **14**, 49—57.

— RAJASILTA, M., HANGELIN, C., HIRVONEN, A., KIISKILÄ, M., & SAARIKARI, V. (1989): Spring abundance, growth and food of 0+ vendace (*Coregonus albula* L.) and whitefish (*C. lavaretus* L. s.l.) in Lake Pyhäjärvi, SW Finland. Finnish Fish. Res. **9**, 221—233.

SIMBERLOFF, D. (1978): Use of rarefaction and related methods in ecology. In: K. L. DICKINSON, J. CAIRNS, Jr., & R. J. LIVINGSTON (eds.), Biological data in water pollution assessment: quantitative and statistical analyses, ASTM STP 652, pp. 150—165.

STENSON, J. A. E., & OSCARSON, H. G. (1985): Crustacean zooplankton in the acidified Lake Gårdsjön system. Ecol. Bull. **37**, 224—231.

STICH, H. B., & LAMPPERT, W. (1981): Predator evasion as an explanation of diurnal vertical migration by zooplankton. Nature **293**, 396—398.

TESSIER, A. J. (1983): Coherence and horizontal movements of patches of *Holopedium gibberum* (Cladocera). Oeceologia (Berlin) **60**, 71—75.

WERNER, E. E., & HALL, D. J. (1988): Ontogenetic habitat shifts in bluegill: the foraging rate — predation risk trade-off. Ecology **69**, 1352—1366.

WINFIELD, I. J., & TOWNSEND, C. R. (1988): Factors affecting prey selection by young bream *Abramis brama* and roach *Rutilus rutilus*: insights provided by parallel studies in laboratory and field. Env. Biol. Fish. **21**, 279—292.

VUORINEN, I. (1986): Selective planktivory — effect on vertical migration and life-cycle parameters of zooplankton. Finnish Mar. Res. **253**, 3—33.

VØLLESTAD, L. A. (1983): Fordeling, vegst og ernaering til årsyngel av mørt, *Rutilus rutilus* i Årungen. Fauna **36**, 18—24.

Address of the authors:
M. WALLS
M. RAJASILTA
J. SARVALA
J. SALO
Laboratory of Ecological Zoology
Department of Biology
University of Turku
SF-20500 Turku
Finland

Limnologica (Berlin) **20** (2), 259—266, Juni 1990

Freshwater Biological Laboratory, University of Copenhagen, Hillerød, Denmark; National Energy Authority, Reykjavik, Iceland

Ecosystem Studies on Temperate Lake Esrom and the Subarctic Lakes Mývatn and Thingvallavatn[1])

By PÉTUR M. JÓNASSON, CLAUS LINDEGAARD, PETER C. DALL, KIRSTEN HAMBURGER & HÁKON ADALSTEINSSON

With 3 Figures

Key words: Lakes; Ecosystems; Biomass; Production; Trophic levels.

Abstract

Ecosystem studies have been carried out on temperate Lake Esrom, Denmark, and the subarctic Icelandic lakes Mývatn and Thingvallavatn for several years in order to describe the energy pathways. Results of biomass and production estimates of phytoplankton, phytobenthos, zooplankton, zoobenthos and fish are summarized and the main energy pathways are outlined, compared and discussed for the three lakes.

1. Introduction

The Freshwater Biological Laboratory at the University of Copenhagen has carried out research on the Lake Esrom ecosystem for 120 years, and has been engaged during the past 20 years in ecosystem studies in the subarctic, Icelandic lakes Mývatn and Thingvallavatn (JÓNASSON 1972, 1977, 1978, 1979a, 1984; JÓNASSON et al. 1977; JÓNASSON & LINDEGAARD 1988).

These latter investigations have enabled comparisons between the structure and functioning of subarctic lake ecosystems and corresponding energy pathways in the temperate Lake Esrom. The results presented here are a brief review of the three ecosystems in terms of biomasses and turnovers within and between the various trophic levels.

[1] Lake Esrom Littoral Research Project, Publ. No. 22. Publication No. 495 from the Freshwater Biological Laboratory, University of Copenhagen, 51 Helsingørsgade, DK-3400 Hillerød, Denmark.

2. Physical descriptions of the lakes and watersheds

The three lakes are shown in Fig. 1 and details of location, morphometric data and physical characteristics of the waterbodies are given in Table 1.

The Icelandic lakes are situated on the Atlantic ridge. This neovolcanic rift zone has produced in Iceland lakes which are fed almost only by ionic rich ground water coming from the surrounding porous lava fields. The Danish Lake Esrom, on the contrary, is located in morai-

Table 1. Physical characteristics of Lake Esrom. Lake Mývatn and Lake Thingvallavatn. Data from EINARSSON (1979), JÓNASSON (1972, 1984a), JÓNASSON et al. (1974), OLAFSSON (1979a), RIST (1979).

Parameter	Lake Esrom	South basin of Lake Mývatn	Lake Thingvallavatn
Latitude	56°00′N	65°35′N	64°12′N
Longitude	12°23′E	17°00′W	21°10′W
Altitude (m a. s. l.)	9	278	101
Age (years)	10,000	2,300	10,000
Subsoil	moraine	lava	lava
Catchment area (km^2)	76	1680	1000
Lake area (km^2)	17.3	29.1	83.0
Maximum depth (m)	22.3	4.2	114
Mean depth (m)	13.5	2.3	34.1
Volume ($10^6 m^3$)	233	68	2800
Discharge ($m^{-3} \cdot s^{-1}$)	0.4	33	100
Retention time	16 years	24 days	330 days
Insolation ($kcal \cdot m^{-2} \cdot yr^{-1}$)	880,000	774,000	745,000
Mean annual air temperature (°C)	8.5	2.2	3.0
Mean air temperature July (°C)	17.8	14.3	12.5
Maximum water temp. (°C)	22.2	18.1	12.3
Mean water temp. (°C)	8.8	4.7	4.2
Degree days (dd)	3200	1700	1550
Ice cover ($days \cdot yr^{-1}$)	57 (0—112)	189 (143—244)	95 (44—132)
Average precipitation ($mm \cdot yr^{-1}$)	602	433	1400
Stratification period (days)	120	polymictic	80—90
Transparency in summer (m)	2—5	0.5	6—15
Water level fluctution (m)	0.3	0.4	0.5

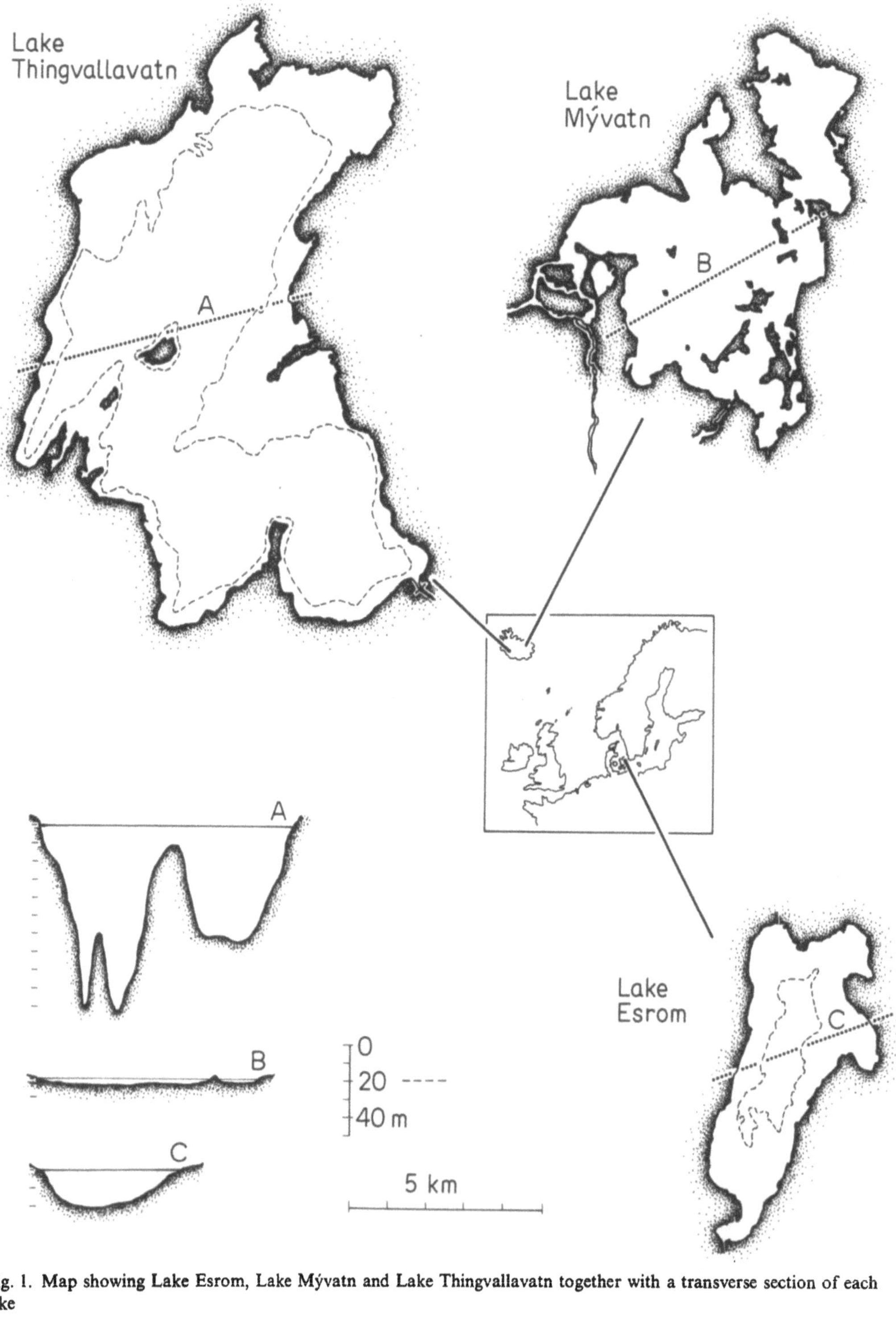

Fig. 1. Map showing Lake Esrom, Lake Mývatn and Lake Thingvallavatn together with a transverse section of each lake

ne surroundings dating back to the time shortly after the latest European glacial period. Its small catchment area (76 km^2), of which more than half is deciduous forest, has protected the lake from severe eutrophication. This, however, also implies that the lake has a retention time of approximate 16 years, compared to which, the Icelandic lakes, and especially the shallow Lake Mývatn (mean depth 2.3 m) have short retention periods of 24 and 330 days for Mývatn and Thingvallavatn, respectively.

The energy input in terms of solar radiation varies only within 15% between the three lakes, but the mean annual temperature regime peaks in Lake Esrom with about 3200 dd (degree days) compared to 1500—1700 dd in case of the subarctic lakes (Table 1).

3. Characteristics of the water quality

It is rather unusual for subarctic or arctic lakes to have water with high ionic content. Lake Mývatn especially, shows conductivity in the range of 145—260 μS · cm^{-2}, values just below the recorded range in Lake Esrom (Table 2). The ionic content of Lake Thingvallavatn is

Table 2. Chemical characteristics of Lake Esrom, Lake Mývatn and Lake Thingvallavatn. Data from JÓNASSON (1972, 1984a), JÓNASSON et al. (1974), LASTEIN (1983), ÓLAFSSON (1979b).

Parameter	Lake Esrom	South basin of Lake Mývatn	Lake Thingvallavatn
pH, winter-summer	8.1—9.2	8.0—10.4	7.0—8.5
Alkalinity (meq $\cdot l^{-1}$)	1.6—2.5	0.9— 1.9	0.5—0.9
Conductivity ($\mu S \cdot cm^{-1}$)	260—320	145—260	65—70
Calcium (mg $Ca^{2+} \cdot l^{-1}$)	50	10	3
Range of ortho-P (μg P $\cdot l^{-1}$)	150—275	2—40	8—24
Range of dissolved inorganic N (μg N $\cdot l^{-1}$)	0—300	8—78	0—36
Range of silicon (μg Si $\cdot l^{-1}$)	100—3000	2200—6300	4400—5900
Input of ortho-P (g P $\cdot m^{-2} \cdot yr^{-1}$)	0.09	1.6	0.9
Input of inorganic N (g N $\cdot m^{-2} \cdot yr^{-1}$)	2.2	0.8	2.4
Input of silicon (g Si $\cdot m^{-2} \cdot yr^{-1}$)	7	240	260
Limiting production factor	Si, N	P in periods	N

65—70 $\mu S \cdot cm^{-2}$, but still high compared to a conductivity of 10 $\mu S \cdot cm^{-2}$ as found in e.g. some Greenlandic (LINDEGAARD 1978) and subarctic Scandinavian lakes (KLOSTER 1978). These differences reflect variations in the specific loadings of calcium in Lake Esrom and of silicon in Lake Mývatn and Lake Thingvallavatn, respectively. Lake Esrom is characterized as a carbonate lake and Lake Mývatn and Lake Thingvallavatn are both softwater lakes. The enormous differences in the content of the nutrients, dissolved inorganic phosphorus and nitrogen, should be emphasized (Table 2).

4. The pelagic producers and consumers

All estimates of biomass and production are given in kcal in order to facilitate comparisons. We assumed that 1 g of Chl **a** equals 30 g C (a very conservative judgement), that carbon makes up 50% of organic matter and, further, that 1 g organic matter measured as dry weight (DW) equals 5 kcal. We further used the following constants in order to describe the zooplankton communities in terms of kcal: 1 g DW = 6 g WW (wet weight), and 1 g DW = 0.9 g AFDW (ash free dry weight).

4.1. Phytoplankton

The chlorophyll **a** content calculated on a volumetric basis for the photosynthetic layer occurs in the ratio 1:10:100 in Lake Thingvallavatn, Lake Mývatn and Lake Esrom, respectively.

The mean annual net phytoplankton production (kcal $\cdot m^{-2} \cdot yr^{-1}$) was estimated at 600, 675 and 1700 in Lake Mývatn for an average of 5 years (1972—1976), Lake Thingvallavatn for an average of 7 years (1975—1982), and Lake Esrom for an average of 15 years (1955—75), respectively (Table 3). The annual range of gross phytoplankton production in Lake Esrom varied in the order of 1600—3100 kcal $\cdot m^{-2}$ during the period 1955—75 (JÓNASSON 1984a), depending primarily on variations in photosynthetic active radiation.

The phytoplankton production curve is bimodal in case of Lake Thingvallavatn and Lake Esrom, whereas in Lake Mývatn it is unimodal showing a peak in gross production during July—August due to nitrogen fixation of blue-green algae (Fig. 2).

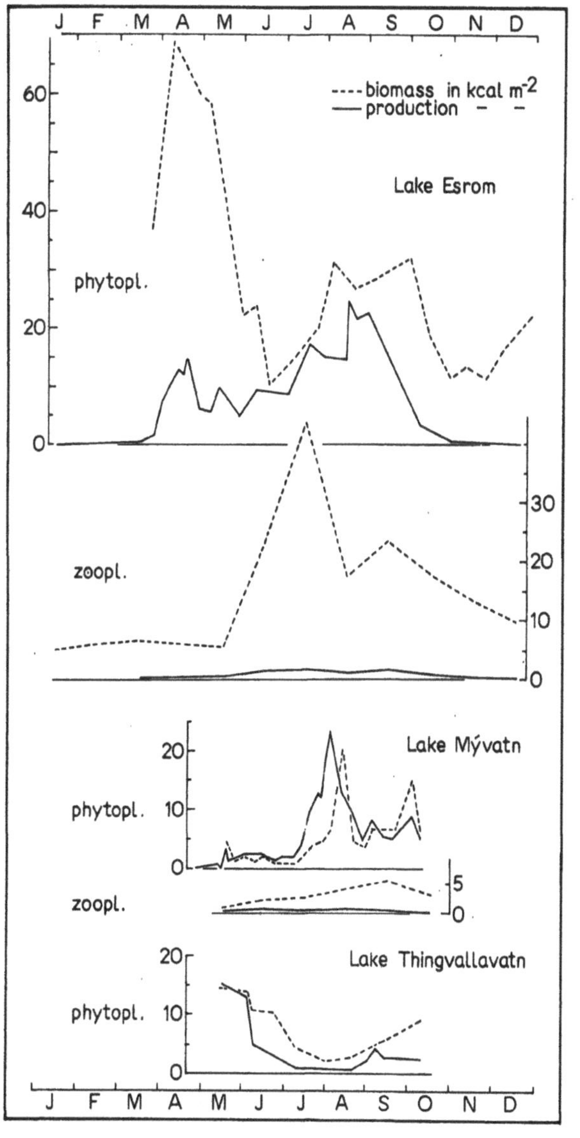

Fig. 2. Annual fluctuation in biomass and production per day of both phytoplankton and zooplankton in Lake Esrom and Lake Mývatn and of phytoplankton alone in Lake Thingvallavatn. Note that the scale is the same for all three lakes

Maximum daily phytoplankton production averaged 15 kcal · m^{-2} in Lake Thingvallavatn, and 25 kcal · m^{-2} in both Lake Mývatn and Lake Esrom. The corresponding photosynthetic depths, estimated as twice the Secchi-depth were about 20, 1, and 5 m, respectively during the periods of maximum primary production.

The differences in net phytoplankton production between the lakes is much more substantial on a volumetric basis (Fig. 3). Net annual phytoplankton production equals 261 kcal · m^{-3} in shallow Lake Mývatn, about 125 kcal · m^{-3} in Lake Esrom and only 20 kcal · m^{-3} in deep Lake Thingvallavatn. Due to the low transparency of 0.5 m in Lake Mývatn during summer, the net production was estimated to be 50% of gross production. The corresponding value in the two other lakes was estimated at 75% (JÓNASSON & LINDEGAARD 1979, 1988).

The composition of the phytoplankton communities shows the same seasonal variations. Diatoms generally dominate during spring and in Lake Mývatn and Lake Esrom blue-green algae increase in biomass and production during mid and late summer. The most striking differences in the successions are partly related to the availability of nutrients. *Melosira islandica* (EHRBG.) O. MÜLL and *M. italica* (EHRBG.) KÜTZ. account for 90% of the plankton in Lake Thingvallavatn from October to June—July. In summer they are reduced to half in the photic layer due to a deficiency of nitrate nitrogen and

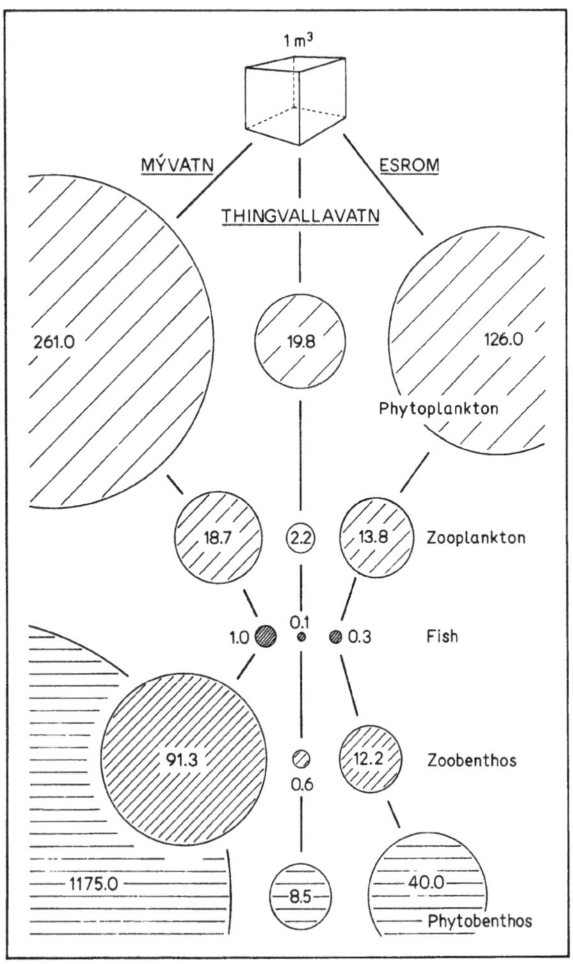

Fig. 3. Diagram showing the annual net production (kcal · m^{-3} · yr^{-1}) of phytoplankton, phytobenthos, zooplankton, zoobenthos and fish in an average cubic meter in Lake Esrom, Lake Mývatn and Lake Thingvallavatn

they occur again at autumn overturn. Small diatoms and Chrysophyceae succeeded the spring maximum and are able to utilize the insignificant amount of nitrate nitrogen during the summer stratification (Fig. 2).

The Lake Mývatn phytoplankton shows no proper bloom of planktonic diatoms during spring. Here the benthic diatoms contribute instead (section 5.1). The peak in biomass and production during July—August is caused mainly by *Anabaena flos-aquae* (LYNGB.) BRÉB. Phosphorus is in short supply during this period (Table 2).

More than 130 phytoplankton species have been recorded in Lake Esrom and their succession described (JÓNASSON & KRISTIANSEN 1967). The seasonal succession of production is well described for 15 years (JÓNASSON 1972, 1984a). The spring bloom of *Asterionella formosa* HABBALL levels off due to shortage of Si and N (JÓNASSON et al. 1974). The late winter and early summer plankton is dominated mostly by green algae until the blue-green algae, with 3—5 common species, build up a bloom during July—August. After the autumn overturn the diatoms dominate again.

4.2. Zooplankton

The zooplankton communities in the three lakes examined differ significantly in both species composition and biomass.

Crustaceans (Cladocera and Copepoda) make up 95—99% of the zooplankton biomass in Lake Esrom (BOSSELMANN 1974). Rotifera peak twice in abundance, during early summer and in autumn and comprise about 5% of total zooplankton biomass (JØRGENSEN 1971; BOSSELMANN 1979a, 1979b). Cladocerans, mainly *Daphnia galeata* SARS, peak in June—July and fade away during autumn (PETERSEN 1983), whereas *Eudiaptomus graciloides* LILLJEBORG and *Cyclops* spp. are present in numbers throughout the year, although showing a prominent peak in June—July. *Eudiaptomus* alone constitutes 92% of the total biomass during January—April (BOSSELMANN 1975a, 1975b).

Zooplankton biomass in Lake Mývatn (Fig. 2) is dominated by herbivorous Rotifera during May—July (c. 60%) and by *Daphnia longispina* (MÜLLER) during August—October (75%). Other frequent species include the omnivorous rotifer *Asplanchna priodonta* GOSSE and the carnivorous *Cyclops* sp. cf. *abyssorum* SARS, both of which are most numerous during summer (ADALSTEINSSON 1979).

Seasonal succession of zooplankton in Lake Thingvallavatn is clear. The plankton is dominated by three species of Crustacea: *Leptodiaptomus minutus* LILLJEBORG, *Cyclops abyssorum* and the most important *Daphnia longispina*. Rotifera are abundant, but their contribution to zooplankton production is small. The estimated production amounted to 100, 43 and 75 kcal · m^{-2} · yr^{-1}, in Lake Esrom, Lake Mývatn and Lake Thingvallavatn, respectively, and thus is in same order of magnitude (Table 3). However, great difference appears, when production is calculated on a volumetric basis (Fig. 3).

4.3. Fish

Fish are the most important carnivores in all three lakes and absolutely dominant in Lake Esrom and Lake Thingvallavatn. But, in Lake Mývatn waterfowl play an important role as top carnivores, since it is the most

Table 3. Net production estimates of Lake Esrom, Lake Mývatn and Lake Thingvallavatn. Values are in kcal · m^{-2} · yr^{-1}. Contributions of the different components are given for an average lake m^{-2}.

Parameter	Lake Esrom	South basin of Lake Mývatn	Lake Thingvallavatn
Primary production of			
phytoplankton	1700	600	675
phytobenthos	350	2700	290
macrophytes + epiphytes	190	500	100 (*Nitella*)
Secondary production of			
zooplankton	186	43	75
littoral zoobenthos	101	—	16
profundal zoobenthos	64*	210	3
fish	4	3	3

* This value is probably reduced to 6 due to a recent decline in profundal fauna.

unique duck lake on the Northern hemisphere (Jónasson 1979b).

Production of fish is calculated from the catch by assuming that the total biomass is five times the average catch and the P/B ratio is 0.5 (Winberg et al. 1972).

Lake Esrom has 8 species of fish. The dominant species is the eel, *Anguilla anguilla* L., which feeds mainly on chironomids (*Chironomus anthracinus* Zett.) during spring and autumn. Other benthic feeders are bream, *Abramis brama* L., tench, *Tinca tinca* L. and pope, *Acerina cernua* L.. The latter is in fact a food competitor to the eel. The plankton feeders are bleak, *Alburnus alburnus* L. and partly perch, *Perca fluviatilis* L.; roach, *Rutilus rutilus* L. feeds both on plants and animals. Pike, *Esox lucius* L. is the piscivore in this community, together with perch. The total average annual catch was 23.7 kg wet weight · ha^{-1} · yr^{-1}. This is a medium harvest for a eutrophic lake. The production was calculated at 4 kcal · · m^{-2} · yr^{-1} (Jónasson 1984a).

Three species of fish occur in Lake Mývatn. Most important is the Arctic charr, *Salvelinus alpinus* L. It has been subject to fishing through centuries. The trout, *Salmo trutta* L., accounted for 14 and 11 % of the total fish catch in 1943—1944, and these figures are apparently close to the mean. The three-spined stickleback, *Gastèrosteus aculeatus* L., is very common, but of no commercial value. The total average production of fish in Lake Mývatn was estimated at c. 3 kcal · m^{-2} · yr^{-1} (Jónasson 1979b).

Lake Thingvallavatn has the same three species as Lake Mývatn, but the proportions are different. Previously the brown trout made up 8—10 % of total catch, but a power plant dam destroyed the main spawning grounds and since then the brown trout has gradually disappeared.

Thingvallavatn is the only lake known to have four different forms of Arctic charr, which fit into different ecological niches: a small pelagic charr feeding on zooplankton, a benthic charr feeding on zoobenthos, a dwarf charr feeding on zoobenthos under stones and in caves, and finally a big piscivorous charr feeding on the pelagic charr and the three-spined stickleback. Yearly catch of Arctic charr in Lake Thingvallavatn is c. 10 kg wet weight · ha^{-1} · yr^{-1} and net production of fish was estimated at 3 kcal · m^{-2} · yr^{-1}.

5. The benthic producers and consumers

The overall estimates of annual production of benthic communities are summarized in Table 3. If we compare the primary producers in each lake with the estimates of total net secondary production per square meter (excluding fish), the result shows rather equal net efficiencies at 7—9 % in the lakes. These reasonable food chain efficiencies, however, cover considerable variations within each of the benthic trophic levels.

5.1. Phytobenthos and macrophytes

The phytobenthos production has been measured on sections of sandy lake shores in lakes similar to Lake Esrom (Hunding & Hargrave 1973), and production of the epilithic communities has been examined in the shallow littoral zone of Lake Esrom (Dall et al. 1984b). Sessile diatoms dominate among tufts of *Cladophora* sp. Total phytobenthos production contributes with 350 kcal · m^{-2} · yr^{-1} for an average square meter of lake (Table 3).

Phytobenthic production of diatoms in Lake Mývatn amounts to 2200 kcal · m^{-2} · yr^{-1} (Olafsson 1979b), while mats of *Cladophora aegagropila* (L.) Rabenh. contribute with an annual production of 500 kcal · m^{-2} (Hunding 1979). The benthic diatoms (*Fragilaria* spp.) are utilized industrially in parts of Lake Mývatn.

Algal biomass is high on the stony substratum in Lake Thingvallavatn, where carpets of, especially *Cladophora glomerata* (L.) Kütz. and *C. aegagropila* down to 10 m depth (Jónsson 1987), contribute to the net production of epilithic algae of c. 1400 kcal · m^{-2} · yr^{-1}, and 290 kcal · m^{-2} · yr^{-1} for an average square meter of lake (Table 3).

The distribution of benthic macrophytes is restricted to depths with sufficient light penetration. Lake Esrom shows a macrophyte community typical of mesotrophic, temperate lakes. The maximum depth of rooted plants is 6—8 m and among species present are 8 species of *Potamogeton*, *Elodea canadensis* Rich., *Ceratophyllum demersum* L., *Myriophyllum spicatum* L., *Chara fragilis* Desvaux, and *C. contraria* Kütz. Reed swamps occur at intervals along the shoreline, but are of minor importance. Total production per average square meter of lake has been estimated at 190 kcal · m^{-2} · yr^{-1} (Table 3).

The macrophyte *Myriophyllum* sp. accounts for a considerable production in the N-basin of Lake Mývatn (Jónasson 1979b; Jónasson & Lindegaard 1988), and its contribution to primary production in the S-basin is estimated to be c. 500 kcal · m^{-2} · yr^{-1}.

In Lake Thingvallavatn the dominating macrophyte is *Nitella opaca* Ag., which builds up a dense vegetation bed from 10 to 25 m and has a net production of c. 500 kcal · m^{-2} · yr^{-1}, which contributes 100 kcal · m^{-2} · yr^{-1} to an average square meter of lake (Table 3).

5.2. Zoobenthos

Large differences between profundal and littoral fauna were found in the dimictic Lake Esrom and Lake Thingvallavatn. Both lakes had a littoral fauna rich in species (c. 40 taxa in Lake Thingvallavatn and c. 100 in Lake Esrom), whereas the profundal fauna was dominated by few species well adapted to either low oxygen concentrations (Lake Esrom) or low energy content in food (Lake Thingvallavatn). In the shallow Lake Mývatn the "profundal" fauna was made up of about 20 taxa characteristic of 95% of the lake area.

Production of the profundal fauna in Lake Esrom during 1954—1960 was about 83 and 17 kcal · m^{-2} · yr^{-1} for the detritivorous and carnivorous species, respectively (Jónasson 1972). During the last 20 years the profundal fauna (especially *Chironomus anthracinus*) was dramatically reduced due to a prolonged anoxic period, and today produces only about 10 kcal · m^{-2} · yr^{-1} (Jónasson 1984b). The profundal zone comprises 64% of the lake area and hence the profundal fauna contributes 64 kcal · m^{-2} · yr^{-1} for an average square meter of lake. This value is probably reduced to 6 due to the recent decline in profundal fauna (Table 3).

The production of several littoral fauna groups in lake Esrom has been calculated: the trichopteran *Tinodes waeneri* L. (Dall et al. 1984a), Chironomidae (Jónsson 1985), and Hirudinea (Dall 1987). The total littoral zoobenthos production was estimated to about 125 kcal · m^{-2} · yr^{-1} by Dall et al. (1984b), but a more recent estimate based on all available results gives 280 kcal · m^{-2} · · yr^{-1}, which is equivalent to 101 kcal · m^{-2} · yr^{-1} for an average square meter of lake (Table 3).

In Lake Mývatn a close relationship between zoobenthos, the mud and *Cladophora* substrate was found. The "profundal" fauna — dominated by the chironomids *Chironomus islandicus* Kieffer and *Tanytarsus gracilentus* (Holmgren) — produced about 200 kcal · m^{-2} · yr^{-1} (Lindegaard & Jónasson 1979).

Zoobenthos showed a clear-cut vertical distribution in Lake Thingvallavatn (Lindegaard 1980). Most interesting is a special Icelandic surf zone community of midges which yield two generations a year and thus a large production. The *Cladophora* fields below 2 m depth feed a large stock of the snail *Lymnaea pereger* L. It has a generation time of two years. *L. pereger* supplies the main food for bottom feeding charr (Snorrason 1982). The enormous *Nitella opaca* belt serves as a sieve for particles, and here *Chironomus islandicus* lives in its 10 cm long tubes as a filtrator between 10 and 25 m depth, and requires 2 years for its development. It has the same lifecycle in Lake Mývatn, but in the shallow lake it is spread all over the bottom, feeding on seston which is available here due to resuspension and sedimentation. In the profundal zone of Lake Thingvallavatn a number of chironomid species are still present, but the fauna is numerically dominated by *Tubifex tubifex* L., which is able to utilize the very low energy content of the profundal sediment.

Secondary benthos production in Lake Thingvallavatn is estimated at 46 kcal · m^{-2} · yr^{-1} in the surf zone, increasing to approximately 75 kcal · m^{-2} · yr^{-1} at 4 m depth, and then decreasing to c. 10 kcal · m^{-2} · yr^{-1} in the *Nitella* zone and to 5 kcal · m^{-2} · yr^{-1} in the profundal zone. On average the littoral and profundal zoobenthos contribute 16 and 3 kcal · m^{-2} · yr^{-1}, respectively, for an average square meter of lake (Table 3).

6. The pathways of energy

Basic evidence for energy flow in lake ecosystems is knowledge of three main factors: edaphic, morphometric and climatic characters of the lake. The edaphic factors determine the nutrient budget of lakes (Likens & Borman 1979). The morphometry of lakes is of importance for production, and for the fate of organic matter in the lake basin. Lakes become polymictic in shallow basins. In deeper basins the lakes are dimictic and a differentiation into a littoral, sublittoral and a profundal zone occurs.

In the case of Lake Mývatn the main energy flow is channelled through the phytobenthos which accounts for 85% of total net primary production (Table 3). This value is similar to that of clear Arctic lakes (Alexander & Barsdate 1971; Hobbie 1973; Kalff & Welch 1974; Rigler et al. 1974; Welch & Kalff 1974; Rigler 1978), where phytobenthos accounts for 80% of total primary production. However, there is little doubt that the share of phytobenthos production in Lake Mývatn is underestimated. The various categories of phytobenthos producers differ greatly in importance as substrate for zoobenthos. With *Cladophora* covering the bottom, the substrate is stabilized, and numerous microhabitats are available for epiphytes and micro- and meiobenthos (cf. Hunding 1979). The vegetation acts as a sieve for resuspended organic matter and plankton, utilized by the benthic filterfeeding herbivores and detritivores. Of the total net primary production 6% is channelled through the zoobenthos in Lake Mývatn.

In shallow, polymictic Lake Mývatn the phytoplankton curve is unimodal, and production is high due to N-fixation. The grazers of the summer maximum are partly the pelagic *Daphnia longispina*, and partly the littoral filtrator, *Eurycercus lamellatus* (Müller), which both are food of the Arctic charr, when the midges have emerged. However, basic for the rich life in Lake Mývatn is the phyto- and zoobenthos, each of which accounts for c. 85% of production in their respective trophic level (Table 3) and therefore are responsible for the rich duck life and the great population of Arctic charr.

In the temperate Lake Esrom with a maximum depth of 22.3 m (mean 13.5 m) it is interesting to note that net production of phytoplankton made up 78% of total primary production, and that production of zoobenthos was equal to that of zooplankton. Thus, a similar proportion of the energy passed through the benthic communities as through the zooplankton (Table 3). The explanation seems to be that the high biomass of phytoplankton in the photosynthetic layer, 10 times as high as that of Lake Myvatn and 100 times that of Lake Thingvallavatn, sediments much faster in Lake Esrom than in Lake Thingvallavatn. In Lake Thingvallavatn sedimentation rate is approximate 60 days for chlorophyll **a** derivates. Due to shorter sedimentation time, approximate 7—10 days, the organic content of the Lake Esrom profundal mud is high, i.e. loss on ignition is up to 37%, but only 15—20% in Lake Thingvallavatn. The high organic content of Lake Esrom mud thus makes it feasible for zoobenthos to occur in great quantities. The high production of zoobenthos also caused a high production of benthic fish, especially the eel, which is also well adapted to low contents of oxygen. The commercial catch data definitely support the view that the benthic fish are dominant in the lake. The morphometry of Lake Esrom — the narrow littoral as compared with the extensive profundal zone — determines the species composition of fish in the lake. During the long summer stratification the benthic feeders have had to move from the profundal and feed in the littoral due to oxygen depletion and to change their feeding behaviour from one ecological pathway to another. But, this is not so any longer, because due to eutrophication, zoobenthos is heavily reduced today in the profundal zone, and has lowered the zoobenthic production to about 45% of total secondary production. Very likely, this will affect the fish stock in the future.

In deep Lake Thingvallavatn (max. depth 114 m, mean depth 34 m) pelagic production is bimodal and dominant (Fig. 2, Table 3). It is evident that the main energy flow passes through the pelagic food web: phytoplankton → zooplankton → pelagic Arctic charr. The proof is that the main harvest of the lake is the pelagic Arctic charr. The carbon dynamics of Lake Thingvallavatn show high utilization of carbon in the pelagic. This is due to a slow sedimentation rate for phytoplankton in a windy and partially stratified lake (LASTEIN 1983). However, it is clear that there is another ecological pathway, which is completely separate from the first one, and that is the phytobenthos-zoobenthos community, especially the zone from 0 to 10 m and to a lesser extent the *Nitella* zone between 10 and 25 m. The food relationship between the zoobenthos and the large benthic charr is evident (MALMQUIST 1988) and realized through centuries by local fishermen.

A comparison of energy pathways in these three lakes shows how tightly the different ecological levels are connected to each other. Fig. 2 shows how phyto- and zooplankton seasonal trends are related to each other in every lake and also that biomass trends of phytoplankton are followed closely by corresponding changes in quantities of zooplankton e.g. from the lowest nutrient status in Lake Thingvallavatn to the highest in Lake Esrom.

Lake Mývatn shows the highest biomass of zoobenthos, one of the highest ever described (LINDEGAARD 1989), and it supports enormous numbers of duck and charr. The deeper the lake the lower the quantities of zoobenthos, because the nutrient value of sediments declines with increasing depth, once a lake is deep enough to prevent frequent upmixing of the surface sediment. In Lake Mývatn the critical depth seems to be c. 2 m, because in the 1—1.5 m deep N-basin, the calorific value of sediment is reduced to less than half found in the deeper S-basin. Lake Mývatn sediment approaches the calorific value of highly eutrophic lakes (JÓNASSON 1979b), while that of Lake Thingvallavatn is very low (LASTEIN 1983). Taking into consideration the previous characteristics of Lake Mývatn phytobenthos being a sieve, it explains the existence of substrate feeding or filtrating chironomids.

The high production of fish in Lake Esrom as compared to the others seems to be explained by the fact that great amounts of energy were channelled through littoral and profundal zoobenthos. Further it should be taken into account that activity and the production periods of fish are considerably longer in Lake Esrom with an ice-cover period of only 57 days as compared to 95 and 190 in the subarctic lakes.

7. Acknowledgements

The authors wish to thank Prof. T. B. REYNOLDSON for valuable suggestions and linguistic correction of the manuscript. We further thank Ms. H. HEEGAARD and Mr. F. PETERSEN for technical assistance.

References

ADALSTEINSSON, H. (1979): Zooplankton and its relation to available food in Lake Mývatn. Oikos **32**, 162—195.

ALEXANDER, V., & BARSDATE, R. J. (1971): Physical limnology, chemistry and plant productivity of a Taiga lake. Int. Rev. ges. Hydrobiol. **56**, 825—872.

BOSSELMANN, S. (1974): The crustacean plankton of Lake Esrom. Arch Hydrobiol. **74**, 18—31.

— (1975a): Population dynamics of *Eudiaptomus graciloides* in Lake Esrom. Arch. Hydrobiol. **75**, 329—346.

— (1975b): Production of *Eudiaptomus graciloides* in Lake Esrom 1970. Arch. Hydrobiol. **76**, 43—64.

— (1979a): Population dynamics of *Keratella cochlearis* in Lake Esrom. Arch. Hydrobiol. **87**, 152—165.

— (1979b): Production of *Keratella cochlearis* in Lake Esrom. Arch. Hydrobiol. **87**, 304—313.

DALL, P. C. (1987): The ecology of the littoral leech fauna (Hirudinea) in Lake Esrom, Denmark. Arch. Hydrobiol. **110**, 259—274.

— HEEGAARD, H., & FULLERTON, A. F. (1984a): Life-history strategies and production of *Tinodes waeneri* (L.) (Trichoptera) in Lake Esrom, Denmark. Hydrobiologia **112**, 93—104.

— LINDEGAARD, C., JÓNSSON, E., JÓNSSON, G., & JÓNASSON, P. M. (1984b): Invertebrate communities and their environment in the exposed zone of Lake Esrom, Denmark. Arch. Hydrobiol. Suppl. **69**, 477—524.

EINARSSON, M. A. (1979): Climatic conditions of the Lake Mývatn area. Oikos **32**, 29—38.

HOBBIE, J. E. (1973): Arctic Limnology: A review. In: BRITTON, M. E. (ed.), Alaskan arctic tundra, pp. 127—168. Tech. paper no. 25. Arctic Inst. of N. Amer.

HUNDING, C. (1979): Oxygen balance of Lake Mývatn, Iceland. Oikos **32**, 139—150.

— & HARGRAVE, B. T. (1973): A comparison of benthic microalgal production measured by C^{14} and oxygen methods. J. Fish. Res. Board Can. **30**, 309—312.

JÓNASSON, P. M. (1972): Ecology and production of the profundal benthos in relation to phytoplankton in Lake Esrom. Oikos Suppl. **14**, 1—148.

— (1977): Lake Esrom research 1867—1977. Folia Limnol. Scand. **17**, 67—89.

— (1978): Zoobenthos of lakes. Verh. Int. Verein. Limnol. **20**, 13—37.

— (ed.) (1979a): Ecology of eutrophic, subarctic Lake Mývatn and the River Laxá. Oikos **32**, 1—308.

— (1979b): The Lake Mývatn ecosystem, Iceland. Oikos **32**, 289—305.

— (1984a): The ecosystem of eutrophic Lake Esrom. Ecosystems of the World **23**, 177—204.

— (1984b): Oxygen demand and long term changes of profundal zoobenthos. Hydrobiologia **115**, 121—126.

— & KRISTIANSEN, J. (1967): Primary and secondary production in Lake Esrom. Growth of *Chironomus anthracinus* in relation to seasonal cycles of phytoplankton and dissolved oxygen. Int. Rev. ges. Hydrobiol. **52**, 163—217.

— LASTEIN, E., & REBSDORF, A. (1974): Production, insolation, and nutrient budget of eutrophic Lake Esrom. Oikos **25**, 255—277.

— ADALSTEINSSON, H., HUNDING, C., LINDEGAARD, C., & OLAFSSON, J. (1977): Limnology of Iceland. Folia Limnol. Scand. **17**, 111—123.

— & LINDEGAARD, C. (1979): Zoobenthos and its contribution to metabolism of shallow lakes. Arch. Hydrobiol. Beih. Ergebn. Limn. **13**, 162—180.

— (1988): Ecosystem studies of North Atlantic Ridge lakes. Verh. Int. Verein. Limnol. **23**, 394—402.

JÓNSSON, E. (1985): Population dynamics and production of Chironomidae (Diptera) at 2 m depth in Lake Esrom, Denmark. Arch. Hydrobiol. Suppl. **70**, 239—278.

JÓNSSON, G. S. (1987): The depth-distribution and biomass of epilithic periphyton in Lake Thingvallavatn, Iceland. Arch. Hydrobiol. **108**, 531—547.

JØRGENSEN, J. (1971): En kvantitativ undersøgelse af zooplanktonet i Esrom Sø med særligt henblik på Rotatoria. MS. thesis, University of Copenhagen. 22 pp.

Kalff, J., & Welch, H. E. (1974): Phytoplankton production in Char lake, a natural polar lake, and in Meretta Lake, a polluted polar lake, Corvallis Island, Northwest Territories. J. Fish. Res. Bd. Can. **31**, 621—636.

Kloster, A. E. (1978): Physical and chemical properties of the waters of Øvre Heimdalen. Holarct. Ecol. **1**, 117—123.

Lastein, E. (1983): Decomposition and sedimentation processes in oligotrophic, subarctic Lake Thingvallavatn, Iceland. Oikos **40**, 103—112.

Likens, G. E., & Borman, F. H. (1979): The role of watershed and airshed in lake metabolism. Arch. Hydrobiol. Beih. Ergebn. Limnol. **13**, 195—211.

Lindegaard, C. (1980): Bathymetric distribution of Chironomidae (Diptera) in the oligotrophic Lake Thingvallavatn, Iceland. In: Murray, D. A. (ed.), Chironomidae, ecology, systematics, cytology and physiology, pp. 225—237. Oxford and New York.

— (1989): Secondary production of zoobenthos in freshwater ecosystems. A review with special reference to Chironomidae (Diptera). Acta Biol. Debr. Oecol. Hung. **3**, 231—240.

— Mæhl, P., Nielsen, B. H. (1979): Zoobenthos of lakes located within and outside the Ilímaussaq intrusion in South Greenland. Verh. Int. Verein. Limnol. **20**, 159—164.

— & Jónasson, P. M. (1979): Abundance, population dynamics and production of zoobenthos in Lake Mývatn, Iceland. Oikos **32**, 202—227.

Malmquist, H. J. (1988): Diet differentiation in polymorphic Arctic charr (*Salvelinus alpinus* (L.)) in Thingvallavatn, Iceland: Diel and seasonal foraging activity. MS. thesis, University of Copenhagen. 51 pp.

Ólafsson, J. (1979a): Physical characteristics of Lake Mývatn and River Laxá. Oikos **32**, 38—66.

— (1979b): The chemistry of Lake Mývatn and River Laxá. Oikos **32**, 82—112.

Petersen, F. (1983): Population dynamics and production of *Daphnia galeata* (Crustacea, Cladocera) in Lake Esrom. Holarct. Ecol. **6**, 285—294.

Rigler, F. H. (1978): Limnology in the high Arctic: a case study of Char Lake. Verh. Int. Verein. Limnol. **20**, 127—140.

— MacCallum, M. E., & Roff, J. C. (1974): Production of zooplankton in Char Lake. J. Fish. Res. Bd. Can. **31**, 637—646.

Rist, S. (1979): Water level fluctuations and ice cover of Lake Mývatn. Oikos **32**, 67—81.

Snorrason, S. (1982): The littoral ecosystem and the ecology of *Lymnaea peregra* in Lake Thingvallavatn, Iceland. Ph. D. thesis, University of Liverpool. 429 pp.

Welch, H. E., & Kalff, J. (1974): Benthic photosynthesis and respiration in Char Lake. J. Fish. Res. Bd. Can. **31**, 609—620.

Winberg, G. G., Babitsky, V. A., Gavrilov, S. I., Gladky, G. V., Zakharenkov, I. S., Kovalevskaya, R. Z., Mikheeva, T. M., Nevyadomskaya, P. S., Ostapenya, A. P., Petrovich, P. G., Potaenko, J. S., & Yakushko, O. F. (1972): Biological productivity of different types of lakes. In: Z. Kajak & A. Hillbricht-Ilkowska (eds.), Productivity problems of freshwaters, pp. 383—404. Krakóv.

Addresses of the authors:

Pétur M. Jónasson
Claus Lindegaard
Peter C. Dall
Kirsten Hamburger
Freshwater Biological Laboratory
University of Copenhagen
51 Helsingørsgade
DK-3400 Hillerød
Denmark

Hákon Adalsteinsson
National Energy Authority
Grensásvegur 9
108 Reykjavik
Iceland

Limnologica (Berlin) **20** (2), 267—271, Juni 1990

Department of Zoology, University of Göteborg, Göteborg, Sweden

Interspecific and Intertrophic Relations in Oligotrophic Water Systems

By JAN A. E. STENSON

With one Figure

Key words: Predator-prey relations; Trophic level interactions; Fish; Invertebrate predators; Zooplankton; Phytoplankton; Water quality.

Abstract

This paper summarizes some of the work on biotic regulation of structure and function of a lake ecosystem performed at the Department of Zoology at Göteborg University. The study, which was experimental in nature, included competitive and predator-prey interactions on the uppermost trophic level, feed-back effects on lower levels, i.e. on grazers and primary producers, and finally interactions between the top levels and abiotic water properties. Fish and invertebrate predators have a significant, specific and predictable impact on the biotic structure, e.g. species and size distribution. Structure and function are intimately connected. Finally there is a clear feed-back influence by biota on the physical and chemical water quality.

Introduction

For many scientists, the research of HRBÁČEK (1958), HRBÁČEK et al. (1961) and BROOKS & DODSON (1965) necessitated a re-evaluation of the accepted knowledge about causal relationships in freshwater ecosystems. The old hierarchical view, with a regulation of structures and processes from below, started to give way for a new view which included pathways of regulation downwards from the uppermost trophic levels. This new information, at the same time both provoking and inspiring, induced a wave of research on biotic interactions in aquatic ecosystems.

At my Institute this work started in the late sixties. The work concentrated, in the beginning, on the study of feed-back mechanisms within and between different trophic levels, with the purpose to estimate the importance of these mechanisms for the structural and functional organisation of lake ecosystems. This entailed studies of competition and predator-prey interactions between groups at the top levels (fish, waterfowl, amphibians); effects of their activities on lower levels, e.g. zooplankton, benthos; effects of grazing on primary producers and production; and finally connections between the top levels and basic abiotic factors commonly used as water quality criteria. More recently other aspects of predator-prey interactions have been included, especially those concerning evolution of anti-predator defense in freshwater invertebrates (STENSON 1981, 1987). Moreover, the duality in influence by grazing zooplankton on phytoplankton, i.e. grazing *per se* and nutrient liberation, is presently being studied in enclosure experiments (ANDERSSON et al. 1988).

An important ultimate objective of the research was to relate theoretical knowledge about regulating mechanisms to practical usage in applied lake management.

Study strategy

The experimental design of the work was based on empirical evidence, from our research group and the literature, of the ability of fish to reduce mean body size and shift species composition of zooplankton communities (e.g. HRBÁČEK 1959; BROOKS & DODSON 1965; STENSON 1972). As fish were shown to prey in a selective manner, with a preference for larger, more conspicuous prey, the effects should be predictable. Thus, by using the fish population as an experimental variable and manipulating its population density and thereby effective predatory impact, it should be possible to study the complex of problems mentioned above.

The work has mainly been carried out in Lake Lilla Stockelidsvatten, an oligotrophic to mesotrophic lake in Bohuslän, about 70 km north of Göteborg, SW Sweden. Several other lakes in the same area have been used as control lakes or for other experimental studies. Detailed information about the lakes is available elsewhere (STENSON 1972, 1976, 1979; STENSON et al. 1978).

The fish population in Lake Lilla Stockelidsvatten (LS in the following text), originally dominated by roach (*Rutilus rutilus* (L.)), was eliminated by means of rotenone treatment in November of 1973. The sequence of changes initiated by this manipulation was assessed in a sampling program including a large number of abiotic and biotic parameters (STENSON et al. 1978). Additional experiments were performed in order to understand, explain and estimate the significance of certain mechanisms. The ultimate goal was to answer a set of basic questions pertaining to regulation and control of plant and animal populations and of functional relationships in freshwater ecosystems.

Results and discussion

The relatively stable situation in the control lakes during the experimental period indicated there was no reason to believe that large scale environmental changes were responsible for the observed changes in the experimental lake. The elimination of the fish population, therefore was the most likely initiator of the sequence of changes in the experimental lake (STENSON et al. 1978; HENRIKSON et al. 1980).

The experimental effects on zooplankton, phytoplankton, primary production and water quality can be described as follows. The zooplankton changed from a community predominated by small cladocerans (e.g. *Bosmina longirostris*) and suspension-feeding rotifers (*Keratella cochlearis*) into a community dominated by larger copepods (e.g. *Eudiaptomus gracilis*) and raptorial rotifers (e.g. *Ascomorpha* spp. and *Asplanchna priodonta*). The biomass and mean body size in both crustaceans and rotifers have increased after the experimental measure (Stenson et al. 1978; Stenson 1982, 1984).

As said above fish predation is the probable keyfactor. Since selective predation by fish has been shown to reduce the mean body size among crustaceans it is evident that the removal of fish would give an opposite response, i.e. an increase in mean size. However, the fact that predation by fish ceased does not explain the drop in numbers of smaller grazers. Instead, a better filtering capacity of larger zooplankters may make these animals more competitively efficient, thus having a negative influence on the relative fitness of smaller species (Dodson 1974). There are also other probable reasons for this development.

Invertebrate zooplanktivores (like *Chaoborus* spp. and the large rotifer *Asplanchna priodonta*) became abundant after the removal of the fish population (Stenson 1978, 1981, 1982). These species are very potent predators, causing significant mortalities in populations of smaller zooplankton, thereby contributing to the size change in zooplankton. The ability of certain invertebrate predators to change their prey environment and to maintain a changed prey structure has been tested and shown to be significant in separate experiments (Nyman et al. 1985; Stenson 1985a; Neill 1985; Stenson, mscr.). Thus, species and size composition of the new pelagic grazer community is probably a consequence of the shift from fish to invertebrate top predators.

The phytoplankton community went through a complete reorganisation during the experiment. The fraction of smaller algae (< 25 μm) decreased, and the dominating species, the desmid *Cosmocladium perissum*, disappeared. Instead there was an addition of algae at the large end of the size spectrum, e.g. *Ceratium hirundinella*. The chlorophyll **a** content of the water decreased significantly, and the limnetic primary production stabilized at a level almost 90% lower than that before fish removal (Stenson et al. 1978; Henrikson et al. 1980; Stenson 1988). Cell size seems to be crucial for productivity. Small cells with a low carbon content tend to have shorter renewal times than larger cells with higher carbon content (Stull et al. 1973; Desortová 1976), a fact which might contribute to the explanation of the drop in primary production of the system.

The change of the phytoplankton assemblage and its production can be attributed to increased grazing pressure from a larger biomass of larger herbivorous zooplankters on the smaller size fraction of algae, leading to a competitive release of larger algae, lowered rates of remineralization via zooplankton — as this process is inversely related to size (Esjmont-Karabin 1984) — and decreased import of nutrients from areas outside the pelagic zone occurring after the fish removal (Andersson et al. 1988). Drainage of nutrients from the pelagic zone is also a result of sinking of fecal material. Copepods produce fecal pellets which sink faster and deeper before the nutrients are released. Cladocerans, on the other hand, produce fecal material which is more quickly regenerated and thus more directly usable for the phytoplankton (Ferrante & Parker 1977; Lehman 1980).

Reasons for the increase of larger algae in Lake LS may be discussed partly along the same lines as conditions for bluegreen algal blooms. In the latter case, large colonial forms appear to be released from competition with green algae and diatoms when large efficient grazers like daphnids are released from intense predation by fish (Shapiro 1980). Thus, neither large dinoflagellates nor blue-greens are effectively grazed by herbivores why they take up nutrients otherwise utilized by their heavily grazed competitors, i.e. smaller plankton algae (Kitchell et al., in press).

The apparent changes in certain abiotic factors can be tied to changes in biota. As the pH in the lake is mainly a function of the carbonic acid system, changes in the carbon dioxide content will influence the equilibrium reactions between the components of the system, thereby changing pH. The decrease in photosynthesis activity implies reduced consumption of carbon dioxide, which dampens the pH fluctuation, resulting in a lower pH level. The transparency increase is a function of reduced particulate matter. The drop in abundance of particles in turn results in a drop in nutrient content of the water. Thus, the reduction of fish changed the lake in an oligotrophic direction (Henrikson et al. 1980).

Very obvious experimental effects occurred among the larger invertebrates. The susceptibility of a prey depends upon the ease with which it can be discovered and captured by the predator. As fish are potent visual predators, large animals, strongly pigmented animals, animals active at daylight and animals which occupy open areas are most susceptible, and should benefit most from the reduction of fish. This was shown to be true. The invertebrate species that increased after the elimination of fish have these properties. Species like *Chaoborus obscuripes* and *Glaenocorisa propingua* are large, relatively darkly pigmented, and active during day in pelagic areas (Henrikson & Oscarson 1978; Stenson 1978). Larvae of the odonate *Leucorrhinia* sp. are also more active during day and less inclined to hide in the bottom sediment than are other species which coexist with fish, etc. (Nilsson 1981; Henrikson 1988).

The experimental manipulations clearly showed the relationship between the fish population and other vertebrate inhabitants or temporary users of the lake ecosystem. Ducklings of mallard (*Anas platyrhynchos*) and ducklings and adults of goldeneye (*Bucephala clangula*) use large invertebrates as food. As fish also feed on the same groups and as fish are more efficient competitors these prey categories will be reduced to levels not exploitable for ducks in lakes with dense fish populations. The removal of fish from Lake LS led to an increase of potential prey which in turn attracted goldeneye ducks to breed in the lake (Eriksson 1976, 1979, 1983; Pehrsson 1979).

Newts (*Triturus* spp.) were not observed in Lake LS before the fish removal. Adult *Triturus vulgaris* were discovered in Lake LS the first time in 1975, 18 months after fish were removed, and the same year newt larvae were found in bottom fauna samples. The population increased to a maximum of approx. 10.000 individuals per ha by 1980. The exclusion of salamander populations from lakes with dense fish stocks probably is a result of predation by fish on the larvae (Aronsson & Stenson, mskr.).

A first test of the validity of our interpretation of the results was the restoration of the fish population in Lake LS. If changes brought about by the elimination of fish were random the accuracy in the predictions about the development after restocking with roach would be low. However, this was not the case. The development followed the predictions concerning structural adaptation to the new top predators, structural feed-back on functional properties, e.g. production etc.

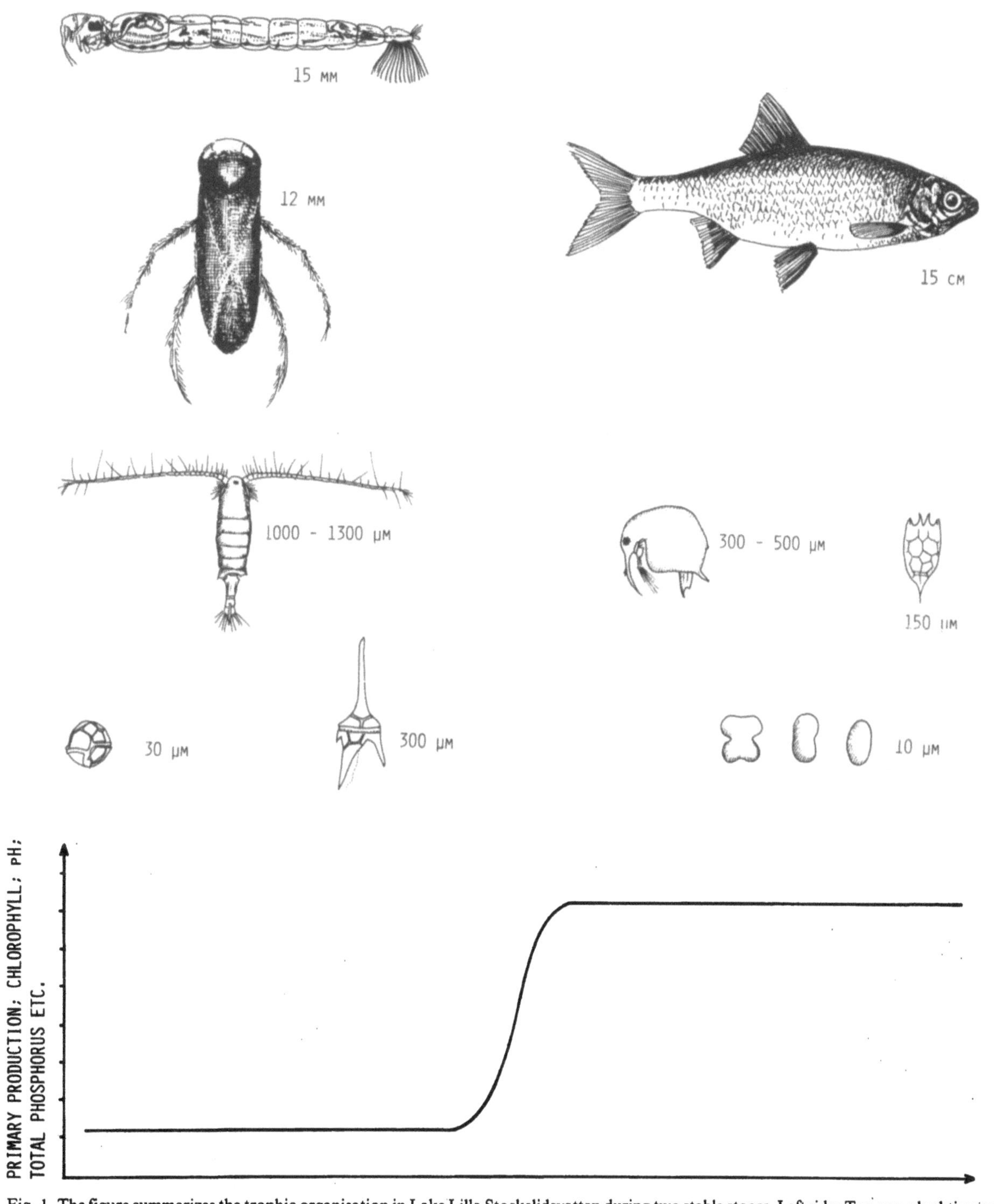

Fig. 1. The figure summarizes the trophic organisation in Lake Lilla Stockelidsvatten during two stable stages. Left side: Top zoo-planktivores are *Chaoborus* spp. and *Glaenocorisa propingua*; dominating zooplankter is *Eudiaptomus gracilis*; dominating phytoplankters are large forms like *Peridinium aciculiferum* and *Ceratium hirundinella*. Right side: Top zooplanktivores are mainly roach (*Rutilus rutilus*) of different size classes; dominating suspension feeders are *Bosmina longirostris* and *Keratella cochlearis*; dominating phytoplankton species is *Cosmocladium perissum*.

The curve below demonstrates factors like primary production, chlorophyll, pH, and total phosphorus during the "oligotrophic" and "eutrophic" stages

In fact, the similarities with the lake before the initial elimination of the roach population were apparent, not only on a general structural and functional level but also on a species level (Stenson 1988) (Fig. 1).

Conclusions

From the results and discussion reviewed above it is possible to draw some conclusions. First, it is evident that predation, by fish as well as by invertebrates, has a significant and specific impact on the biotic structure in terms of species composition and size distribution. Second, there are certain important feed-back relations between structural and functional components of the lake ecosystem. Third, there is a clear feed-back influence by the biotic structure and activity on basic abiotic properties of the water. The knowledge of these relations should make the system more predictable. Furthermore, support from other studies in other types of lakes enables us to go from interpretations of site-specific conditions and reactions to generalities which make the results useful not only from the theoretical point of view but also from an applied perspective.

Final remarks

As mentioned in the introduction, one important purpose of the research was to tie theoretical knowledge about regulating mechanisms to practical use. Experiences from the original field experiment have proven to be very useful in several connections:

The feed-back relations between fish and water quality criteria can be used in the interpretations of trophic changes in natural waters. The knowledge of these mechanisms can, furthermore, suggest manipulative measures, so-called food chain manipulations, as corrective treatments for, e.g., eutrophicated lakes (KITCHELL et al. 1979 and in press; CARPENTER et al. 1987; MILLS et al. 1987; BENNDORF 1988).

The efficiency of fish in reducing populations of larger invertebrates implies a competitive exclusion of non-piscivorous, invertebrate-feeding duck species from lakes with dense fish populations, knowledge relevant in practical waterfowl management (ERIKSSON 1983).

In order to understand the acidification process, and to find adequate corrective measures, knowledge about these regulating mechanisms is important in order to discriminate between toxic effects caused by the chemical change of the environment and indirect effects caused by altered predator-prey relations (ERIKSSON et al. 1980; STENSON 1985b; STENSON & ERIKSSON 1988).

Acknowledgements

The work within this project has been supported by grants from the Swedish Natural Science Research Council, the National Swedish Environment Protection Board and the National Board of Fisheries.

References

ANDERSSON, G., GRANELI, W., & STENSON, J. A. E. (1988): The influence of animals on phosphorus cycling in lake ecosystems. Hydrobiologia **170**, 267—284.

BENNDORF, J. (ed.) (1988): Biomanipulation. *Limnologia* **19** (1), 1—116.

BROOKS, J. L., & DODSON, S. I. (1965): Predation, body size, and composition of plankton. Science **150**, 28—35.

CARPENTER, S. R., KITCHELL, J. F., HODGSON, J. R., COCHRAN, P. A., ELSER, J. J., ELSER, M. M., LODGE, D. M., KRETCHMER, D., HE, X., & VON ENDE, C. N. (1987): Regulation of lake primary productivity by food web structure. Ecology **68**, 1863—1876.

DESORTOVÁ, B. (1976): Productivity of individual algal species in natural phytoplankton assemblage determined by means of autoradiography. Arch. Hydrobiol./Suppl. **49**, 415—449.

DODSON, S. I. (1974): Zooplankton competition and predation: An experimental test of the size-efficiency hypothesis. Ecology **55**, 605—613.

ERIKSSON, M. O. G. (1976): Food and feeding habits of downy Goldeneye *Bucephala clangula* (L.) ducklings. Ornis. Scand. **7**, 159—169.

— (1979): Competition between freshwater fish and Goldeneyes *Bucephala clangula* (L.) for common prey. Oecologia (Berlin) **41**, 99—107.

— (1983): The role of fish in the selection of lakes by nonpiscivorous ducks: Mallard, Teal and Goldeneye. Wildfowl **34**, 27—32.

— HENRIKSON, L., NILSSON, B. I., NYMAN, H. G., OSCARSON, H. G., STENSON, J. A. E., & LARSSON, P. K. (1980): Predator-prey relations inportant for biotic changes in acidified lakes. Ambio **9**, 248—249.

ESJMONT-KARABIN, J. (1984): Phosphorus and nitrogen excretion by lake zooplankton (Rotifers and Crustaceans) in relationship to individual body weights of the animals, ambient temperature and presence or absence of food. Ekol. Pol. **32**, 3—42.

FERRANTE, J. G., & PARKER, J. I. (1977): Transport of diatom frustrules by copepod fecal pellets to the sediment of Lake Michigan. Limnol. Oceanogr. **22**, 92—98.

HENRIKSON, B. I. (1988): The absence of antipredator behaviour in the larva of *Leucorrhinia dubia* (Odonata) and the consequences for their distribution. Oikos **51**, 179—183.

HENRIKSON, L., & OSCARSON, H. G. (1978): Fish predation limiting the abundance and distribution of *Glaenocorisa propinqua* (Hemiptera). Oikos **31**, 102—105.

— NYMAN, H. G., OSCARSON, H. G., & STENSON, J. A. E. (1980): Trophic changes, without changes in the external nutrient loading. Hydrobiologia **68**, 257—263.

HRBÁČEK, J. (1959): Density of the fish population as a factor influencing the distribution and speciation of the species in the genus *Daphnia*. XVth International Congress of Zoology, London. pp. 794—796.

— DVOŘÁKOVÁ, M., KOŘÍNEK, V., & PROCHÁZKOVÁ, L. (1961): Demonstration of the effect of the fish stock on the species composition of zooplankton and the intensity of metabolism of the whole plankton association. Verh. Int. Verein. Limnol. **14**, 192—195.

KITCHELL, J. F., O'NEILL, R. V., WEBB, D., GALEPP, G. W., BARTELL, S. M., KOONCE, J. F., & AUSMUS, B. S. (1979): Consumer regulation of nutrient cycling. BioScience **29**, 28—34.

— HENDERSON, H. F., GRYGIEREK, E., HRBÁČEK, J., KERR, S. R., PEDINI, M., PETR, T., SHAPIRO, J., STEIN, R. A. Jr., STENSON, J. A. E., & ZARET, T.: Expert consultation on the management of lakes by food-chain manipulation. Food and Agricultural Organization of the United Nations (in press).

LEHMAN, J. T. (1980): Release and cycling of nutrients between planktonic algae and herbivores. Limnol. Oceanogr. **25**, 620—632.

MILLS, E. L., FORNEY, J. L., & WAGNER, K. J. (1987): Fish predation and its cascading effect on the Oneida Lake food chain. In: KERFOOT, W. C., & SIH, A. (eds.), Predation, direct and indirect impacts on aquatic communities, pp. 118—131. New England, Hanover and London.

NEILL, W. E. (1985): The effects of herbivore competition upon the dynamics of *Chaoborus* predation. Arch. Hydrobiol., Beih. Ergebn. Limnol. **21**, 483—491.

NILSSON, B-I. (1981). Susceptibility of some odonate larvae to fish predation. Verh. Int. Verein. Limnol. **21**, 1612—1615.

NYMAN, H. G., OSCARSON, H. G., & STENSON, J. A. E. (1985): Impact of invertebrate predators on the zooplankton composition in acid forest lakes. Ecol. Bull. (Stockholm) **37**, 239—243.

PEHRSSON, O. (1979): Feeding behaviour, feeding habitat utilization, and feeding efficiency of Mallard ducklings (*Anas platyrhynchos* L.) as guided by a domestic duck. Swedish Wildl. Res. **10**, 191—218.

SHAPIRO, J. (1980): The importance of trophic-level interactions to the abundance and species composition of algae in lakes. In: J. BARICA & L. R. MUR (eds.), Hypertonic ecosystems (Developm. in Hydrobiol. **2**), pp. 105—116. The Hague.

STENSON, J. A. E. (1972): Fish predation effects on the species composition of the zooplankton community in eight small forest lakes. Rep. Inst. Fresw. Res. Drottningholm **52**, 132—148.

— (1976): Significance of predator influence on composition of *Bosmina* spp. populations. Limnol. Oceanogr. **21**, 814—822.

— (1978): Differential predation from fish on two species of *Chaoborus* (Diptera, Chaoboridae). Oikos **31**, 98—101.

— (1979): Predator-prey relations between fish and invertebrate prey in some forest lakes. Rep. Inst. Freshw. Res. Drottningholm **59**, 166—182.

— (1981): The role of predation in the evolution of morphology, behaviour and life history of two species of *Chaoborus*. Oikos **37**, 323—327.
— (1982): Fish impact on rotifer community structure. Hydrobiologia **87**, 57—64.
— (1984): Interactions between pelagic metazoan and protozoan zooplankton, an experimental study. Hydrobiologia **111**, 107—112.
— (1985a): Strategies for prey-selection and anti-predator defence in lake ecosystems. Rep. Inst. Freshw. Res. Drottningholm **62**, 150—160.
— (1985b): Biotic structures and relations in the acidified Lake Gårdsjön system. Ecol. Bull. (Stockholm) **37**, 319—326.
— (1987): Variation in capsule size of *Holopedium gibberum* (ZADDACH): A response to invertebrate predation. Ecology **68**, 928—934.
— (1988): Animal structure and primary production, an experimental study. In: G. BALVAY (ed.); Eutrophication and lake restoration. Water quality and biological impacts, pp. 161—169. Thonon-les-Bains.
— & ERIKSSON, M. O. G. (1988): Ecological mechanisms important for the biotic changes in acidified lakes in Scandinavia. Arch. Environ. Contam. Toxicol. **17**, in press.
— BOHLIN, T., HENRIKSON, L., NILSSON, B-I., NYMAN, H. G., OSCARSON, H. G., & LARSSON, P. (1978): Effects of fish removal from a small lake. Verh. Int. Verein. Limnol. **20**, 794—801.
STULL, E. A., AMAZAGA, E. DE, & GOLDMAN, C. R. (1973): The contribution of individual species of algaee to primary productivity of Castle lake, California. Verh. Int. Verein. Limnol. **18**, 1776—1783.

Address of the author:

JAN A. E. STENSON
Department of Zoology
University of Göteborg
P. O. Box 25059
S-400 31 Göteborg
Sweden

Limnologica (Berlin) **20** (2), 273—278, Juni 1990

Department of Biology, Åbo Akademi, Åbo, Finland

Limnological Changes in Two Recently Isolated Lakes

By TORE LINDHOLM & JOHN ERIKSSON

With 8 Figures

Key words: Brackish-water lakes; Eutrophication; Toxic cyanobacteria; *Oscillatoria agardhii*.

Abstract

Two brackish-water lakes, the shallow Markusbölefjärden and the stratified, meromictic Långsjön, both about 1.5 km² and situated at sea level in Central Åland, SW Finland, were isolated from the sea in 1971—72 to serve as freshwater reservoirs. The salinity decreased rapidly in both lakes and in 1974 Långsjön turned holomictic. Both lakes became increasingly eutrophic and blooms of cyanobacteria frequently caused water quality problems. In 1984 and 1985 both lakes were studied with regard to nutrient loading and limnological conditions. Fish and bird kills and toxic cyanobacteria (above all *Oscillatoria agardhii*) occurred in both lakes. Very high total phosphorus values were observed during the summers, indicating a large internal loading. The phytoplankton growth was phosphorus-limited in the epilimnion of Lake Långsjön, whereas nitrogen and iron limitation occurred in Lake Markusbölefjärden. Both lakes had high chlorophyll **a** levels in epilimnion (25—250 µg/l) as well as high and fluctuating pH values (7.0 to 9.5). One reason for the eutrophication is the increased water renewal time after the isolation, another is the storage of nutrient-rich water in spring. Moreover, both lakes have large populations of roach and bream, which may influence the nutrient status of the lakes. During the 1980s, efforts have been made to reduce the external and internal loading in order to improve the water quality and to prevent toxic cyanobacterial blooms.

Introduction

In the Finnish and Swedish archipelagos new lakes still arise when semi-enclosed basins gradually become isolated from the Baltic Sea due to land uplift (WILLÉN 1962; BAGGE & TULKKI 1967; LINDHOLM 1982). The postglacial land uplift is about 5 mm/year in the Åland area. This lake formation — which is seldom mentioned in limnology textbooks — is speeded up by reed growth and sedimentation along shallow sill areas. In many cases small canals have been made to permit passage of boats from the lakes to the sea e.g. for fishing. Outlet ditches have also been enlarged in order to drain farmland around the lakes. Numerous archipelago basins have been isolated or nearly isolated by road embankments. Finally, a great number of coastal and archipelago basins have been abruptly isolated to form freshwater reservoirs (e.g. LINDHOLM 1975, 1982; SEVOLA 1975; AHLBÄCK 1978; VOGT 1978). The Åland area is especially rich in semi-enclosed coastal waters (some of which show meromictic features) as well as recently isolated lakes. Biologists working at Husö Biological Station (see Fig. 1) have studied many lakes and coastal basins since the station was established in 1959 (for references, see e.g. LINDHOLM 1982). The limnological features and recent changes in two of the largest lakes on Åland, Markusbölefjärden and Långsjön, will be presented here. Together the lakes form the most important drinking water reservoir on the Åland Island.

Materials and methods

Water samples were taken with a Ruttner sampler at the deepest points in the lakes. Secchi depth was measured with a 20 cm diameter white disc. Light was measured with a LI-COR 188 radiometer. Dissolved oxygen, pH and conductivity were analyzed according to GOLTERMAN et al. (1978). The chlorophyll **a** samples (500 ml) were filtered on Whatman GF/C filters and the filters were stored deep-frozen for 4—5 months. Chlorophyll **a** was analyzed by acetone extraction as described by MARKER et al. (1980). Total phosphorus (TP) and total nitrogen (TN) were analyzed according to GRASSHOFF (1976) and KOROLEFF (1979). The species composition of fresh water and bloom samples was studied with an inverted microscope. Dense phytoplankton samples were taken with a plankton net (25 µm) for microscopical examination and toxicity tests. Total alkaline phosphatase activity (APA) was determined as described by GAGE & GORHAM (1985).

Study area

The lakes Markusbölefjärden and Långsjön (Fig. 1) are situated in the Åland mainland about 15 km N from the town of Mariehamn. The local climate is characterized by mild autumns, moderate winters, late springs and relatively dry summers. The lakes are usually ice-covered from the middle of December to the end of April. The bedrock of Åland consists of rapakivi granite, which is often bare or has merely a thin moraine cover. The actual study area is, however, characterized by fertile clay and moraine soils with inclusions of lime. The drainage areas of both lakes are varied, including forests, farmland, lakes and shore meadows (Table 1). Although the local population is relatively small and there is no big industry, the lakes are greatly influenced by man (see also next chapter). General data about the lakes are presented in Table 1. More details are given

Table 1. General data about the lakes Markusbölefjärden and Långsjön. The lakes are interconnected and the water level is the same in both, but their catchment and drainage areas are treated separately in this table. Data from HELMINEN (1978) and LINDHOLM (1982)

	Markusbölefjärden	Långsjön
Area (ha)	156	143
Maximum length (km)	3.7	4.5
Shoreline length (km)	9.8	11.6
Maximum depth (m)	9	18
Volume (10^6 m³)	7	9
Catchment area (km²)	13.8	16.3
Lake % of drainage area	5.0	0.3
Farmland % of drainage area	21.2	14.0
Estimated retention time (years)	2.5	2.8

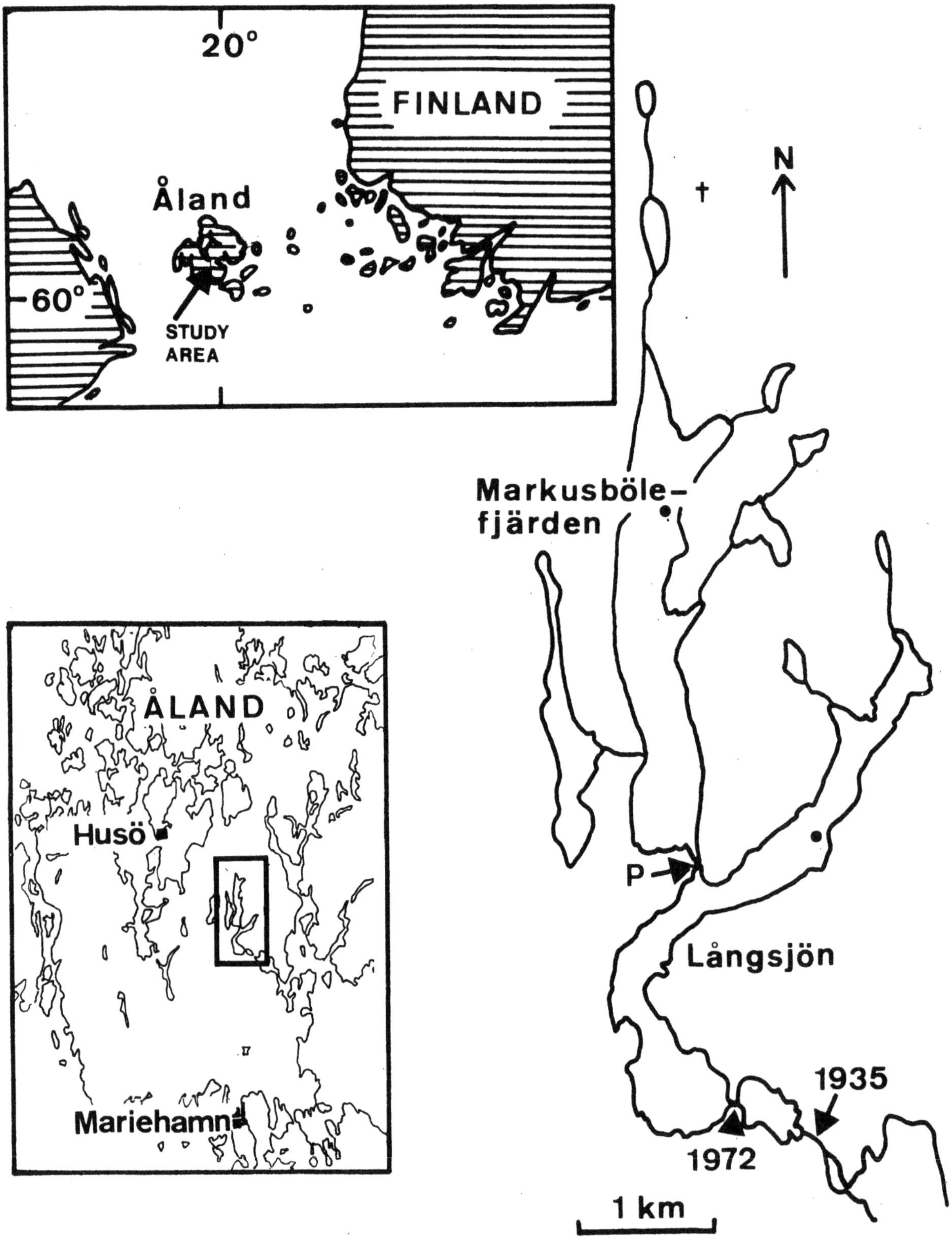

Fig. 1. Study area. The canal was made in 1935 and the lake system was finally isolated in 1972. The pump station of the water company is marked with P, the church of Finström with a cross

in HELMINEN (1978), KARLSSON et al. (1981), LINDHOLM (1982) and ERIKSSON & LINDHOLM (1985).

Influence by man

In the Middle Ages the straits of that time (the lake system of today) were open waters permitting boats to pass from the sea even to the church of the Finström parish (Fig. 1). In the 17th century the inlets even became official channels ("water-ways of the Crown"). For hundreds of years the main post road from Sweden to Finland passed over the outlet of Lake Långsjön. It is likely that some kind of bridge was used. Thus, there was some human influence on the lake chain already hundreds of years ago.

During the last centuries both lakes were isolated

from the sea due to land uplift and the diatom frustules of the sediments indicate that the water was fresh (RÄSÄNEN 1983). In 1935 a canal was built from Lake Långsjön to the sea (the nearby archipelago inlet; Fig. 1). The lakes became brackish (1—3 permille) and Lake Långsjön turned meromictic (WIKGREN 1965; LINDHOLM 1975, 1982; RÄSÄNEN 1983). In the late 1960s there was a need to find additional freshwater supplies for the town of Mariehamn and the surrounding areas (approx. 14000 inhabitants), whereby the lakes Markusbölefjärden—Långsjön were chosen. In 1971 and 1972 the lakes were isolated and their water levels were regulated. In 1974, when the water was still slightly brackish, the lakes were already used as drinking water reservoirs. During and after the refreshening (Fig. 2) the lakes became more and more eutrophic, and considerable taste and odour problems were encountered. Occasionally, copper sulphate was used to reduce the phytoplankton growth. Later some aerators were installed to improve the oxygen conditions of the lakes. Measures have also been taken to reduce pollution from farms and villages around the lakes.

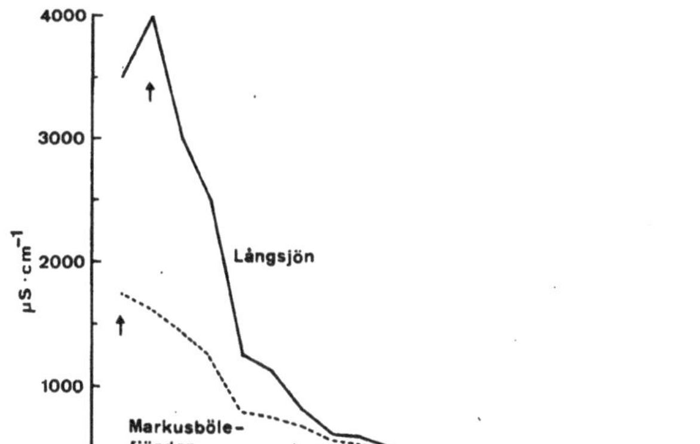

Fig. 2. Conductivity at 1 m in the lakes Långsjön and Markusbölefjärden 1971—1988. The refreshening was particularly fast during the rainy year 1974. Some of the data obtained from Ålands Vatten AB, Mariehamn

Lake Markusbölefjärden

Before the isolation in 1971 Lake Markusbölefjärden was eutrophic, but showed no permanent stratification (Fig. 3). The salinity was about 1 permille (about 1800 µS/cm, Fig. 2) and the flora and fauna did not differ much from the flora and fauna of near-by eutrophic lakes. Extensive blooms of various cyanobacteria were

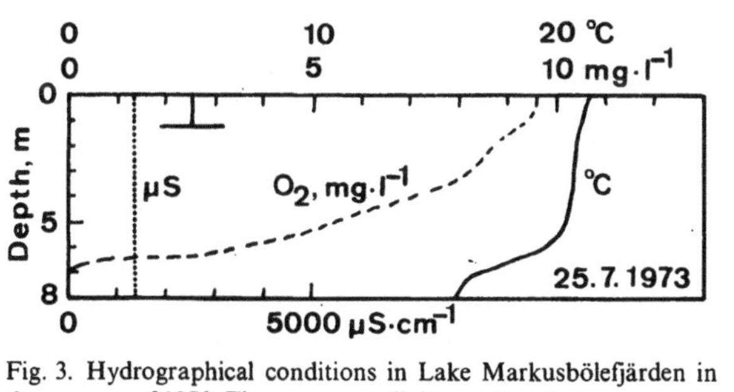

Fig. 3. Hydrographical conditions in Lake Markusbölefjärden in the summer of 1973. There was no salinity gradient but a temporary thermocline and poor oxygen conditions

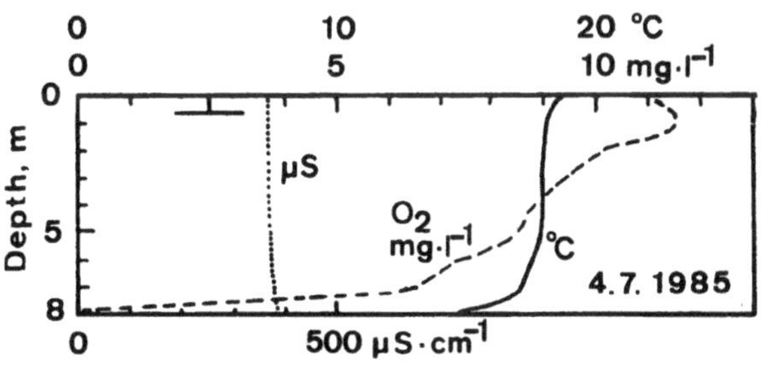

Fig. 4. Hydrographical conditions in Lake Markusbölefjärden in the summer of 1985 when the water was fresh. Note the oxygen depletion near the bottom and the low Secchi disc transparency

observed both before and after the isolation. After the isolation the lake has become even more eutrophic, probably due to:

— increased water renewal time
— continuous external loading (pollution)
— internal loading due to periods of anoxia
— internal loading due to high pH and large areas of epilimnetic bottom
— nitrogen fixation by cyanobacteria.

Lake Markusbölefjärden is exposed to winds from most directions and the thermal stratification is only temporary (a few weeks). However, even during short stagnation periods the hypolimnion becomes anoxic (Figs. 3 and 4). Consequently, there is a considerable release of nutrients from the hypolimnetic sediments. This internal loading is intermittently mixed with the surface water. The pH of the epilimnion may also be high enough to cause a release of phosphorus from the large areas of epilimnetic sediments. This combined internal loading results in summer total phosphorus

Table 2. Total phosphorus (TP) and total nitrogen (TN) at various depths in the lakes Markusbölefjärden (Ma) and Långsjön (Lå) during 1984 (µg/l). Dates in brackets. Selected profiles from ERIKSSON & LINDHOLM (1985)

		(25.5)		(20.7)		(27.8)		(18.10)	
		TP	TN	TP	TN	TP	TN	TP	TN
Ma	0 m	63	1020	192	2210	200	1710	134	2080
	4 m	48	790	187	2180	183	1690	133	1420
	8 m	71	680	196	2070	173	1450	131	1290
		(17.5)		(23.7)		(30.8)		(19.10)	
Lå	0 m	47	890	73	1210	70	1300	67	1020
	5 m	51	820	56	1890	51	1240	—	—
	10 m	39	900	126	1550	254	2010	68	1070
	15 m	60	1400	207	1240	376	2260	68	1440

Table 3. Chemical and biological characteristics of the euphotic zone (0—3 m) in Lake Långsjön and Lake Markusbölefjärden in 1985. Numbers indicate mean values ± SD for June to August

	Långsjön	Markusbölefjärden
TP (µg/l)	55 ± 9	122 ± 64
TN (µg/l)	1048 ± 109	1334 ± 315
pH	8.4 ± 0.3	8.7 ± 0.4
Chl. a (µg/l)	41.2 ± 10.0	96.7 ± 69.7
TN/TP	21.5 ± 4.3	12.8 ± 4.4
Cellular phosphorus (mg/g dry weight)	3.73 ± 0.90	5.38 ± 0.73
Alkaline phosphatase activity (nmol · mg dry weight^{-1} · h^{-1})	25.83 ± 10.73	4.57 ± 0.74

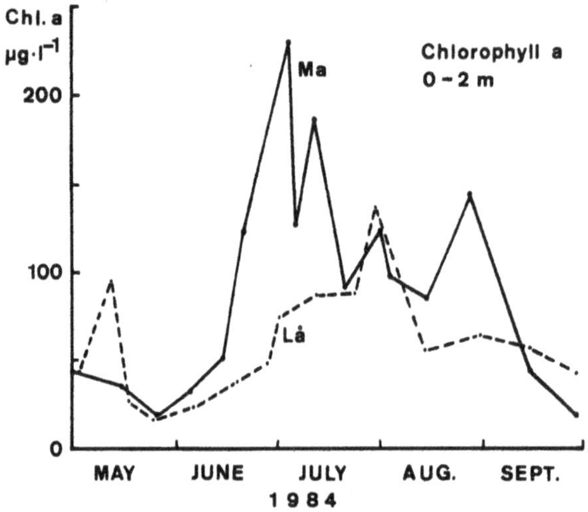

Fig. 5. Chlorophyll a in the topmost 2 m layer (mean of 2 or 3 measurements) in the lakes Markusbölefjärden and Långsjön during the summer of 1984. The phytoplankton biomass peaks occurred at different times

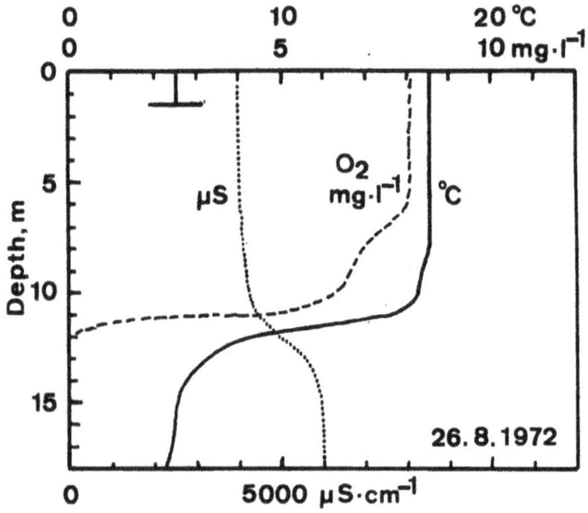

Fig. 6. Hydrographical conditions in Lake Långsjön in August 1973 (after a storm) when the lake was still meromictic

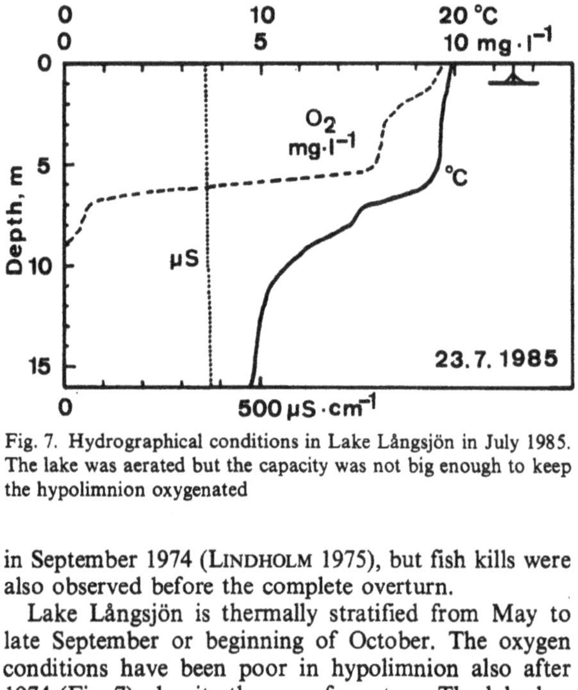

Fig. 7. Hydrographical conditions in Lake Långsjön in July 1985. The lake was aerated but the capacity was not big enough to keep the hypolimnion oxygenated

(TP) values up to about 200 µg/l (Table 2). The Secchi depth of Lake Markusbölefjärden has been about 0.5 m during the phytoplankton production peaks in high summer. The chlorophyll a values of the epilimnion have often been 100—200 µg/l (Fig. 5). Due to the high TP levels, the mean summer TN/TP ratio is 12.9 and the actual ratios often below 10 (Table 3). The low TN/TP ratios suggest that nitrogen rather than phosphorus is the main limiting factor for phytoplankton growth in Lake Markusbölefjärden. This assumption is further supported by the high cellular phosphorus (CP) content and the extremely low APA observed in the lake (Table 3; HEALEY & HENDZEL 1980; GAGE & GORHAM 1985). Experimental work on nutrient deficiency (ERIKSSON & LINDHOLM, unpublished) has confirmed that nitrogen and at times also iron limited the phytoplankton growth in Lake Markusbölefjärden during the summers of 1984 and 1985. Conspicuous blooms of several *Anabaena* species and *Microcystis aeruginosa* have occurred. The phytoplankton flora of Lake Markusbölefjärden has at times also been rich in green algae and euglenoids.

The fish population of Lake Markusbölefjärden is large and dominated by roach (*Rutilus rutilus*) and bream (*Abramis brama*), which may have a negative influence on the water quality (BRAABRAND et al. 1984; PENCZAK & TATRAI 1985). In the summers of 1984 and 1985 partial fish kills occurred during periods of dense cyanobacterial blooms (several species). A toxic strain of *Oscillatoria agardhii* was occasionally present, but it is not known whether it contributed to the fish kills.

Lake Långsjön

Before the isolation in 1972, Lake Långsjön was meromictic with a halocline between 5 and 10 m (WIKGREN 1965; LINDHOLM 1975). Accordingly, the oxygen conditions were very poor and the deep water strongly reducing. In 1972 the halocline was situated at 12 m (Fig. 6) and in 1974, two years after the isolation, Lake Långsjön turned holomictic (LINDHOLM 1975, 1982). At the same time a large volume of stagnant, anoxic water, rich in nutrients, entered the euphotic zone and caused heavy algal blooms. There was a big fish kill in September 1974 (LINDHOLM 1975), but fish kills were also observed before the complete overturn.

Lake Långsjön is thermally stratified from May to late September or beginning of October. The oxygen conditions have been poor in hypolimnion also after 1974 (Fig. 7), despite the use of aerators. The lake has a great internal loading (Table 2), but due to the thermal stratification this loading is only gradually released during the autumnal overturn (usually in September). The TP values for epilimnion have been 40—70 µg/l. The summer Secchi depth has usually been between 1 and 2 m, and occasionally as low as 0.6 m. In 1984 the summer chlorophyll a values fluctuated between 20 and more than 100 µg/l (Fig. 5). The TN/TP ratios and the APA have been significantly higher and the CP significantly lower than in Lake Markusbölefjärden (Table 3). These values indicate that phosphorus is the main limiting factor for phytoplankton growth. However, the values are not high and low enough, respectively, to indicate absolute P deficiency. In fact, experimental studies have shown that nitrogen is the secondary growth limiting factor in Lake Långsjön (ERIKSSON & LINDHOLM, unpublished). Conspicuous surface blooms of cyanobacteria have occurred (*Anabaena* spp. in June and July, *Aphanizomenon flos-aquae* and *Microcystis aeruginosa* in late summer), but not always at the same time as in Lake Markusbölefjärden (Fig. 5). Hepatotoxic strains of *Oscillatoria agardhii* have been present and sometimes dominant among the phytoplankton. A toxic red form of *O. agardhii* occurred under the ice in March 1984. During 1984 and 1985 fish and bird kills were observed in July and August (LINDHOLM & ERIKSSON 1985; ERIKSSON et al. 1986). The fish fauna is similar to that of Lake Markusbölefjärden.

Discussion and conclusions

Isolation of brackish archipelago inlets due to land uplift or human measures cause gradual or even fast limnological changes in the recently isolated basins (WILLÉN 1962; BAGGE & TULKKI 1967; LINDHOLM 1982). In most cases there have been no long-term ecological studies and therefore there is only a limited amount of published information on this kind of lake formation. The lakes Markusbölefjärden and Långsjön represent late, although disturbed stages in this ecological and geological succession. Another lake system, Östra and Västra Kyrksundet, with a quite similar development, has been presented in other papers (LINDHOLM 1982; LINDHOLM et al. 1989; BONSDORFF & STORBERG, this volume).

Transient stages in lake formation have many interesting features of their own (LINDHOLM 1982; LINDHOLM & WEPPLING 1987). Semi-isolated brackish archipelago inlets have very few marine organisms. However, when the basins become freshwater lakes, considerable limnological and biological changes occur. Some important limnological changes, from meromixis to holomixis, as well as the increased time for water renewal, were mentioned above (see also LINDHOLM 1982). In coastal inlets open to the sea there is a considerable transport of nutrient-rich freshwater (meltwater) already under the ice in early spring. In reservoirs, where there is a need to store the spring runoff, nutrients are also trapped (Fig. 8). Due to increased plankton production and increased sedimentation, the rate of sedimentation will increase and result in anoxia in the sediments and in the hypolimnion of deep reservoirs. Thus, many mechanisms contribute to the eutrophication observed in recently isolated lakes (LINDHOLM 1982).

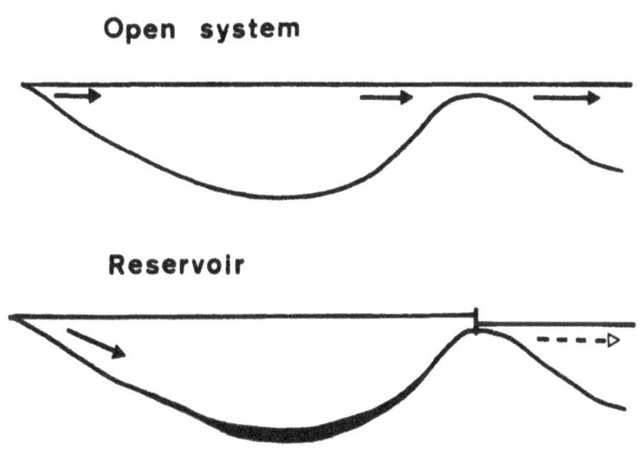

Fig. 8. Schematic figure showing one important hydrological change when a coastal inlet is isolated from the sea. In an open system the nutrient-rich melt-water goes as a surface stream over the brackish water (under the ice in early spring). In a closed system, especially if there is a need to fill the reservoir, the nutrient-rich waters are trapped. The dark bottom illustrates the increased sedimentation

From the limnological point of view Markusbölefjärden is a hypertrophic lake with a large internal loading. Artificial aeration has been used to improve the oxygen conditions and measures have been taken to reduce the external loading. In spite of this, dense blooms of cyanobacteria (especially *Microcystis aeruginosa*) occur and the water quality has been poor during the summers. The unpredictable stratification and mixing periods during the summer greatly influence the phytoplankton production and make management difficult. Lake Långsjön is a stratified, eutrophic lake with blooms of toxic and non-toxic cyanobacteria. During the 1980s several measures have been taken to reduce the external and internal loading. Two aerators have been employed, but their capacity has been too small to keep the deep waters well aerated. As the lakes are both needed to provide raw water, further efforts are necessary to improve their water quality.

Although the two lakes are interconnected, have a similar history and are both affected by farmland pollution, their different hydrographical characteristics are clearly expressed in their biological features. The nutrient status, the factors limiting phytoplankton growth, biomass (expressed as chlorophyll **a**), as well as succession of species composition are markedly different in these two parts of the lake system. Lake Långsjön is predictable with regard to mixing, and less rich in "mobile" phosphorus than Lake Markusbölefjärden and is therefore a more promising object with regard to restoration.

Acknowledgements

This study was supported by grants from the Academy of Finland (Research Council for the Environmental Sciences) and the Maj and Tor Nessling Foundation.

References

AHLBÄCK, O. (1978): Vattenvårdsproblem i Pargas sötvattenbassäng. Skärgård **2**, 28—31 (in Swedish).

BAGGE, P., & TULKKI, P. (1967): Studies on the hydrography and biota of recently isolated lakes. Merentutkimuslait. julk. **223**, 13—34.

BRAABRAND, A., FAAFENG, F., KÄLLQVIST, T., & NILSSEN, J. P. (1984): Can iron defecation from fish influence phytoplankton production and biomass in eutrophic lakes? Limnol. Oceanogr. **29**, 1330—1334.

ERIKSSON, J. E., & LINDHOLM, T. (1985): Intern och extern belastning i sjöarna Markusbölefjärden och Långsjön (Internal and external nutrient loading in Lake Markusbölefjärden and Lake Långsjön). Husö Biol. Stat. Medd. **25**, 3—15 (in Swedish, English summary).

— MERILUOTO, J. A. O., & LINDHOLM, T. (1986): Can cyanobacterial peptide toxins accumulate in aquatic food chains? In: F. MEGUSAR & M. GANTAR (eds.), Perspectives in Microbial Ecology. Proc. IV Int. Symp. Microbial Ecology, pp. 655—658. Ljubljana.

GAGE, M. A., & GORHAM, E. (1985). Alkaline phosphatase activity and cellular phosphorus as an index of the phosphorus status of phytoplankton in Minnesota Lakes. Freshwater Biol. **15**, 227—233.

GOLTERMAN, H. L., CLYMO, R. S., & OHNSTAD, M. A. M. (1978): Methods for Physical and Chemical Analysis of Fresh Waters. IBP Handbook No. 8. Oxford.

GRASSHOFF, K. (1976): Methods of Seawater Analysis. Weinheim.

HEALEY, F. P., & HENDZEL, L. L. (1980): Physiological indicators of nutrient deficiency in lake phytoplankton. Can. J. Fish. Aquat. Sci. **37**, 442—453.

HELMINEN, O. (1978): Tillrinningen till Markusbölefjärden och Långsjön 1976 (The water discharge into Lake Markusbölefjärden and Lake Långsjön (Åland) in 1976). Husö Biol. Stat. Medd. **20**, 5—30 (in Swedish, English summary).

KARLSSON, O., LINDHOLM, T., & STORBERG, K-E. (1981): Tillståndet i Markusbölefjärden och Långsjön under 1970-talet (Water quality in the Markusbölefjärden-Långsjön lake chain in the 1970's). Husö Biol. Stat. Medd. **23**, 14—29 (in Swedish, English summary).

KOROLEFF, F. (1979) Meriveden yleisimmät kemialliset analyysimenetelmät (Methods for chemical analysis of sea water). Meri **7**, 1—60 (in Finnish, English abstract).

LINDHOLM, T. (1975): Utvecklingen mot holomixi i Långsjön, en meromiktisk kustsjö på Åland, efter isoleringen från havet år 1972. Vesitalous **16**, 20—23 (in Swedish).

— (1982): Dynamics of hydrography and primary production in three stratified coastal lakes on Åland (SW Finland). Acta Acad. Aboensis, Ser. **B**, **42**, 1—75.

— & ERIKSSON, J. E. (1985): Problemalger och fiskdöd i åländska vattentäkter. Ympäristö ja Terveys 1/1985, 41—44 (in Swedish).

— — & MERILUOTO, J. A. O. (1989): Toxic cyanobacteria and water quality problems: Examples from a eutrophic lake on Åland, SW Finland. Water Research, **23**, 481—486.

— & WEPPLING, K. (1987): Blooms of phototrophic bacteria and phytoplankton in a small brackish lake on Åland, SW Finland. Acta Acad. Aboensis Ser. B, **47** (2), 45—53.

MARKER, A. R., NUTSCH, E. A., RAI, H., & RIEMANN, B. (1980): The measurement of photosynthetic pigments in fresh waters and standardization of methods: Conclusions and recommendations. Ergeb. Limnol. **14**, 91—106.

PENCZAK, T., & TATRAI, I. (1985): Contribution of bream, *Abramis brama* (L.), to the nutrient dynamics of Lake Balaton. Hydrobiologia **126**, 59—64.

RÄSÄNEN, M. (1983): Ahvenanmaan järvialtaissa 1900-luvulla tapahtuneiden veden suolaisuuspitoisuusmuutosten vaikutus altaiden limnologiaan ja piilevälajistoon. Turun Yliopiston maaperägeologian osaston Julk. **49**, 1—45 (in Finnish).

SEVOLA, P. (1975): Uudenkaupungin makeavesialtaan ekologinen tutkimus. Lounais-Suomen Vesiensuojeluyhdistys r. y. Julk. **25**, 1—86 (in Finnish).

WIKGREN, B-J. (1965): Salt vatten i insjöar. Husö Biol. Stat. Medd. **8**, 1—37 (in Swedish).

WILLÉN, T. (1962): The Utål Lake Chain Central Sweden and its phytoplankton. Oikos Suppl. **5**, 1—156.

VOGT, H. (1978): An ecological and environmental survey of the humic man-made lakes in Finland. Aqua Fennica **8**, 18—24.

Address of the authors:

TORE LINDHOLM
Department of Biology
Åbo Akademi
Porthansg. 3
SF — 20500 Åbo
Finland

Limnologica (Berlin) **20** (2), 279—284, Juni 1990

Department of Biology and Husö Biological Station, Åbo Akademi, Åbo, Finland;
National Board of Water and Environment, Vaasa District Office, Vaasa, Finland

Ecological Changes in a Formerly Meromictic Coastal Lake

By ERIK BONSDORFF & KARL-ERIK STORBERG

With 7 Figures

Key words: Meromixis; Hydrography; Zoobenthos, Fish community; Recovery potential; Ecological changes.

Abstract

In 1932 Lake Västra Kyrksundet on Åland, SW Finland was reconnected with the Baltic Sea through an artificial canal. The water level was lowered 0.5 m, and brackish (about 5 ppt) water entered the lake. From 1932 to 1979 the lake was meromictic with a permanent halocline at 6.5—8 m, with complete oxygen deficiency below that. No zoobenthos existed below the littoral, and fish production was dependent on secondary production above the halocline. In 1979 the halocline was broken artificially through oxygen bubbling of the bottom-near water and by cutting the connection with the sea.

Hydrography, zoobenthos and fish were studied in relation to the recent changes in the lake. The meromictic conditions diminished rapidly, and by 1981 the water was fully limnic. As the bottom water was reoxygenated, sediment-bound nutrients were released increasing the pelagic primary production. Thus the transparency was dramatically reduced after 1979. Zoobenthos first appeared below the former halocline in September 1979, thereafter showing an annual successional pattern dominated by chironomid larvae. The fish community was also altered, as the typically benthic-feeding bream and ruffe gained in importance on behalf of roach and perch. Although the hydrographical conditions stabilized since 1979, the benthic and fish communities showed great variations within and between years.

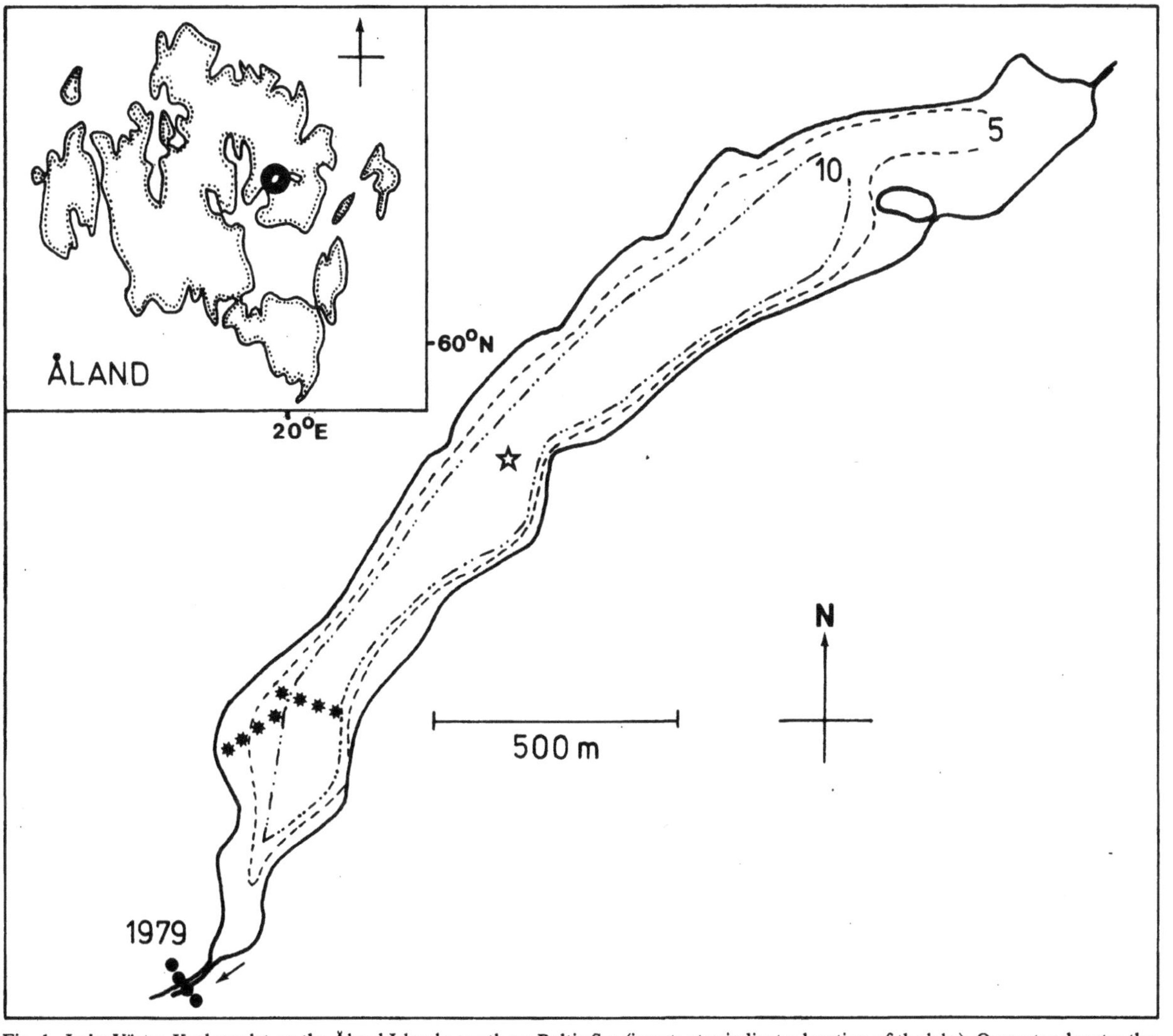

Fig. 1. Lake Västra Kyrksundet on the Åland Islands, northern Baltic Sea (insert; star indicates location of the lake). Open star denotes the principal sampling site for hydrography and zoobenthos, and the black stars mark the area of gill net fishing. In the SW end of the lake the small dam built in 1979 is indicated

Introduction

On the Åland Islands (SW Finland) almost all lakes are relatively small and have evolved from sea inlets through land uplift (WIKGREN 1965; LINDHOLM 1982). The process is still ongoing, and new coastal lakes are constantly being formed. During the last century this development has been altered by human impact in many cases (LINDHOLM 1975a, b). Lake Västra Kyrksundet (Fig. 1), which is a fjordlike inlet at sea level, was reconnected with the nearby Baltic Sea through an artificial canal in 1932. Thus the water level was lowered about 0.5 m, and brackish water (4.5—5.5 ppt S) entered the lake. From 1932 until 1979, when the connection to the sea was again cut, Lake Västra Kyrksundet was meromictic with a permanent halocline at 6.5—8 m (max. depth in the lake is 18 m), and permanent oxygen deficiency below that (Fig. 2). During the spring 1979 the halocline was broken through oxygen bubbling of the bottom-near water, and by the construction of a small dam in the canal to the sea (Fig. 1). Thus the ecosystem was artificially brought back to the developmental stage of before 1932 (WIKGREN 1965; STORBERG 1980, 1982; LINDHOLM 1982).

The aims of the present study were to study the hydrographical changes from the meromictic to fully limnic conditions in relation to the recovery (establishment, succession and dynamics) of the zoobenthos and fish community. This study covers the years 1975—1985. The couplings between the abiotic and biotic components of the lake were compared to theoretical successional and competitive models (CONNELL & SLATYER 1977; BAZZAZ & SIPE 1987; LAMPERT 1987).

Material and methods

Lake Västra Kyrksundet has previously been thoroughly investigated by, e. g., LINDHOLM (1982), who also gave a general description of coastal meromictic lakes. The basic morphometric features of the lake are: Length 2.5 km, width 0.3 km, surface area 59.5 ha, shore line 5.9 km, maximum depth 18 m (mean: 8.5 m), volume 5×10^6 m³, and drainage area 40.8 km².

The hydrography and basic chemistry was assessed as described by LINDHOLM (1982), covering mainly the primary parameters (temperature, conductivity, pH, oxygen content, alkalinity, °dH, chemical oxygen demand). The level of pelagic primary production was assessed only by measuring the transparency (Secchi disk depth). Zoobenthos was sampled at a standard station at 14—18 m depth using an Ekman-Birge box grab (17 × 17 cm). A reference station at 6.5—8 m was sampled occasionally to register the initial recruitment below the former halocline. 5 replicate samples (sieved on a 0.5 mm mesh screen, and stored individually in 4% buffered formaldehyde) were taken at each sampling date, and species composition, abundance and biomass were measured for each individual sample. The fish community was monitored using a Finnish standard gill net set, with 8 nets of 1.8 × 30 m (12, 15, 20, 25, 35, 45, 60 and 75 mm meshes knot to knot). The nets were set at dusk (standardized to 18.00) and taken at dawn (standardized to 06.00) the following morning. The selectivity of gill nets has been reviewed by HAMLEY (1975).

For the zoobenthos all results are given per m², and for the fish community as catch per effort (i. e. one nights fishing with the entire set of gill nets).

Results

The mechanically induced changes of Lake Västra Kyrksundet showed an immediate response in several hydrographical and chemical parameters. The bottom water was reoxygenated, and no H_2S-layer was recorded during the study between 1979 and 1985 (Fig. 2). As a consequence of the dam disconnecting the influx of brackish water, the conductivity of the lake water was rapidly reduced (from clearly brackish to purely limnic), and within two years a steady state with limnic conditions was reached (Fig. 3). Simultaneously the alkalinity was reduced from 1.1 to < 0.7 mval. The hardness of the water (°dH) showed a similar pattern (20.7 in 1975, 12.9 in 1980, and 6.1 in 1981), and chemical oxygen demand declined from > 40 mg/l $KMnO_4$ before 1979 to 20—25 mg/l in 1979—1981. Thus the overall hydrographical conditions in Lake Västra Kyrksundet showed markedly different features after the mechanical changes of the lake in 1979 (summarized in Fig. 3), with the exception of the pH-conditions, which remained at the same level. The inter-annual thermal conditions and oxygen content for a model year (in this case 1982; Fig. 4) clearly demonstrate the radical change of the hydrography of the lake. The permanent stratification (as described for 1975—76 by LINDHOLM 1982) is gone, and the graphs illustrate the gradual changes from the spring situation throughout summer until a complete circulation of the water mass in the fall.

The changes in hydrography (including the reoxygenation of the sediments, with massive releases of nutrients etc.) greatly increased the pelagic primary production, illustrated as a marked reduction in transparency throughout the productive season (Fig. 5). This enhanced state of eutrophication has been dominated by cyanobacteria of the genera *Oscillatoria*, *Aphanizomenon* and *Anabaena* (LINDHOLM et al. 1989).

The zoobenthos below the depth of the former halocline (6.5—8 m) showed the first signs of recruitment in September 1979, about half a year after the lake was re-isolated from the sea. The only immigrants observed, however, were larval chironomids (*Chironomus plumosus*-type). Above the former halocline on an average 9—11 species/genera, totalling 500—1000 ind. per m² with a biomass of about 15 g wwt per m², were recorded already before 1979 and throughout the study. At 14—18 m both total numerical abundance and biomass reached similar levels in 1980, and have shown seasonal fluctuations governed by the hydrographical changes ever since (Table 1). The species composition did, for natural reasons, not reach the level of the shallow sublittoral (2—6 m), and from 1979—85 a total of 8 species/genera were recorded (1—4 spp. per sampling occasion). The community at 2—6 m was dominated by insect larvae, but oligochaetes, gastropods and crustaceans were also important. At the deeper station only insect larve and water mites have been recorded since 1979. Table 1 summarizes the annual seasonality of the zoobenthos in relation to hydrography for October 1981 through October 1982 (the amplitudes of variation

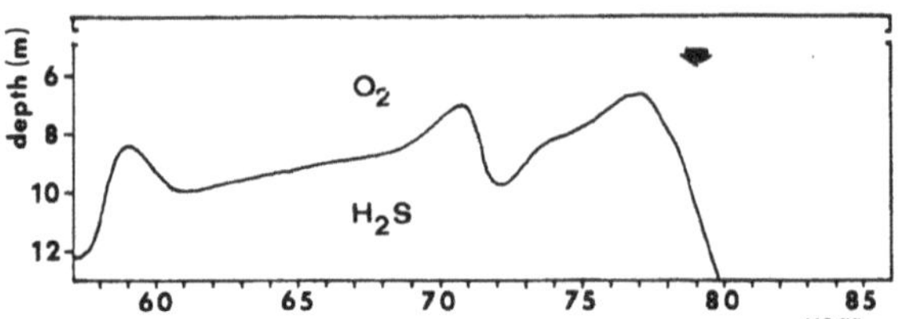

Fig. 2. The O_2/H_2S-boundary layer in the free water body of Lake Västra Kyrksundet before and after isolation from the sea in 1979 (indicated by an arrow). The data before 1975 are from WIKGREN (1965) and LINDHOLM (1982)

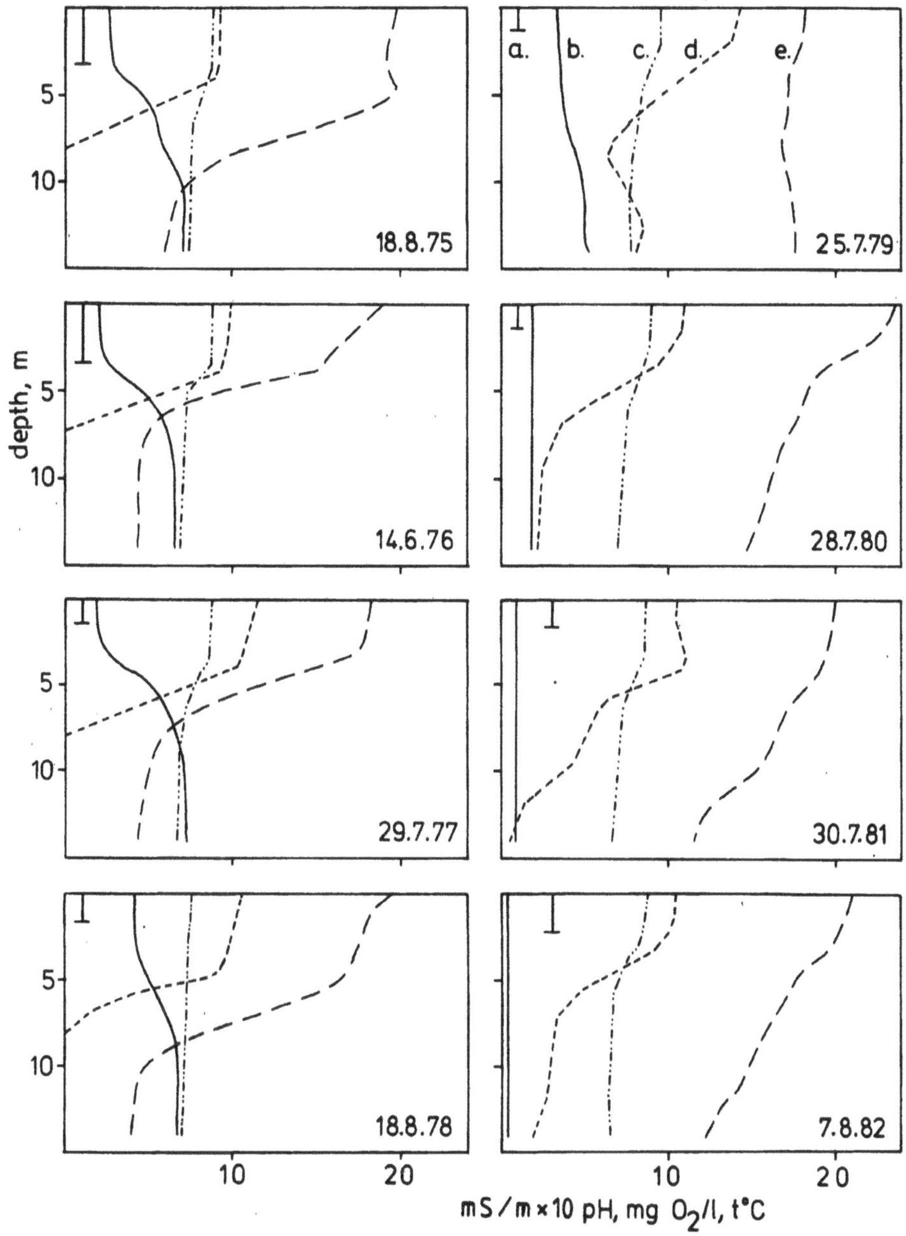

Fig. 3. Basic hydrographical summer features of Lake Västra Kyrksundet before (1975—78) and after (1979—82) isolation from the ambient (Baltic Sea. The sampling dates are indicated in the graphs. a = transparency (Secchi disk depth; m); b = conductivity (mS/m); c = pH; d = oxygen content (mg/l); e = temperature (°C)

Table 1. An annual cycle of the zoobenthic community characteristics in Lake Västra Kyrksundet (October 1981—October 1982) in relation to the basic hydrographical features of the bottom-near (14—18 m) water. S = number of species; A = total abundance (ind/m^2); B = total biomass (g wwt/m^2)

Date:	11.10 1981	22.05 1982	13.06 1982	23.06 1982	06.07 1982	19.07 1982	03.08 1982	23.08 1982	15.10 1982
t °C	11.8	7.1	7.5	7.5	8.4	10.7	12.3	14.5	11.4
pH	7.21	6.81	6.59	6.47	6.55	6.50	6.56	6.58	7.21
O_2 (mg/l)	9.28	8.58	5.81	3.60	3.04	2.19	1.54	1.20	9.95
S	3	4	4	4	2	1	2	2	4
A	1391	671	831	464	332	332	429	374	706
B	11.9	14.5	18.4	10.2	8.1	7.5	11.7	12.1	12.4

within the zoobenthic community remained the same also in 1983—85).

The fish community of Lake Västra Kyrksundet showed large variations throughout the investigation (Fig. 6), but the physical change in the environment in 1979 induced some major changes in the dominance relationships between species. Roach (*Rutilus rutilus*) and perch (*Perca fluviatilis*) totally dominated the standing stock until 1980, when bream (*Abramis brama*) and ruffe (*Acerina cernua*) increased in importance in terms of biomass (Fig. 6). Although the roach became less dominant in terms of biomass after 1979, the population structure (size frequency distribution) was not affected (Fig. 7), and thus it has been able to (with increasing levels of eutrophication) regain in importance until 1988 (Husö Biological Station; unpublished results). One reason for the slow "ecological return" of the roach may be the altered food chains, which was demonstrated in a reduced growth potential after 1979 (in August 1976 the length-weight relationship could be given as $y = 99.957 - 15.968\,x + 0.781\,x^2$, $r^2 = 0.982^{***}$, and in August 1985 as $y = 46.382 - 8.302\,x + 0.48\,x^2$; $r^2 = 0.962^{***}$). Thus the length-weight curve for the popula-

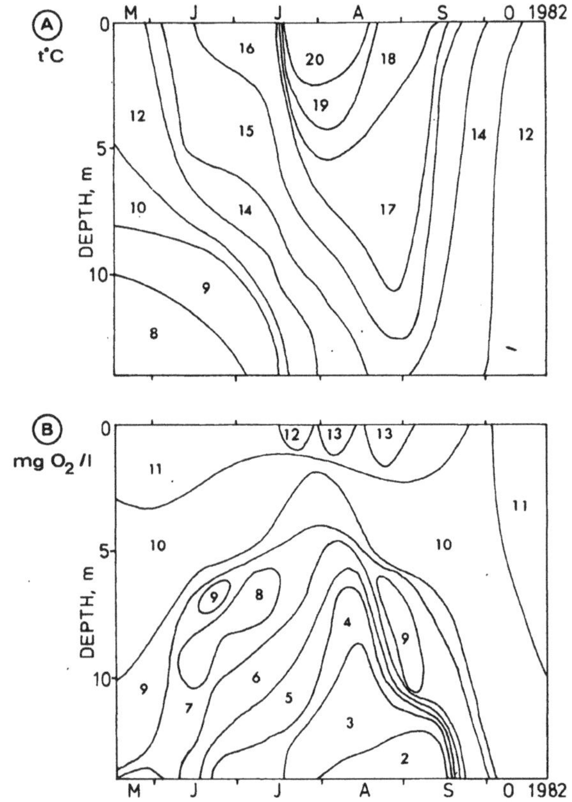

Fig. 4. The temperature (A) and oxygen (B) regimes of Lake Västra Kyrksundet exemplified by the productive season in 1982, three years after isolation from the sea (i.e. after changing the formerly meromictic conditions of the lake)

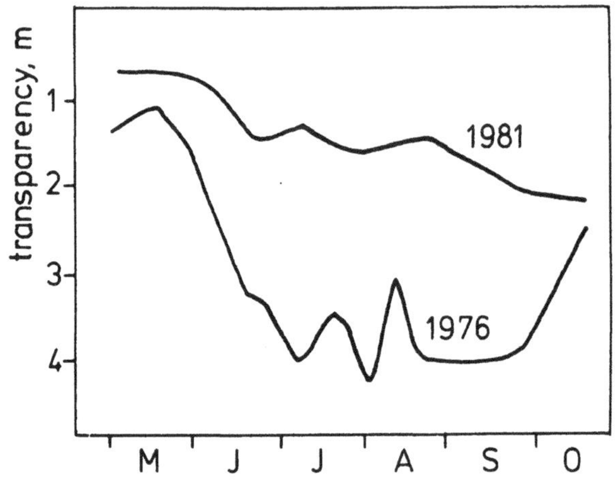

Fig. 5. The transparency (Secchi disk depth; m) during the productive season of Lake Västra Kyrksundet before (1976) and after (1981) isolation from the sea

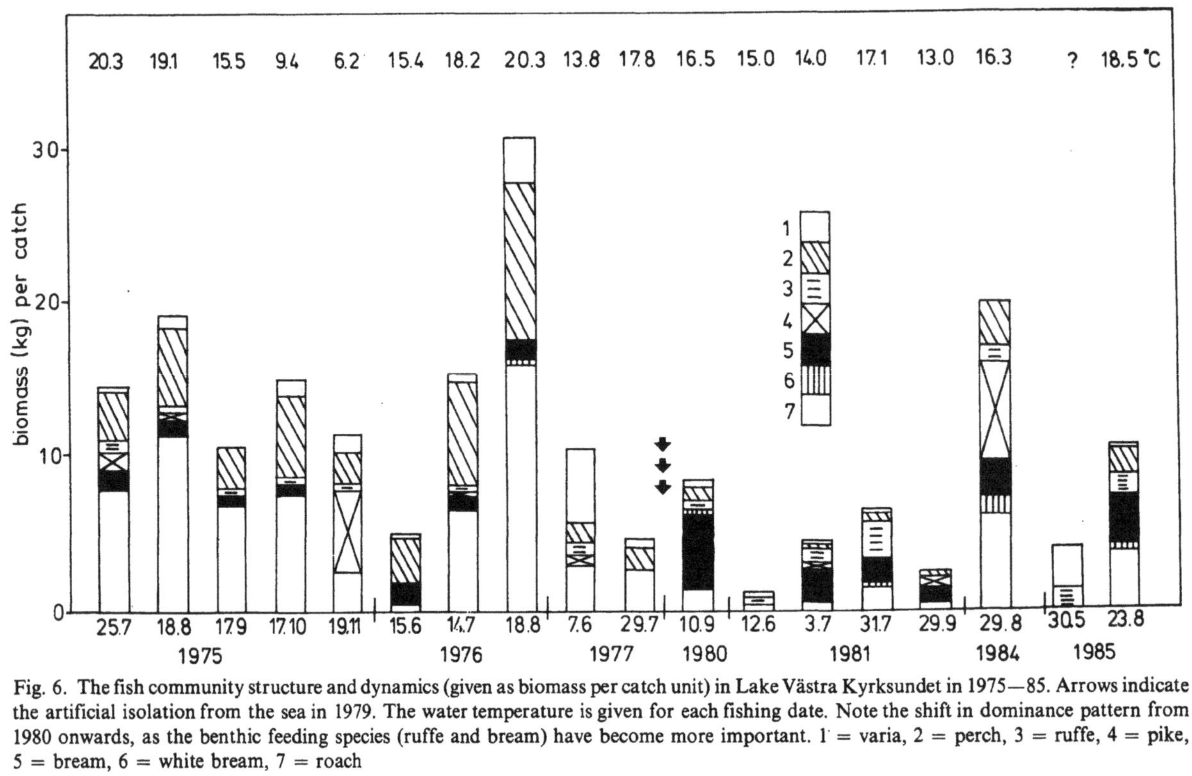

Fig. 6. The fish community structure and dynamics (given as biomass per catch unit) in Lake Västra Kyrksundet in 1975—85. Arrows indicate the artificial isolation from the sea in 1979. The water temperature is given for each fishing date. Note the shift in dominance pattern from 1980 onwards, as the benthic feeding species (ruffe and bream) have become more important. 1 = varia, 2 = perch, 3 = ruffe, 4 = pike, 5 = bream, 6 = white bream, 7 = roach

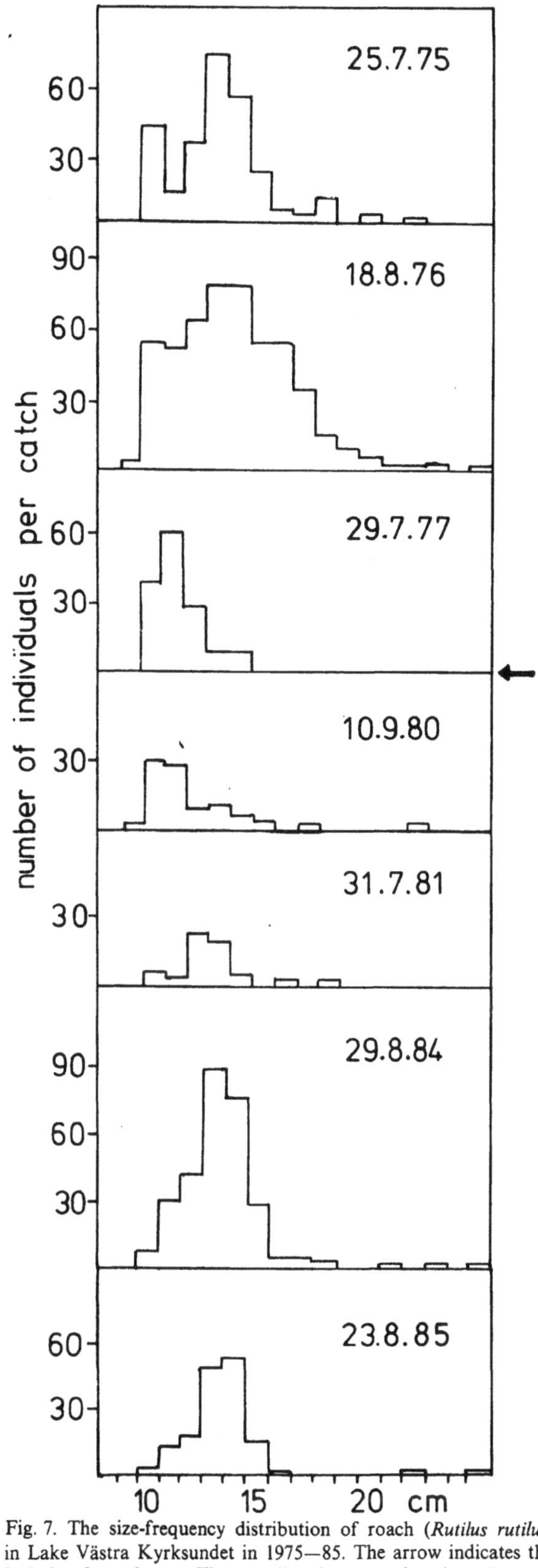

Fig. 7. The size-frequency distribution of roach (*Rutilus rutilus*) in Lake Västra Kyrksundet in 1975—85. The arrow indicates the isolation from the sea. The sampling dates are given in each graph

tion prior to 1979 indicated higher condition values than a few years after the isolation process.

Discussion

The rapid changes in hydrography (Figs. 2 and 3) after the isolation of the lake from the connection with the Baltic Sea, correspond with the data of LINDHOLM (1975a, b, 1982) from the same area. Since 1985, however, the dam between the lake and the canal to the surrounding archipelago, seems to have leaked occasionally, as slightly increased conductivity (25—40 mS/m), oxygen deficiency in August—September, and an increased transparency (1.9—3.1 m) were recorded in 1988 (Husö Biological Station; unpublished data). Thus it seems as though the lake rapidly (in about two years) developed towards purely limnic hydrographical conditions, no equilibrium point has been reached, and the water body is still sensitive to even small influxes of brackish water.

The recolonization of the deeper parts of the lake by benthic invertebrates demands oxic conditions not only in the water, but also in the surface sediment (MCLACHLAN & CANTRELL 1976). The reoxygenation of the water/sediment interface took about 5 months (the black, H_2S-containing topmost layer of the sediment gradually became brown, non-smelling, and the oxygen-content remained well above 0). The first chironomid larvae appeared in the zoobenthic samples in September 1979, which is partly explained by the improved conditions, but also by the mesh size of the sieve used (0.5 mm; the smallest larvae possibly escaped through the sieve), and the timing of the recruitment of chironomid eggs from the surface (JONASSON & KRISTIANSEN 1967). This will occur at the earliest in late May, when the first flying adults have hatched from the lake (WIEDERHOLM et al. 1977; PAASIVIRTA & KOSKENNIEMI 1987). HARGRAVE (1972) studied the oxygen demand of profundal sediments in eutrophic lakes, and found that an anoxic sediment core demanded about 60 days of constant oxygen supply (< 10 °C) before a steady state in oxygen demand was reached. The process of reoxygenation of sediments is also highly temperature-dependent (GRANÉLI 1978), possibly explaining the almost 5 month delay in Lake Västra Kyrksundet, as temperatures in the bottom water rose to around 15 °C, greatly increasing the oxygen consumption in the sediment. Consequently, in Lake Västra Kyrksundet the first immigrants were *Chironomus plumosus*-larvae, known to resuspend the surface-sediment, and also to withstand low oxygen concentrations (GRANÉLI 1979).

Several authors have described the early benthic colonization of lakes and reservoirs (e.g. PATERSON & FERNANDO 1970; FERRARIS & WILHM 1977; PAASIVIRTA & KOSKENNIEMI 1980, 1987, CASPERS 1983; VOSHELL & SIMMONS 1984), and it seems obvious that seasonality in both hydrography and food availability governs the annual cycle, as described in Table 1, with peak abundances in late summer or during the fall (BEATTIE 1982; JONSSON 1985). In markedly stratified waters, the oxygen content of the bottom waters may well be the determining factor, altering the seasonal cycle (JONASSON & KRISTIANSEN 1967). Finally, biotic couplings must also be taken into account, as fish predation seems to play an important role in regulating the standing stock of limnic invertebrates (HAYNE & BALL 1956; LAMMENS 1982).

The fish assemblage of Lake Västra Kyrksundet changed markedly in terms of dominance patterns (Fig. 6). This is partly due to the restriction in area caused by the dam, but mainly to the much increased local feeding area, as the lake was reoxygenated. The species that mainly benefited from this were the benthic feeding fish, viz. bream (*Abramis brama*) and ruffe (*Acerina cernua*). LAMMENS (1982) demonstrated the opposite process for bream, as its competitors for food expanded their feeding habitats. BREGAZZI & KENNEDY (1984) also showed that perch responded negatively to increasing eutrophication, and PERSSON (1983, 1986) has shown that the growth rate of roach suffers from increased competition for food (as in this case by bream). The reduced length-weight relationships recorded in the present study also correspond with the reverse pattern described for roach in a limnic ecosystem on Åland artificially influenced by highly nutritive brackish water (WIKLUND et al. 1984).

In recent years several attempts have been made to generalize the recovery patterns of various aquatic environments after physical or chemical disturbances, and one general conclusion has been that, although rapid, the early stages of ecosystem recovery and succession are vulnerable to repeated disturbances, and generally show large amplitudes of variation within the system (CONNELL & SLATYER 1977; LAMPERT 1987). The present study basically confirms that the initial recovery is rapid both for the abiotic factors (hydrography, sediment properties) and the biotic components at both population- and community level, but that the succession towards a stable ecosystem demands longer time intervals.

Acknowledgements

The work was conducted from Husö Biological Station (Åbo Akademi), which is thanked for providing both good working facilities and equipment. Mr. C.-S. ÖSTERMAN helped with much of the field- and laboratory work, Dr. E. M. BLOMQVIST and Dr. E. KOSKENNIEMI commented on the manuscript, which is gratefully acknowledged. The style of the presentation was also improved by Prof. B. PEJLER and an anonymous referee. Financial support was given by the Provincial Government of Åland (through Husö Biological Station), the Maj & Tor Nessling foundation, and the Academy of Finland.

References

BAZZAZ, F. A., & SIPE, T. W. (1987): Physical ecology, disturbance, and ecosystem recovery. In: E.-D. SCHULZE & H. ZWÖLFER (eds.). Ecol. Stud. **61**, 203—227.

BEATTIE, D. M. (1982): Distribution and production of the larval chironomid populations in Tjeukemeer. Hydrobiologia (The Hague) **95**, 287—306.

BREGAZZI, P. R., & KENNEDY, C. R. (1982): The responses of a perch, *Perca fluviatilis* L., population to eutrophication and associated changes in fish fauna in a small lake. J. Fish. Biol. **20**, 21—31.

CASPERS, N. (1983): Sukzessionsanalyse des Makrozoobenthos eines neu angelegten stehenden Gewässers. Arch. Hydrobiol., Suppl. **65**, 300—370.

CONNELL, J. H., & SLATYER, R. O. (1977): Mechanisms of succession in natural communities and their role in community stability and organization. Am. Nat. **111**, 1119—1144.

FERRARIS, C., & WILHM, J. (1977): Distribution of benthic macroinvertebrates in an artificially destratified reservoir. Hydrobiologia (The Hague) **54**, 169—176.

GRANÉLI, W. (1978): Sediment oxygen uptake in South Swedish lakes. Oikos **30**, 7—16.

— (1979): The influence of *Chironomus plumosus* larvae on the exchange of dissolved substances between sediment and water. Hydrobiologia (The Hague) **66**, 149—159.

HAMLEY, J. M. (1975): Review of gillnet selectivity. J. Fish. Res. Bd. Can. **32**, 1943—1969.

HARGRAVE, B. T. (1972): Oxidation-reduction potentials, oxygen concentration and oxygen uptake of profundal sediments in a eutrophic lake. Oikos **23**, 167—177.

HAYNE, D. W., & BALL, R. C. (1956): Benthic productivity as influenced by fish predation. Limnol. Oceanogr. **1**, 162—175.

JONASSON, P. M., & KRISTIANSEN, J. (1967): Primary and secondary production in Lake Esrom. Growth of *Chironomus anthracinus* in relation to seasonal cycles of phytoplankton and dissolved oxygen. Int. Rev. ges. Hydrobiol. **52**, 163—217.

JONSSON, E. (1985): Population dynamics and production of Chironomidae (Diptera) at 2 m depth in Lake Esrom, Denmark. Arch. Hydrobiol., Suppl. **70**, 239—278.

LAMMENS, E. H. R. R. (1982): Growth, condition and gonad development of bream (*Abramis brama* L.) in relation to its feeding conditions in Tjeukemeer. Hydrobiologia (The Hague) **95**, 311—320.

LAMPERT, W. (1987): Predictability in lake ecosystems: the role of biotic interactions. In: E.-D. SCHULZE & H. ZWÖLFER (eds.), Ecol. Stud. **61**, 333—346.

LINDHOLM, T. (1975a): Coastal meromictic lakes on Åland (SW Finland). Aqua Fennica **1975**, 24—40.

— (1975b): Långsjön och Markusbölefjärden före och efter isoleringen från havet. Husö Biol. Stat. Medd. **16**, 28—49.

— (1982): Dynamics of hydrography and primary production in three stratified coastal lakes on Åland (SW Finland). Acta Acad. Aboensis, Ser. B. **42**, 1—75.

— ERIKSSON, J. E., & MERILUOTO, J. A. (1989): Toxic cyanobacteria and water quality problems — examples from a eutrophic lake on Åland, SW Finland. Water Res. **23**, 481—486.

MCLACHLAN, A., & CANTRELL, M. A. (1976): Sediment development and its influence on the distribution and tube structure of *Chironomus plumosus* L. (Chironomidae, Diptera) in a new impoundment. Freshwater Biol. **6**, 437—443.

PAASIVIRTA, L., & KOSKENNIEMI, E. (1980): The Chironomidae (Diptera) in two polyhumic reservoirs in western Finland. In: D. A. MURRAY (ed.), Chironomidae: Ecology, systematics, cytology and physiology, pp. 233—238.

— (1987): The chironomid (Diptera) fauna in a Finnish reservoir during its first four years. Entomol. scand. Suppl. **29**, 239—246.

PATERSON, C. G., & FERNANDO, C. H. (1970): Benthic fauna colonization of a new reservoir with particular reference to the Chironomidae. J. Fish. Res. Bd. Can. **27**, 213—222.

PERSSON, L. (1983): Food consumption and the significance of detritus and algae to intraspecific competition in roach *Rutilus rutilus* in a shallow eutrophic lake. Oikos **41**, 118—125.

— (1986): Effects of reduced interspecific competition on resource utilization in perch (*Perca fluviatilis*). Ecology **67**, 355—364.

STORBERG, K.-E. (1980): Fiskbeståndet i fem åländska kustsjöar (Inre Verkviken, Kyrksunden, Långsjön och Markusbölefjärden). Rekommendationer och åtgärdsförslag. Forskn. rapp. till Ålands Landskapsstyrelse. Husö Biol. Stat. N. S. **14**, 1—26.

— (1982): Fiskbeståndet i Västra Kyrksundet åren 1975—1981. Forskn. rapp. till Ålands Landskapsstyrelse. Husö Biol. Stat. N. S. **26**, 1—10.

VOSHELL, Jr., J. R., & SIMMONS, G. M. (1984): Colonization and succession of benthic macroinvertebrates in a new reservoir. Hydrobiologia (The Hague) **112**, 27—39.

WIEDERHOLM, T., DANELL, K., & SJÖBERG, K. (1977): Emergence of chironomids from a small man-made lake in northern Sweden. Norw. J. Entomol. **24**, 99—105.

WIKGREN, B.-J. (1965): Salt vatten i insjöar. Husö Biol. Stat. Medd. **8**, 1—36.

WIKLUND, T., WISTBACKA, B., & KOIVISTO, V. (1984): Brackish water invading a lake system (Åland Islands, N. Baltic Sea). Ophelia, Suppl. **3**, 259—263.

Address of the authors:

ERIK BONSDORFF
Department of Biology and Husö Biological Station
Åbo Akademi
SF — 20500 Åbo
Finland

KARL-ERIK STORBERG
National Board of Water and Environment
Vaasa District Office
PB 262
SF — 65101 Vaasa
Finland

Limnologica (Berlin) **20** (2) 285—289, Juni 1990

Norwegian Institute for Water Research (NIVA), Oslo, Norway

Lake Mjøsa, Recipient and Ecosystem

By HANS HOLTAN

With 8 Figures

Key words: Eutrophication; Phosphorus; Algal bloom; Water flow; Pollutional load.

Abstract

Up to about 1950, Lake Mjøsa, the largest lake in Norway, was a harmonious oligotrophic lake. At that time, it was not considered that a large and deep lake like Mjøsa, could be affected by pollution. The lake was therefore used as recipient for untreated crude sewerage water from municipalities, industry and agricultural activities. As a result the lake became more and more eutrophic. The activities related to the lake (drinking water, bathing, recreation, fishing etc.) became seriously affected due to the algal bloom. In 1975 and 1976 the water could hardly be used as drinking water.

Due to the rapid deterioration of the water quality, the national authorities initiated in 1976 the "Save Mjøsa Campaign". In the following years, from 1976 to 1980, the pollutional load was reduced considerably. In spite of this campaign, the ecological balance has up to now not been fully reestablished. Still further action in reducing the pollutional load is required.

Introduction

During the summer seasons of 1975 and especially of 1976 the bluegreen alga *Oscillatoria bornetii* fa. *tenuis* made a dramatic appearance in the largest lake in Norway — Lake Mjøsa. This alga produces organic substances (geosmin) which gives the water a very unpleasant taste and odour. The substances passed through the water works, and the tapwater could hardly be used for drinking. The problems of taste and odour increased even further when the water was chlorinated. These incoveniences affected a population of approx. 200.000 persons along the lake and downstream the watercourse.

On the basis of the results available at the Norwegian Institute for Water Research (NIVA), from long-term studies which started in 1966, one concluded that the reasons for these happenings were clear: Too much municipal waste water, industrial pollution and polluted water from agricultural activities were released into the lake. Because of the dramatic events, the authorities started a "Save Mjøsa Campaign", and in the following years one worked hard for reducing the pollutional load to the lake. Since that time the pollution situation in the lake has been under observation by a monitoring programme initiated by The Norwegian State Pollution Control Authority. NIVA has been responsible for the research activities.

Lake Mjøsa Basin

During a very long geological time and several glacier periods, Lake Mjøsa has been dug out along some permian faults (SKJESETH 1963). When the last glacier melted, the moraine material dammed up the lake a little more. The outlet river, Vorma, has now excavated its way through this moraine. These formation processes resulted in a long, narrow and deep basin — a fjord lake basin. From north to south the length of the lake is 117 km, but the largest width is only 9 km (ØSTREM 1984). With a maximum depth of 449 m, Lake Mjøsa is Norway's and Europe's fourth deepest lake. The deepest point is about 328 m below the present sea level. The mean depth is 153 m. The shoreline is mostly steep and rocky, and macro-vegetation does not exist in the littoral zone.

Mjøsa is a reservoir for hydro-electric power production, and the lake is dammed up 3.61 m which is equivalent to a reservoir approx. 2.3 per cent of the lake volume.

Catchment Area

At the outlet (Vorma), the catchment area (Fig. 1) is 16.420 km^2 or about 45 times the surface area (362 km^2) of the lake. Sixty to seventy per cent of this area consists

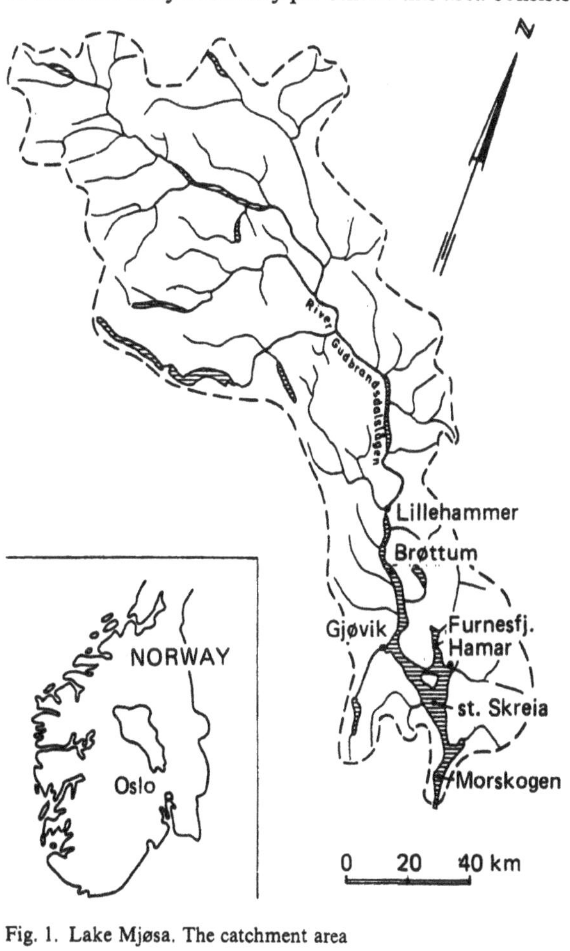

Fig. 1. Lake Mjøsa. The catchment area

of mountainous regions with gabbroic and granitic bedrocs and many glaciers. Parts of the famous mountain chains, Jotunheimen, Dovre and Rondane, belong to these regions, including the highest mountain peak in Norway, Galdhøpiggen (2469 m).

The drainage water from the mountainous regions is carried to the lake by the river Gudbrandsdalslågen which is the main tributary. In the summer time the river water is very silty because of the glacier-eroded substances. Along the river Gudbrandsdalslågen and its main tributaries there are agricultural areas and some smaller communities. At the present time about 50,000 persons are living in the valley system. The steep valley sides are covered with forest.

Around the central zone of the lake there are sedimentary and metamorphic rocks of cambro-silurian age and some calcareous rock types. This district is one of Norway's best agricultural areas (about 18 per cent of the total area).

Approximately 150,000 persons are living in the immediate surroundings of the lake. The most important population densities are found in the towns of Lillehammer, Gjøvik and Hamar, but many persons are also living in smaller communities. Approx. 75,000 inhabitants live in the rural districts.

The wood pulp industry, food industry and metallurgic industry are among the larger industries around Mjøsa.

Hydrological Conditions

Approximately 75 per cent of the water masses in Mjøsa flow through the river Gudbrandsdalslågen which has a mean annual flow of 250 m^3/s. The river is regulated for hydro-electric power production, and the mean winter water flow has thus increased from 64 to 104 m^3/s, while the mean summer flow has decreased from 509—451 m^3/s.

The snow-melting processes in the low-land areas are taking place during April—May, and sometimes these processes can cause so-called low-land flow. Later on, during June—July, when snow and ice in the mountains are melting, high flow occurs, which is especially heavy during periods of warm and rainy weather. Approximately 60 per cent of the annual water flow takes place during this period. Flow-peaks up to 2000 m^3/s are not unusual at that time.

Water Circulation

The general temperature variations in Lake Mjøsa are in agreement with what is usual in temperate lakes (SCHIØTZ 1887; HUITFELDT-KAAS 1905; HOLTAN 1973). Down to approximately 200 m below surface, the vernal circulation period lasts from the latter half of April to the end of June, while the autumnal circulation lasts from October to January/February. During some winters the central zones are even icefree.

The deep waters (below 200 m) appear to be unstable during the icebound period in winter. From Fig. 2 it can be seen that a stable thermal stratification is not established until June, and this lasts right up until autumn, and winter cooling starts in the deeper water. Below 200 m the temperature of the water always ranges between 3.4 and 4.0 °C.

The water from the main tributary, river Gudbrandsdalslågen, flows through the whole lake along the right-hand shoreline (Coriolis forces). The wind, however, which normally is blowing along the longitudinal direction of the lake, generates currents and internal waves (seiches).

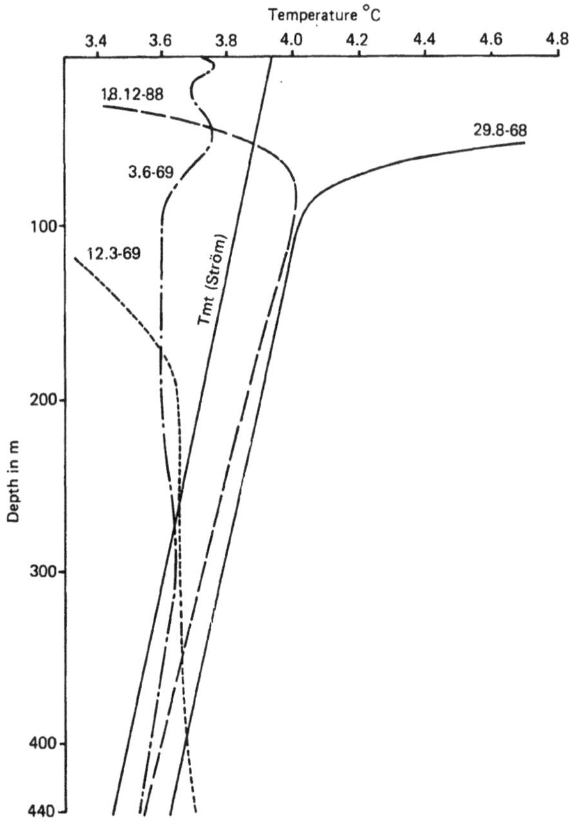

Fig. 2. The temperature distribution of Lake Mjøsa 1968 to 1969. STRØM's temperature curve for maximum density is given (STRØM 1945)

Water circulation in Lake Mjøsa is thus a complex system involving wind currents, currents caused by internal seiches, and gravitation currents. The surface water temperature can change from one day to another due to this complex system.

In the summer the epilimnic water is approximately 10 per cent of the lake volume. When the thermocline is established in late June, the water from the tributaries flows through the lake in the upper layer. This is important to the biological production in the lake.

Water Chemistry

The water in Lake Mjøsa is soft and has a pH of approx. 7 (Table 1). The main chemical components have been fairly stable during the last 20 years. The sulphate content has probably increased and the hydrogen carbonate decreased a little. The concentration of heavy metals and aluminium in the water is fairly low. Originally Lake Mjøsa had much lower values of nutrients than what appears from the table.

Eutrophication of Lake Mjøsa

During the first half of this century Lake Mjøsa was an almost unpolluted (oligotrophic) lake (HOLMBOE 1900; HUITFELDT-KAAS 1906, 1916, 1946; STRØM 1930; BRAARUD et al. 1928; LINDSTRØM et al. 1973), but the produc-

Table 1. Mean chemical values of the water quality in Lake Mjøsa, 1986.

pH	Cond (mS/m)	Ca^{++} (mg/l)	Mg^{++} (mg/l)	Na^{+} (mg/l)	K^{+} (mg/l)	SO_4^{--} (mg/l)	Cl^{-} (mg/l)	HCO_3^{-} (mg/l)	Tot P (μg/l)	PO_4-P (μg/l)	Tot N (μg/l)	NO_3-N (μg/l)	Chl **a** (μg/l)
6.9	4.3	5.4	0.7	1.0	0.7	9.9	1.3	11.8	6.1	3.1	550	415	2.6

tion capacity was greater than in some other similar fjordlakes in Norway. Species of the diatom group, e.g. *Asterionella formosa* and even *Melosira granulata*, were important. Some years, in the middle of the summer, a bloom of the green alga *Botryococcus braunii* was reported.

During the period until the late 1940s there seem to have been only small changes in the biomass and in the algal composition. However, from about 1930 the diatom *Asterionella formosa* and the bluegreen *Anabaena flos-aquae* have been increasingly more abundant. After 1950 the algal biomass increased from year to year. *A. formosa* and *A. flos-aquae* were the dominating species. Along the shorelines the attached green alga *Ulothrix zonata* became more abundant. During the 1960s the algal biomass increased considerably. Many of the clean-water species disappeared and diatoms such as *Fragilaria crotonensis*, *Diatoma elongatum*, *Stephanodiscus* spp. and *Melosira granulata* var. *angustissima* became more and more dominating. *Anabaena flos-aquae* formed water blooms during warm and calm weather.

This development continued during the 1960s and 1970s. The situation culminated in 1975 and especially in 1976, when the bluegreen alga *Oscillatoria bornetii* fa. *tenuis* made a dramatic appearance and caused the water to be unsuitable for use (drinking, bathing and so on) (HOLTAN 1979).

"Save Mjøsa Campaign"

Because of these dramatic events, the authorities were convinced about the seriousness of the situation and started a "Save Mjøsa Campaign".

Before the "Save Mjøsa Campaign" started in the 1970s, all municipal and industrial waste water, as well as the drainage water from agricultural areas etc., was flowing untreated into the lake or its tributaries (HOLTAN 1979). During the summer period, after the thermocline was established in June, the polluted water affected mainly the epilimnion. E.g. 40 per cent of the annual phosphorus load was supplied to surface layer of the lake during the production season (HOLTAN 1981). In this way the concentrations of nutrients, e.g. phosphorus which is the key element for the eutrophic situation in Lake Mjøsa, were very high in the epilimnion. From time to time phosphorus concentrations of 20 to 25 μg P/l were measured in the surface layer. The mean concentration of total phosphorus in the entire lake was approx. 10 μg P/l in the middle of the 1970s.

During the "Campaign" period, 42 new sewage treatment plants for retention of phosphorus were built and about 280 km new sewer pipes were installed (HOLTAN 1981). As a result of all this work, the total supply of phosphorus to Lake Mjøsa was reduced from approx. 310 tons in 1976 to approx. 225 tons in 1980.

When the "Campaign"-period was ended in 1980, the algal production and biomass had decreased considerably. The mean chlorophyll **a** content, which was measured to about 5 μg Chl **a**/l as a mean during the summer season in 1976, was reduced to only half of this value (2.5 μg Chl **a**/l) in 1980. The value of primary production, which was 570 mg C/m^2 · day in 1976, was only 110 mg C/m^2 · day in 1980. The Secchi depth during the production periods had increased from 4 meters in 1976 to about 6 meters in 1980.

Present Situation

As can be seen from the graphs (Figs. 3, 4, 5 and 6), the reduction of phosphorus has been of great importance to the primary production and algal biomass in the lake.

After 1980 (KJELLBERG 1982, 1986, 1988) the values of primary production, algal biomass, and Secchi depth have varied from year to year. The primary production values seeming to have an increasing trend. The algal community has in the late seasons been dominated by the same groups as in the 1950s and 1960s, viz. diatoms (mostly *Asterionella formosa*, *Tabellaria fenestrata*, *Fragilaria crotonensis*), while Cryptophyceae (*Cryptomonas* spp., *Katablepharis ovalis* and *Rhodomonas lacustris*) have been dominating in spring and early summer.

The bluegreen alga *Oscillatoria bornetii* fa. *tenuis*, which dominated the algal communities in the middle

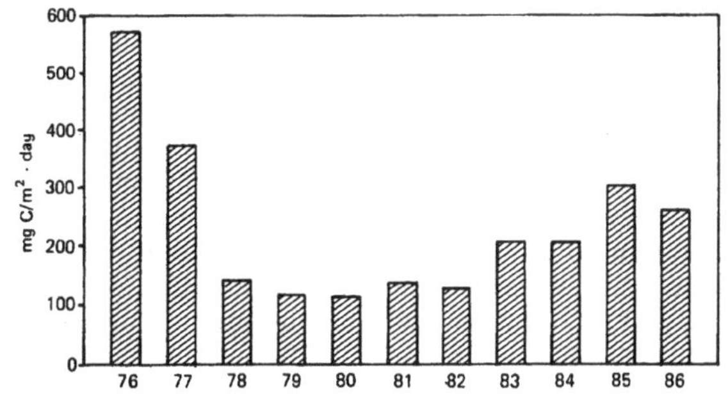

Fig. 3. Mean values for primary production during summer (15/5—31/10) in the central parts of the lake

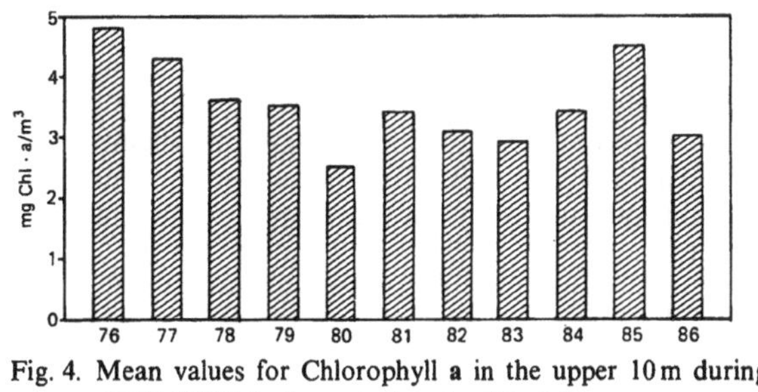

Fig. 4. Mean values for Chlorophyll **a** in the upper 10 m during summer (15/5-31/10) in the central parts of the lake

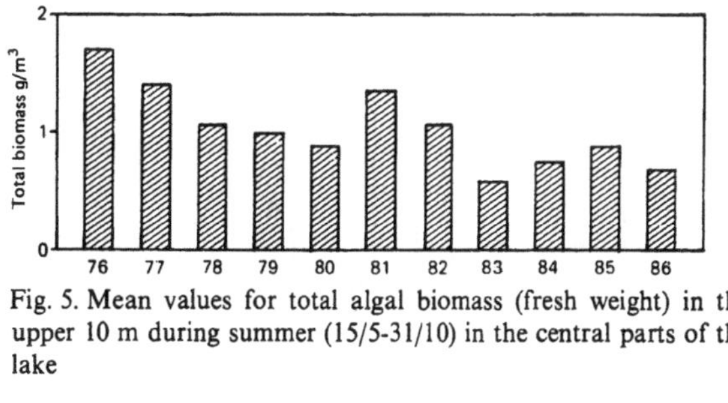

Fig. 5. Mean values for total algal biomass (fresh weight) in the upper 10 m during summer (15/5-31/10) in the central parts of the lake

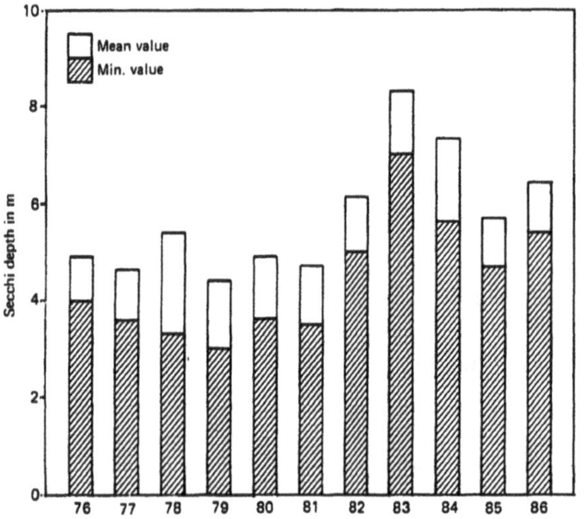

Fig. 6. Mean values for Secchi depth transparency in the central part of the lake from the end of June to the beginning of September

of the 1970s, has only been present in small quantities in the 1980s (Fig. 7). *Anabaena flos-aquae* has been present every year in the surface layer during the late summer. This alga is very inconvenient for bathing activities, but the algal volume compared with the volume of other algal species is small.

Discussion

The phosphorus concentration in Lake Mjøsa has always been low compared with other eutrophic lakes. VOLLENWEIDER (1976) considers 10 µg P/l to be a safe lower concentration limit for phosphorus. In Lake Mjøsa and many other Norwegian lakes it seems to be necessary from a eutrophic point of view to keep the concentrations down to a far lower level. Experiences show that it would be more realistic to set a lower limit at 7 µg P/l. This empirical value also seems to agree well with tolerable concentration values which have been arrived at through experiments (TRAAEN 1975). The eutrophic development in a lake is not only dependent on the phosphorus load. A number of other biological, physical and chemical factors are also involved.

In Lake Mjøsa, the low silica concentrations in late summer may influence the algal successions during this season since this component is limiting for the growth of the diatoms (LUND 1965; SCHELSKE & STOERMER 1972; KILHAM 1971; TITMAN 1976). These authors maintain that the increased addition rate of phosphate relative to silicate causes an increasingly large number of diatoms to become silicon-limited, driving down the silica concentrations and thus eliminating species that had formerly been dominant because they were superior competitors for phosphate. As even more phosphate is added relative to other nutrients, those other nutrients will start limiting other species in the lake (KILHAM & TILMAN 1977).

The diatoms *Asterionella formosa*, *Fragilaria crotonensis* and *Tabellaria fenestrata* are normally the dominating algae in Lake Mjøsa. However, in 1975 and 1976 the bluegreen alga *Oscillatoria bornetii* fa. *tenuis* was dominating in the late summer. The silicon content decreased parallelly with the diatom production and during the bluegreen period the silica concentration was very low (about 0). In the vernal circulation period the silica concentration is about 2 mg SiO_2/l.

Variations in the climatic conditions have also to be taken into consideration when we discuss eutrophication problems. In the middle of the 1970s, when the bluegreen alga (*Oscillatoria*) bloomed in Lake Mjøsa, the weather during the production season was relatively warm and dry. Till then municipal raw sewage and industrial waste water (point sources) from the surroundings were transported untreated into the surface layer (epilimnion) of the lake. Consequently the basis of a heavy algal bloom was optimal. During the "Campaign" period, from 1976 to 1980, the point sources of e.g. phosphorus were heavily reduced. Therefore, today fertilizers and manure from agricultural activities are of greater importance compared with other sources. During summers with rainy weather, nutrients from the agricultural areas are to a greater extent washed into the lake and its tributaries than during dry weather periods. Consequently, the primary production and algal biomass are now greater in years with cold and rainy weather than in years with dry weather during the production periods.

In spite of the "Lake Mjøsa Campaign", the water quality in the surface layers of the lake is not satisfactory for bathing and other recreational activities. Moreover, the primary production seems to have increased a little in the later years, but the bluegreens are only present in small amounts. The authorities are now very apprehensive about the development and afraid of a new algal bloom. They have therefore now worked out a plan for further efforts against pollutional activities around the lake. This includes further rehabilitation of municipal sewage plants, efforts against agricultural drainage, and construction of more treatment plants against industrial wastes.

In the deep water strata the water quality seems to have improved (Fig. 8). The oxygen situation is on the same level as before; saturation values of ca. 80 per cent or more near the bottom. The crustacean plankton,

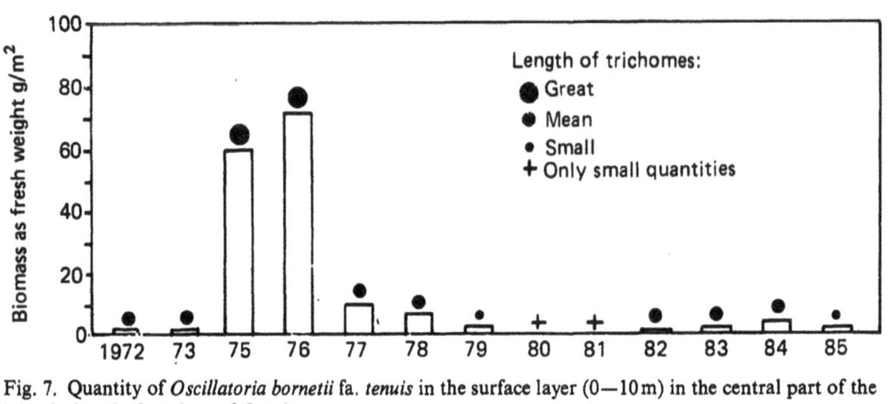

Fig. 7. Quantity of *Oscillatoria bornetii* fa. *tenuis* in the surface layer (0—10 m) in the central part of the lake during the last days of October

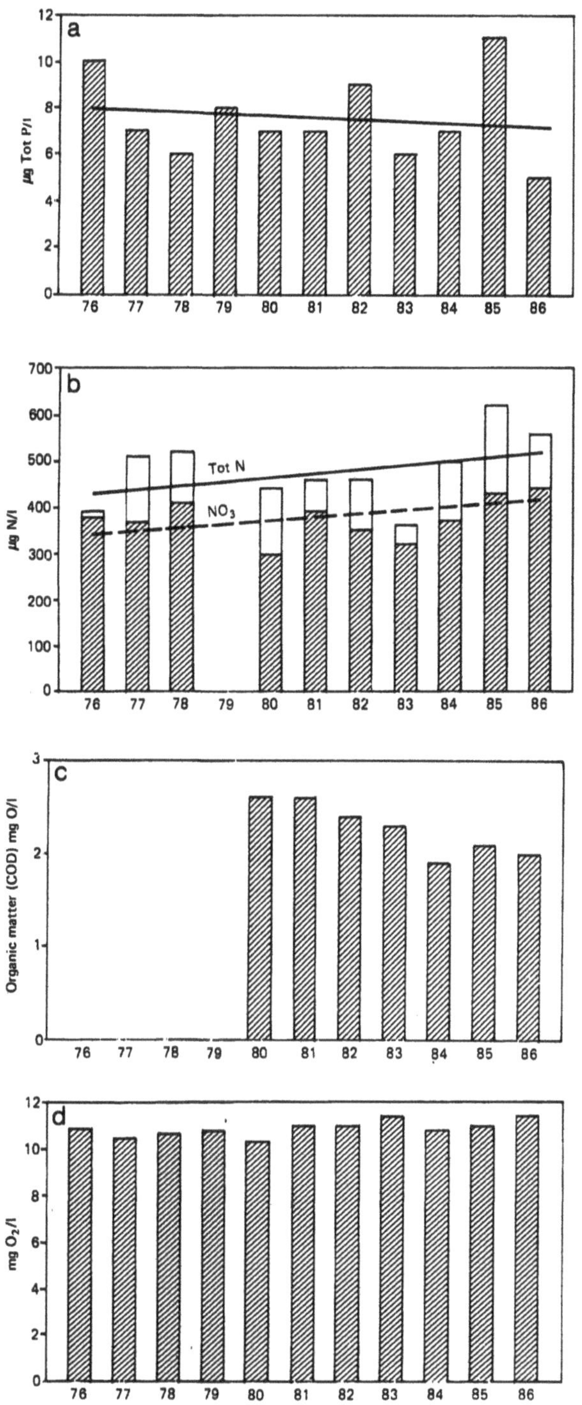

Fig. 8. Mean values for total phosphorus (a), total nitrogen and nitrate (b), organic matter (c) and oxygen (d) in the depth (200—400 m) in the central part of the lake during the vernal circulation period. (May)

bottom fauna and fish communities have not changed during the last years (HOLTAN 1979). Therefore it is believed that Lake Mjøsa can be at an ecological steady state if the concentrations of nutrients are kept on the same, or preferably a somewhat lower level than today.

Acknowledgement

The author would like to thank Dr. GØSTA KJELLBERG at the Norwegian Institute for Water Research, who has been the scientific leader for the water management programme running in Lake Mjøsa.

Literature

BRAARUD, T., FØYN, B., & GRAN, H. H. (1928): Biologische Untersuchungen in einigen Seen des östlichen Norwegens. August—September 1927. Avh. Det Norske Vidensk. — Akad., Oslo. Mat. Nat. V. sk. kl. nr. **2**, 1—37.

HOLMBOE, J. (1900): Undersøgelser over norske ferskvandsdiatoméer. Arch. Mat. Naturv. **22**, 1—72.

HOLTAN, H. (1973): Temperature distribution and water circulation in Lake Mjøsa. Symposium on Hydrology of Lakes, Helsinki 1973. Proceedings. U.S., Int. Ass. of Hydr. Science **1973**, 42—48.

— (1979): The Lake Mjøsa Story. Arch. Hydrobiol., Beih. **13**, 242—258.

— (1981): Eutrophication of Lake Mjøsa and its recovery. Wat. Qual. Bull., Vol. 6, No. 4, 1981.

HUITFELDT-KAAS, H. (1905): Temperaturmessungen in dem See Mjøsen und in drei anderen tiefen norwegischen Seen. Arc. Mat. Nat. **BXXVII**.

— (1906): Planktonundersøgelser i Norske Vande. Christiania.

— (1916): Mjøsens fisker og fiskerier. Det Kgl. Norske Videnskabers Selskabs Skrifter Nr. **2**, 1—157.

— (1946): The plankton in Mjøsa. Nytt Mag. Naturvid. **85**, 161—221.

KILHAM, P. (1971): A hypothesis concerning silica and the freshwater planktonic diatoms. Limnol. Oceanogr. **16**, 10—18.

— & TILMAN, D. (1977): The importance of resource competition and nutrient gradients for phytoplankton ecology. Arch. Hydrobiol., Beih. **13**, 110—119.

KJELLBERG, G. (1982): Overvåking av Mjøsa. Bakgrunnsdata, historikk og videreføring. Del. A. NIVA-rapport nr. **54/82**, 104 pp.

— (1986): Overvåking av Mjøsa. Sammendrag, trender og kommentarer 1976—85, NIVA-rapport nr. **241/86**, 80 pp.

— (1988): Tiltaksorientert overvåking i 1987 av Mjøsa. NIVA-rapport nr. **320/88**, 59 pp.

LINDSTRØM, E. A., SKULBERG, R., & SKULBERG, O. M. (1973): Observations on planktonic diatoms in the lake-river system Lake Mjøsa — Lake Øyeren — River Glåma, Norway. Norwegian J. Bot. **20**, 183—195.

LUND, J. W. G. (1965): The ecology of the freshwater phytoplankton. Biol. Rev. **40**, 231—293.

SCHELSKE, C. L., & STOERMER, E. F. (1972): Phosphorus, silica and eutrophication of Lake Michigan. Amer. Soc. Limnol. Oceanogr. Spec. Symp. **1**, 157—171.

SCHIØTZ, O. E. (1887): Resultatet af en del undersøgelser over temperaturforholdene i dypet av Mjøsen. Forh. Skand. Naturf. Møte **13**.

SKJESETH, S. (1963): Contributions to the geology of the Mjøsa districts and the classical sparragmite area in Southern Norway. Norges Geologiske Undersøkelse nr. 220, 129 pp.

STRØM, K. M. (1930): Limniological observations on Norwegian Lakes. Arch. Hydrobiol. **21**, 97—124.

— (1945): The temperature of maximum density in fresh waters. Geofys. Publ. **16**, No. 8, 145 pp.

TITMAN, D. (1976): Ecological competition between algae: Experimental confirmation of resource-based competition theory. Science **192**, 463—465.

TRAAEN, T. (1976): Vassdragsbiologi, Virkninger av rensetekniske tiltak. Prosjektkomiteen for rensing av avløpsvann (PRA) nr. **13**, 40 pp.

VOLLENWEIDER, R. A. (1976): Advances in defining critical loading levels for Phosphorus in Lake Eutrophication. Mem. Ist. Ital. Idrobiol. **33**, 53—83.

ØSTREM, G., FLAKSTAD, N., & SANTHA, J. M. (1984): Dybdekart over norske innsjøer. Norwegian Water Resources and Electricity Board 1984, 128 pp.

Address of the author:

HANS HOLTAN
Norwegian Institute for Water Research
P. O. Box 69 Korsvoll
N-0808 Oslo
Norway

Limnologica (Berlin) **20** (2), 291—306, Juni 1990

Institute of Limnology, University of Uppsala, Uppsala, Sweden

Acidification and Phytoplankton Development in Some West-Swedish Lakes 1966—1983

By GRETA MORLING & TORBJÖRN WILLÉN

With 9 Figures

Key words: Phytoplankton; Water chemistry; Acidification; Phytoplankton biomass; Chlorophyll; Liming.

Abstract

During the period 1966—1983 biological and chemical changes were analysed in some acidified lakes and in a non-acidified lake in the area of Lygners Vider, Western Sweden. Three different phases of the acidification process could be distinguished: a "normal" phase (1966—1969), the real acidification phase (1970—1977), and thirdly, the stabilized or slightly recovering phase (1978—1983). The acidification did not markedly change the number and composition of phytoplankton, but corresponding phases of the phytoplankton development and abundance could be followed in detail. Several cyanophyceans, some chrysophyceans and chlorophyceans and a few other algae disappeared after phase I. During phase III a smaller recolonization was observed. Among the 191 taxa of net phytoplankton observed in the lakes, not less than 101 belonged to Conjugatophyceae. Although most of them were recorded during the whole period, some examples are given of taxa only occurring during phases I and III. No marked changes in chlorophyll **a** concentration were recorded. The somewhat divergent phytoplankton development of the non-acidified Lake Änketjärn is discussed.

1. Introduction

The acidification of soil and water in Sweden has been the subject of comprehensive studies and documentation during recent decades. The acidification course in a lake often is illustrated in the shape of a process running through three phases (Monitor 1986). During phase I the pH value is rather high and stable. At a mean alkalinity value of 0.05 mequ $\cdot l^{-1}$ a marked decrease of pH is recorded: the lake has reached the unstable phase II. Further acid additions to the lake give an alkalinity value of zero resulting in stabilized pH values at a level below 5. In the present study three phases also are distinguished but which are not quite identical with those just described: phase III implies a slight tendency to "deacidification" in some lakes (FORSBERG & MORLING 1987).

An investigation of phytoplankton and physical/chemical conditions was started in 1966/1967 in a number of oligotrophic lakes in western Sweden, originally planned as a study of the ecological conditions of different *Ceratium* species (MORLING 1979). After some years, however, the chemical conditions of the lakes drastically changed, with increasing values of sulphate and decreasing pH values (MORLING 1981). At the same time marked changes in phytoplankton composition were recorded. These obvious and rapid changes led to further investigations of both the physical/chemical conditions and the phytoplankton development. In some lakes these analyses were made continuously from 1966/1967 to 1982/1983.

The aim of the whole investigation is to study the acidification process in some west-Swedish lakes. In this paper the main stress is laid on water-chemical changes and their consequences for phytoplankton composition and frequency. In another paper the zooplankton development is discussed (MORLING & PEJLER 1990, this volume).

2. Material and Methods

The lakes investigated are situated on Lygners Vider plateau in western Sweden (Figs. 1 and 2). Background data on environment, climate, lake morphology and water chemistry were supplied earlier (MORLING 1981).

As a rule, samples for phytoplankton analyses were collected once a month and sometimes — during periods of marked changes in the phytoplankton composition — every fortnight during the growing period from April until October. Samples were also taken during the period November—January in years when the lakes did not freeze.

Among the fourteen lakes investigated some were paid special attention by frequent samplings, viz. lakes Stora Hålevatten, Oxsjön, Lilla Svansjön, Stora Svansjön and Änketjärn.

All samplings were performed from a row-boat in the pelagic area of the lakes. Samples for species identification were taken by net (mesh size 63 μm) from the surface layer. For quantitative analysis, water samples making up a total volume of 20 l have been collected. In most cases, this water volume was concentrated by pouring through the plankton net. Thus, the whole study is based on net plankton material.

Species identification was performed on living samples. For later and more thorough analyses all samples were preserved with IKI solution or formaldehyde. Quantitative analysis (UTERMÖHL 1958) was made on samples preserved with IKI solution + acetic acid.

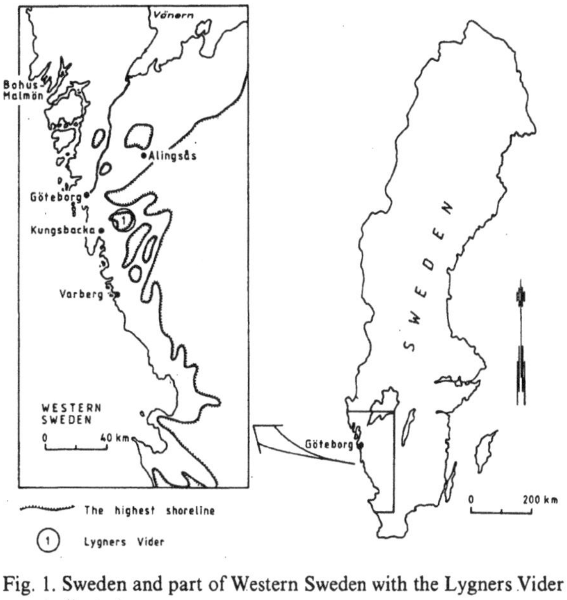

Fig. 1. Sweden and part of Western Sweden with the Lygners Vider area indicated

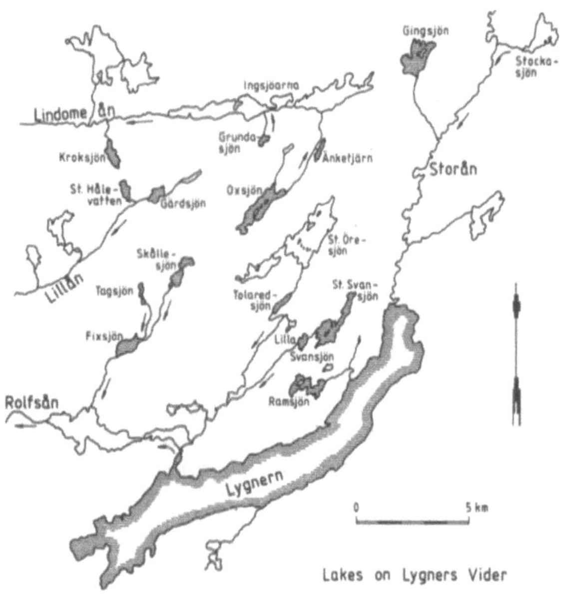

Fig. 2. Detailed map of Lygners Vider with the lakes investigated named and shaded

These samples were allowed to sediment in 10 ml (occasionally in 1 or 100 ml) KOLKWITZ-UTERMÖHL chambers for 6 h. Counting was performed by using an inverted plankton microscope. Depending on the frequency of the species, sometimes several chambers of the same sample were counted to provide more reliable mean values of the number per litre (cells or colonies).

The systematic grouping of phytoplankton (Appendix 1) follows FOTT (1971) with a few exceptions. Statistical tests of the changes of some common taxa between the different phases have been performed using the Kruskal-Wallis and Mann-Whitney tests (probability level 0.10).

The programme of this investigation did not include estimations of the total phytoplankton biomass based on number and volume of different species. Some chlorophyll a analyses, however, were performed to provide a measure of the biomass and its variation. These chlorophyll analyses according to TOLSTOY (1969) were performed two to three times per lake and season.

With one exception, all lakes investigated have been limed after 1979. Only Lake Änketjärn is regarded as a reference lake and here the liming has been postponed.

3. Results

3.1. The three phases

During the first period of this investigation the chemical and biological conditions were rather unchanged in the lakes (Fig. 3). In about 1970, however, rapid water chemical changes occurred: increasing values of sulphate, alkalinity decreasing to zero followed by decreasing pH values. After that the concentration levels were stabilized; some tendencies, however, for further changes were seen. At the end of the investigation period higher pH values were recorded in several lakes. These pH changes also resulted in secondary chemical changes as well as changes in the phytoplankton composition (cf. MORLING 1981; MORLING et al. 1985).

Three different periods or phases may be distinguished, mainly based on variations of pH values; naturally, the limits are rather imprecise.

	Phase I 1966—69	Phase II 1970—77	Phase III 1978—83
The Lygners Vider lakes, pH	6.8	6.8—4.6	4.6—4.8
Lake Änketjärn, pH	6.8	6.8—5.5	5.5—6.4

Lake Änketjärn is the only lake to show clearly deviating values; the conditions of the lake have been discussed separately by MORLING et al. (1985) and FORSBERG & MORLING (1987) where more detailed physical and chemical data are given.

3.2. Physical and chemical conditions

Changes of the physical and chemical conditions of the lakes have previously been discussed (MORLING 1981). As background information to this paper only one diagram is given in which pH values from the period 1980—1983 are included (Fig. 3).

Lake Oxsjön has been chosen as representative for the changes of different physical and chemical factors during the three phases mentioned (Table 1).

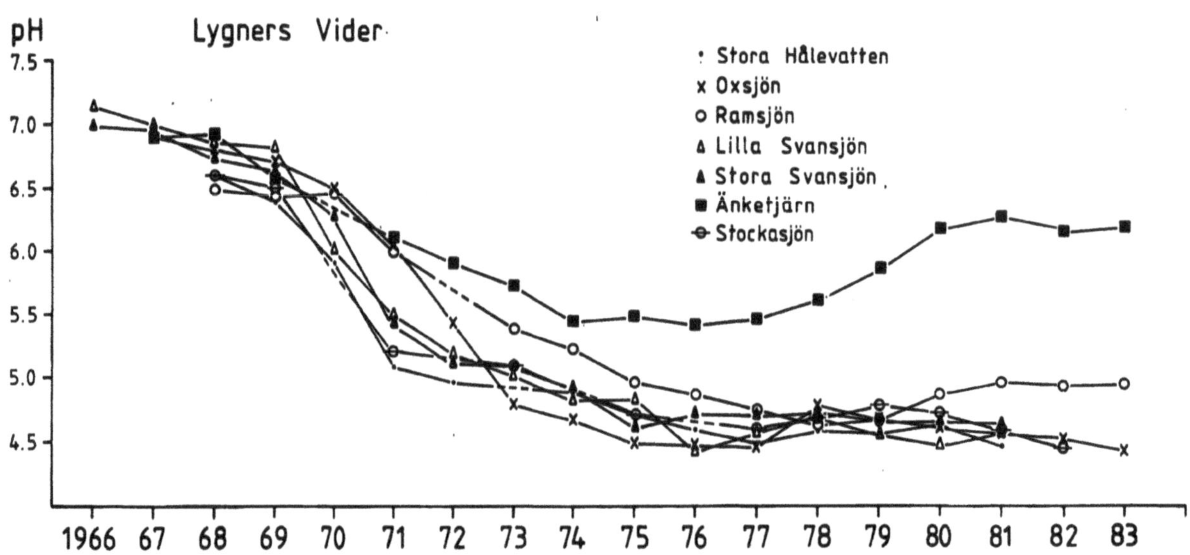

Fig. 3. Annual variation of pH in the lakes most regularly investigated. Means for May—October

Table 1. Lake Oxsjön. Physical and chemical data (minimum and maximum values) from the three acidification phases

	Phase I	Phase II	Phase III
pH	6.6—6.8	6.5—4.5	4.7—4.8
Conductivity ($\mu S_{20°}$)	53—59	69—82	66—75
Ca ($mg \cdot l^{-1}$)	2	2.2—2.6	2.2—2.8
Mg ($mg \cdot l^{-1}$)	—	1.4—1.5	1.1—1.7
Na ($mg \cdot l^{-1}$)	—	6.6—7.1	5.2—6.7
K ($mg \cdot l^{-1}$)	—	1	0.7—1.1
HCO_3 ($mg \cdot l^{-1}$)	0.5	0.2—0	0
SO_4 ($mg \cdot l^{-1}$)	5.5—6.5	7.8—14.5	13.0—10.4
Cl ($mg \cdot l^{-1}$)	9.6—10.6	11.0—13.5	10.4—11.0
NO_3 ($\mu g \cdot l^{-1}$)	100	150—300	150—280
PO_4 ($\mu g \cdot l^{-1}$)	2—4	3—5	4—6
Fe ($\mu g \cdot l^{-1}$)	—	60—310	50—200
Mn ($\mu g \cdot l^{-1}$)	150	160—360	180—250
Al ($\mu g \cdot l^{-1}$)	—	—	230—480
Colour ($mg\ Pt \cdot l^{-1}$)	7—15	7—20	10—25
Secchi depth (m)	6.3—7.2	5.0—7.3	3.9—6.3

3.3. Phytoplankton

3.3.1. List of species

All phytoplankton taxa — some desmids excluded — observed during the investigation period, 1966—1983, are listed in Appendix 1. Special notes are made for taxa recorded during the whole period as well as those only observed during phase I or phase III. Frequently occurring species belonging to Desmidiales are included in Appendix 1; species of rare occurrence are enumerated in Appendix 2 (p. 303—306).

3.3.2. Chlorophyll a

Some chlorophyll **a** analyses have been performed and the results are given in Table 2. With few exceptions, all chlorophyll values are lower than 2 $mg \cdot m^{-3}$. Seasonally, the highest values were consistently recorded during July and August. As a rule, the values from May were somewhat higher than those from April, but only in Lake Oxsjön could a marked spring maximum be distinguished as a result of a strong development of diatoms, mainly *Tabellaria*. The extremely high value, 6.6 $mg \cdot m^{-3}$, from Lake Lilla Svansjön was caused by a large number of *Mougeotia* filaments occurring in the pelagial during August and September.

With the exception of the small seasonal changes mentioned, it is not possible to perceive any real difference in chlorophyll concentration between the three different phases.

3.3.3. Phytoplankton development

Cyanophyceae

Cyanophyceae seem to be the algal group most sensitive to decreasing pH values. Already at the start of the investigation several taxa showed a low frequency, e.g. the *Anabaena* spp. At pH 5.5—6.0 the frequency was reduced and at still lower pH the species disappeared. In Lake Änketjärn, *Anabaena spiroides* had a rather high frequency from the start (about 20 trichomes $\cdot l^{-1}$). During the period 1974—1980 it was not recorded at all, but in 1981 and 1982 it reappeared (Fig. 4). The requirement of the cyanophyceans for a relatively high pH values is illustrated by a continuous occurrence of several taxa in Lake Änketjärn, a lake which had a higher pH value

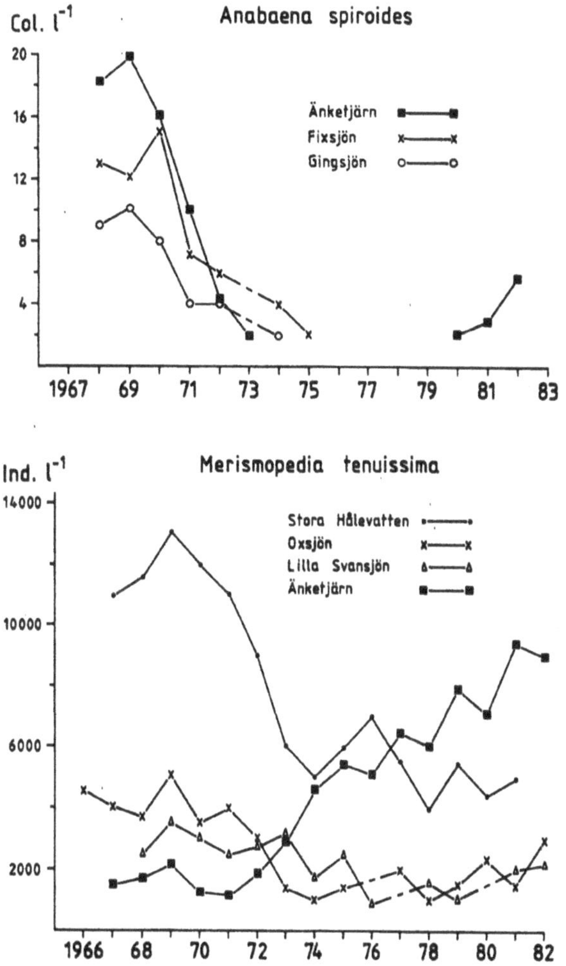

Fig. 4. Annual variation of *Anabaena spiroides* (means for May—September) and *Merismopedia tenuissima* (means for May—September) in some lakes of Lygners Vider

Table 2. Chlorophyll **a** values ($mg \cdot m^{-3}$) from five lakes during 1967—1980; means from six months, April—September

	Chlorophyll a ($mg \cdot m^{-3}$)				
	1967	1969	1973	1975	1980
L. St. Hålevatten	—	0.8—1.2	—	—	0.9—1.5
L. Oxsjön	0.8—1.2	—	1.0—3.1	0.6—1.5	0.5—1.2
L. Ramsjön	1.0—1.8	—	0.8—1.4	0.9—1.6	1.1—1.5
L. Lilla Svansjön	0.6—1.7	0.8—1.8	0.6—6.6	—	—
L. Änketjärn	—	1.0—1.8	0.7—1.9	1.1—2.0	—

during the whole period of investigation than other Lygners Vider lakes (Fig. 3).

Most tolerant to decreasing pH values are the two *Merismopedia* spp. observed, *M. punctata* and *M. tenuissima*, which are found during the whole period even if a marked lower frequency was observed in several lakes at pH values of about 5 (Fig. 4). Significant changes (p 0.10 Mann-Whitney) of *M. tenuissima* in Lake Stora Hålevatten were found between phase I and II and phase II and III, respectively, between phase I and II in Lake Oxsjön, and between phase II and III in Lake Änketjärn. — On some occasions, total individual numbers of about $10{,}000 \cdot l^{-1}$ were recorded in lakes Fixsjön and Skållesjön. No clear tendency of dominance by one of the two *Merismopedia* spp. has been observed in any lake.

Chrysophyceae

Mallomonas caudata has always been more frequent than *M. tonsurata* in the lakes; both taxa reached their highest cell numbers in August. *Mallomonas caudata* was observed in three lakes in very varying numbers: in Lake Gingsjön 400—600 ind. $\cdot l^{-1}$, in Lake Gärdsjön 3,000—13,000 ind. $\cdot l^{-1}$ and in Lake Änketjärn 2,000—3,000 ind. $\cdot l^{-1}$. No tendency for decreasing cell numbers correlated to lower pH values was evident.

As a rule, *Synura sphagnicola* in lakes Gärdsjön and Änketjärn as well as *Synura uvella* in Lake Fixsjön were recorded sporadically in single specimens. In some years (1969, 1975 and 1981), however, *S. sphagnicola* was more frequent during a longer period.

Dinobryon spp. frequently occurred in most lakes and reached their highest cell numbers during July. One example is given in Fig. 5 where the frequency of four species from the period 1966—1981 is illustrated.

Especially the two varieties *alpinum* and *palustre* of *D. cylindricum* are two taxa counted in large numbers. In 1973 and 1974 they reached peak values in Lake Oxsjön with about 10,000 cells $\cdot l^{-1}$ of var. *alpinum* and about 30,000 cells $\cdot l^{-1}$ of var. *palustre* to compare with about 4,000 cells $\cdot l^{-1}$ in most other years investigated. In these years the water had a greyish-rose colour in Lake Oxsjön with a transparency value of 3.5 m instead of the normal 7 m.

D. cylindricum and *D. sertularia* occurred in most lakes at low frequency; sometimes only single specimens were observed. *D. divergens* was also continuously recorded but always in small quantities.

D. bavaricum was continuously observed during the period in lakes Ramsjön and Änketjärn. At the start, rather high cell numbers were recorded from both lakes, about 2,500 cells $\cdot l^{-1}$. In Lake Ramsjön the number constantly decreased to reach a level of about 500 cells $\cdot l^{-1}$ at the end of the study. In Lake Änketjärn the number also decreased during phase II but increased in 1979—1982 to 2.500—3.500 cells $\cdot l^{-1}$.

D. pediforme was only observed in lakes Oxsjön and Stora Hålevatten during the later part of the investigation period: in L. Oxsjön from 1973 (pH 4.75) and in L. Stora Hålevatten from 1974 (pH 4.8).

Stichogloea doederleinii (Fig. 5) occurred in Lake Änketjärn during the whole investigation period and showed significant changes beween phase I and II and phase II and III, respectively. In three other lakes, however, the species disappeared in phase II at a pH value of 4.8—5.0: in Lake Oxsjön in 1973, in Lake Ramsjön in 1976 and in Lake Tagsjön in 1971. In Lake Tolaredsjön, however, there was a remarkable occurrence of single specimens in several samples during the period 1977—1981 at pH 4.9—5.0.

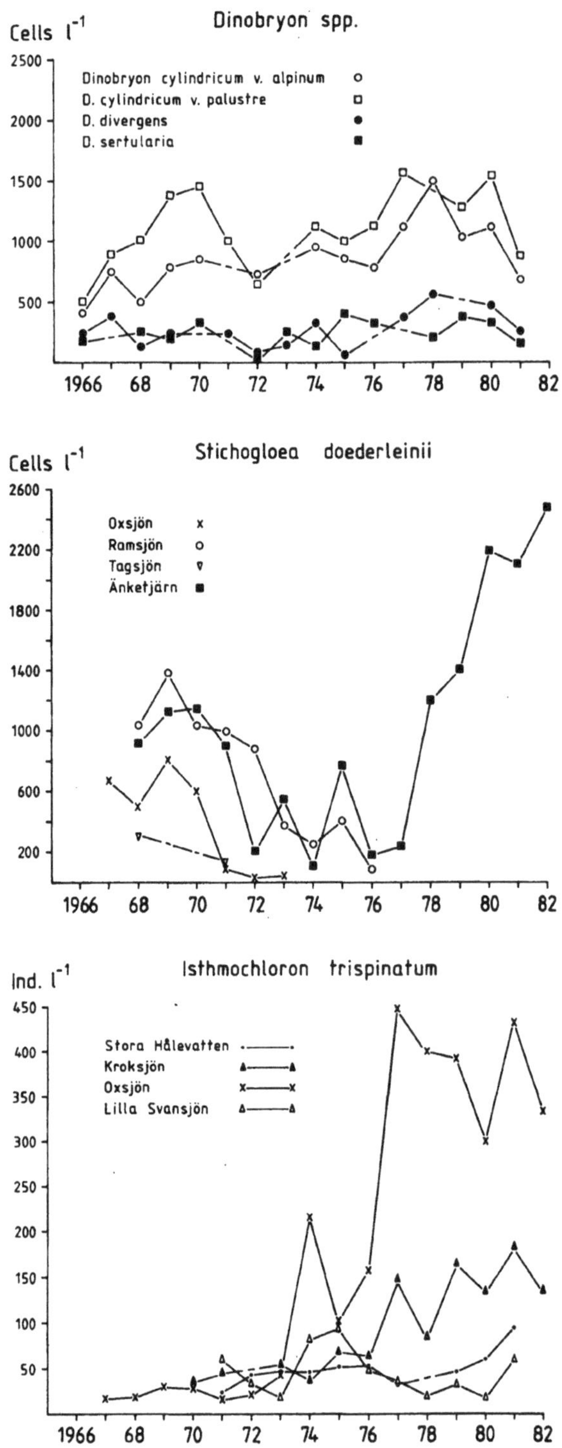

Fig. 5. Annual variation of *Dinobryon* spp. in Lake Lilla Svansjön (means for May—September), *Stichogloea doederleinii* (means for June—September), and *Isthmochloron trispinatum* (means for April—June) in some lakes of Lygners Vider

Xanthophyceae

Isthmochloron trispinatum (Fig. 5) showed a similar development in the five lakes where it was observed. In Lake Stora Svansjön — not shown in the figure — its occurrence was nearly identical with that in Lake Lilla Svansjön. The species reached peak development in April and May favoured by rather low water temperatures. *Isthmochloron trispinatum* is one good example of taxa increasing in numbers during phase III. Significant changes between phase I and II and phase II and III, respectively, could be confirmed for Lake Oxsjön and between phase II and III for Lake Kroksjön.

Bacillariophyceae: Centrales
Centric diatoms generally had a low abundance in the lakes investigated. *Melosira islandica* spp. *helvetica* and *M. italica v. valida*, however, occurred rather frequently and it was possible to evaluate the variations during the period. The species were most abundant at the beginning, decreased in numbers at sinking pH values and could be indicated as "single" at the end of the period. *M. islandica* spp. *helvetica* showed peak values in spring and in September/October.

Cyclotella spp. also were more abundant at the beginning of the period to decrease obviously at pH 5.5

Rhizosolenia longiseta was only observed in Lake Gingsjön with an irregularly decreasing frequency showing a lower maximum in spring and a higher one in September with cell numbers of 2,000—4,000 $\cdot l^{-1}$.

Bacillariophyceae: Pennales
Asterionella formosa was observed occasionally in a few colonies in most lakes. Only in Lake Oxsjön was a rather high frequency recorded during the whole period: in the autumns of 1978 (November) and 1979 (October) peak values of about 2,000 col. $\cdot l^{-1}$ were observed (Fig. 6).

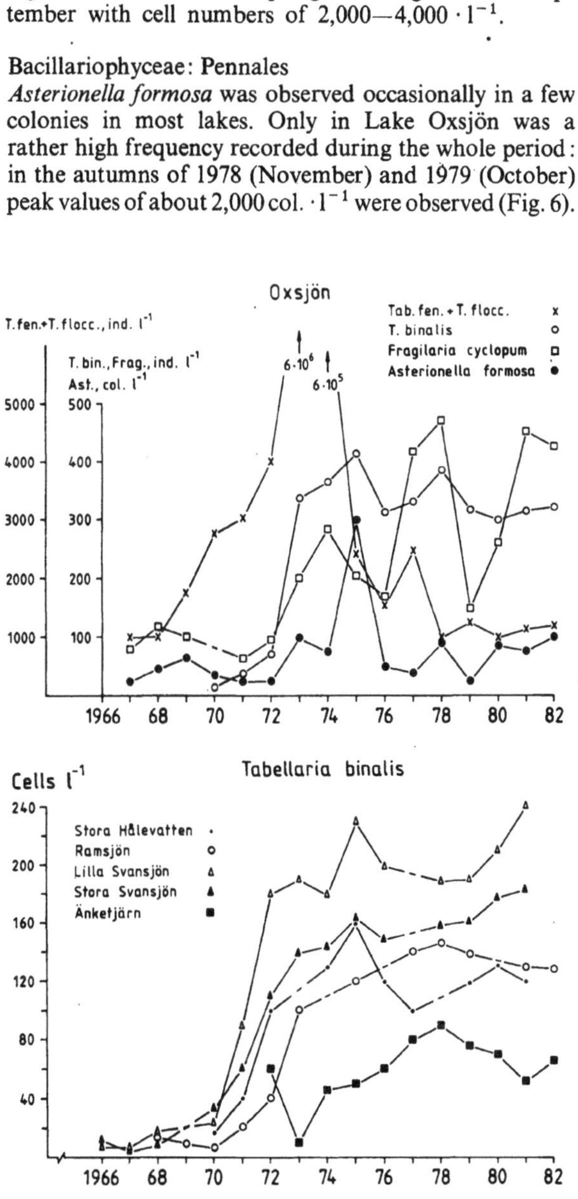

Fig. 6. Annual variation of some diatoms in Lake Oxsjön (means for May—September) and of *Tabellaria binalis* in some lakes of Lygners Vider

Fragilaria cyclopum (Fig. 6), observed in five lakes, showed a higher frequency after 1972 at pH values of 5.2—5.9. In Lake Oxsjön the species formed "bundles" of 5—10 individuals on the *Tabellaria* colonies in 1973 and 1974 (see below).

Frustulia rhomboides as well as *F. saxonica* were common in about half of the lakes. The highest numbers were recorded in Lake Lilla Svansjön with 2,000—3,000 cells $\cdot l^{-1}$ at the end of the investigated period.

Stauroneis anceps was recorded at pH values of 5.0—5.4.

Stenopterobia intermedia occurred in five lakes during phase I and increased in number during the first years of phase II. After 1973 it disappeared.

Tabellaria binalis (Fig. 6) is an example of taxa increasing in numbers at decreasing pH values (phases II and III). Only in lakes Lilla and Stora Svansjön was it observed during phase I. In the last mentioned lake significant changes were verified between phase I and II and between phase II and III, respectively; in Lake Lilla Svansjön between phase II and III. The frequency was rather constant during the growth period except for a small spring maximum. The species is also included in Fig. 6 showing the development of different diatoms in Lake Oxsjön.

Tabellaria fenestrata and *T. flocculosa* were common in all lakes and had a similar development. As a rule, *T. fenestrata* predominated with about 70—80% of the total cell number of the both species. The species were observed in rather high numbers during the whole growing period with a marked maximum in May/June. At the beginning and the end of the investigation the frequency was about the same, but during phase II considerable variations occurred.

In Lake Oxsjön an enormous development of *T. fenestrata* and *T. flocculosa* was observed in May and early June in 1973 and 1974 (Fig. 6). The concentration of Si immediately before the mass development was 2.9 and 2.2 mg $\cdot l^{-1}$, respectively; pH 4.6. In September the corresponding values were 0.6 and 1.1, respectively; pH 4.5. The algal colonies grew like a fur cover on sticks

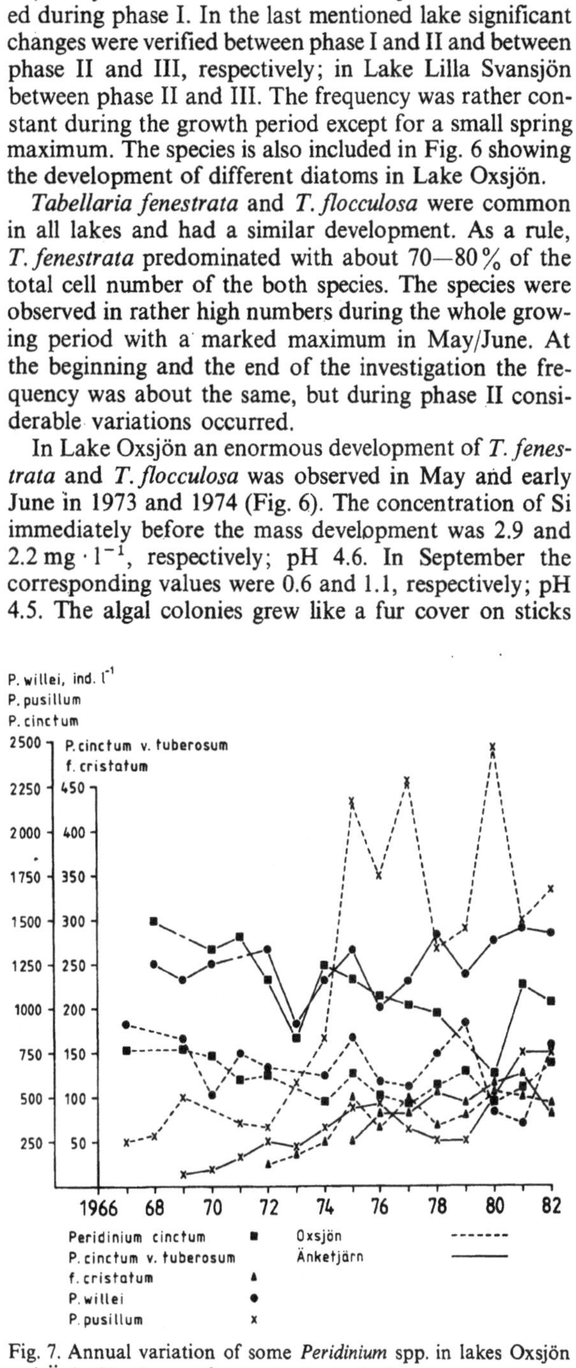

Fig. 7. Annual variation of some *Peridinium* spp. in lakes Oxsjön and Änketjärn (means for April—September)

and straws like one-metre long veils from branches hanging down in the water. The colonies filled the water down to a depth of about 3 metres. In June the growth diminished and had disappeared in August.

Dinophyceae

Peridinium spp. frequently occurred in the Lygners Vider lakes (Fig. 7). In most cases there was a tendency for higher cell numbers in relation to decreasing pH values.

Peridinium bipes sparsely occurred in most lakes; only single specimens were recorded during the period 1977—1979.

Peridinium cinctum generally increased in numbers during phases II and III. In Lake Änketjärn, however, there was a significant reduction. In Lake Oxsjön a significant change was found between phase I and II (Fig. 7).

Peridinium cinctum v. *tuberosum* f. *cristatum* — not earlier recorded from Sweden was observed in five lakes only during phases II and III with a tendency for frequencies to be higher in lakes with decreasing pH values.

Peridinium pusillum showed a marked increase in most Fixsjön while *P. palustre* occurred in single specimens in five lakes.

Peridinium pussilum showed a marked increase in most lakes at pH 5.0—5.5 exemplified by the development in Lake Oxsjön (Fig. 7) where a significant change was found between phase I and II. In Lake Änketjärn, however, rather few individuals occurred during the whole period; a significant increase was verified between phase II and III.

Peridinium willei was recorded from all lakes. In Lake Änketjärn it frequently occurred during the investigation period but in most other lakes the individual numbers decreased during phases II and III.

The *Peridinium* species reached their maximum development in spring and early summer. The reproduction (with a "*Gymnodinium*" phase) of *P. cinctum*, *P. cinctum* v. *tuberosum* f. *cristatum* and *P. willei* occurred in late April and in May at a water temperature of 10—11 °C. The naked "*Gymnodinium*" cell, making itself free from the theca, measures 30—40 µm in diameter. The corresponding reproduction of *P. pusillum* occurred later — in June — at a temperature of about 18 °C. The "*Gymnodinium*" cell measures 10—15 µm in diameter.

Ceratium carolinianum also belongs to the group of phytoplankton occurring in all lakes. During phase II a marked decrease was observed (Fig. 8) but after 1973 a clear recovery was recorded in spite of the fact that the pH values were low or even decreasing. A significant change between phase I and II was confirmed for Lake Stora Hålevatten and Lake Änketjärn. The species started its development slowly in spring and it was not until the water temperature reached 17 °C that noteworthy cell numbers were recorded. Maximum values were reached in August and September. Resting cysts were formed in October at 8—10 °C. In Lake Oxsjön, however, single living specimens were observed in late January 1975, temperature 2.4 °C (cf. Morling 1979).

Ceratium cornutum was only recorded from Lake Änketjärn where it disappeared at a pH value of 5.5 (Fig. 8). The species seemed to demand a water that was richer in nutrients than that required by other *Ceratium* spp.

Ceratium hirundinella was observed in most lakes with a rather high frequency during phase I — except in Lake Änketjärn. During phases II and III, however, the cell numbers decreased and at the end of phase III only single specimens were sporadically recorded, e.g., in Lake

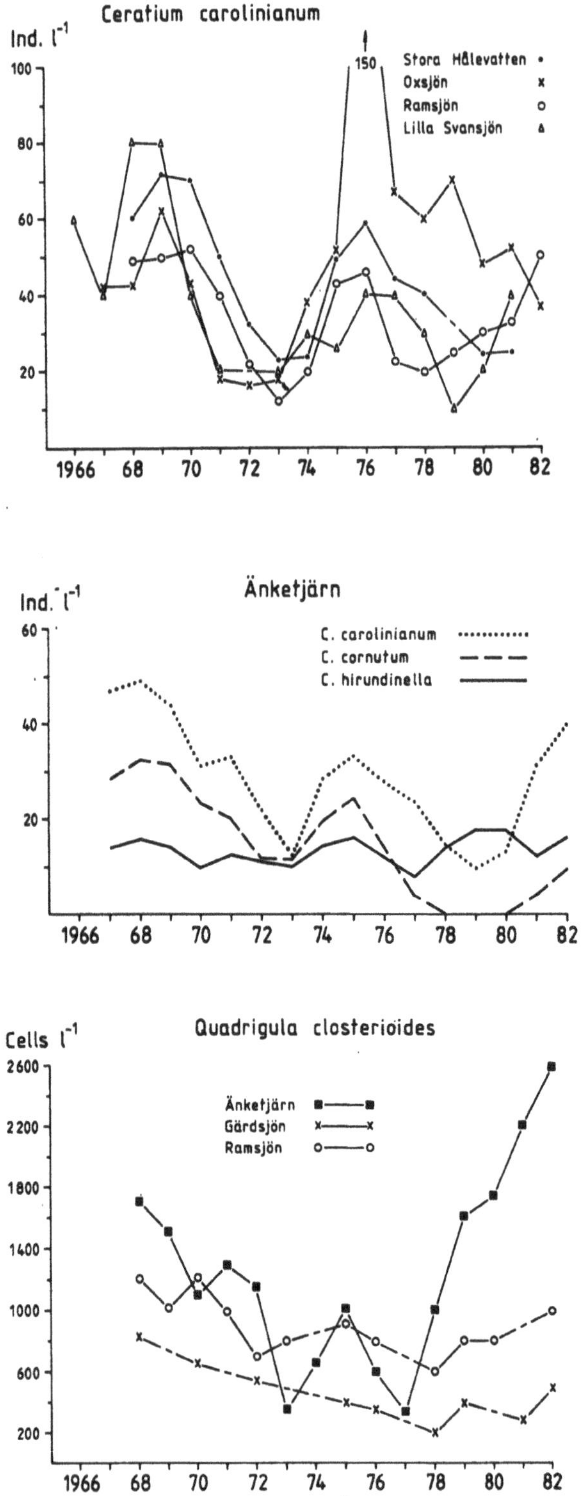

Fig. 8. Annual variation of *Ceratium carolinianum* (means for June—September), *Ceratium* spp. in Lake Änketjärn (means for July—August), and *Quadrigula closterioides* (means for May—September) in some lakes of Lygners Vider

Lilla Svansjön 1975, pH 4.8 and 1978, pH 4.7; Lake Oxsjön 1979, pH 4.8. In Lake Änketjärn, moderate numbers of the species were recorded during the whole period (Fig. 8). Between phase II and III a significant change was verified.

Chlorophyceae: Volvocales and Tetrasporales

Only a few taxa belonging to these orders occurred sporadically in the lakes and always in small quantities.

Volvox aureus was only observed during phase I, and *Pseudosphaerocystis lacustris*, most common in spring, during phase III.

Chlorophyceae: Chlorococcales
This order was represented by several taxa; the individual numbers, however, were low during the whole period or had a slight tendency to decrease in frequency as pH values decreased.

Ankistrodesmus falcatus was only observed in Lake Änketjärn with 10—20 cells l^{-1}, but was slightly more frequent during phases I and III.

Botryococcus braunii occurred in all lakes with varying frequencies: in Lake Oxsjön about 250 col. $\cdot l^{-1}$, in Lake Tolaredsjön about 100 col. $\cdot l^{-1}$ and in Lake Änketjärn about 300—500 col. $\cdot l^{-1}$, etc. No clear tendency for changes to occur with decreasing pH values was seen, but in lakes Stora Hålevatten, Oxsjön, Ramsjön and Lilla Svansjön a smaller increase in the number of colonies was recorded during phase III. The species was most abundant in August and September; in some cases 40—50 col. $\cdot l^{-1}$ were observed in late autumn, 7 °C.

Oocystis lacustris occurred with 500—1,000 cells $\cdot l^{-1}$ during the whole period.

Pediastrum boryanum was most abundant among the *Pediastrum* spp. 20—30 coenobia per litre were recorded from Lake Änketjärn where it was most frequent. In most other lakes about 10 coen. $\cdot l^{-1}$ were counted during phase I; the species became rarer during phases II and III.

Quadrigula closterioides was observed in four lakes (Fig. 8) showing a slightly decreasing frequency except in Lake Änketjärn where a peak value was reached in 1982. In this lake, significant changes were found between phase I and II between phase II and III, respectively.

Chlorophyceae: Ulotrichales
Binuclearia tatrana was observed in ten lakes with highest frequency during phase III. In lakes Fixsjön, Skållesjön and Oxsjön rather few trichomes, 40—50 µm in length, occurred; in Lake Lilla Svansjön the highest frequency was recorded, trichome length 100—200 µm.

Conjugatophyceae: Zygnemales
Sterile trichomes of *Mougeotia* were recorded in all lakes but there were only few occasions when it was possible to examine and determine fertile specimens. *Mougeotia* sp. was very frequent in lakes Lilla and Stora Svansjön; it was common during phase I, increased considerably during phase II and also frequently occurred during phase III. In lakes Oxsjön and Änketjärn, however, only a few tiny filaments were found during the whole period. Some other comments will be added in the following chapter.

Conjugatophyceae: Desmidiales
Only a few frequently occurring desmids are included in Appendix 1 but in Appendix 2 other taxa observed are enumerated. As a rule, few or single specimens were found. Altogether about 125 different taxa are recorded; 94 of them occur in Lake Änketjärn and 26 taxa only in this lake.

Arthrodesmus trispinatus occurred sporadically during phase I but increased considerably in numbers during phases II and III and was counted in about 200 ind. $\cdot l^{-1}$ at the end of the investigation.

Bambusina brebissonii was found in most lakes with 5—10 trichomes per litre during phase I. Minimum values were observed in Lake Skållesjön, while 16—18 trichomes $\cdot l^{-1}$ were recorded from Lake Lilla Svansjön during all phases.

Closterium was represented, on average, by three different species in most lakes; in Lake Änketjärn 16 taxa occurred. The low frequency of only 3—5 ind. $\cdot l^{-1}$ does not permit an evaluation of any relationship to pH fluctuations. Only four species were continuously counted in 8—12 ind. $\cdot l^{-1}$, viz. *Closterium dianae*, *C. intermedium*, *C. kuetzingii* and *C. setaceum*.

Cosmarium, as a rule, was represented by three different species in each lake; in Lake Änketjärn, however, there were 12 species. The frequencies of *Cosmarium* spp. were higher — 10—25 ind. $\cdot l^{-1}$ — than that of *Closterium* spp. The number of the four taxa *Cosmarium contractum* v. *ellipsoideum*, *C. contractum* v. *minutum*. *C. depressum* and *C. subtumidum* was higher during the whole period and at the end of the investigation mean values of 120, 90, 400 and 100 ind. $\cdot l^{-1}$ were recorded. *C. depressum* was very frequent in single samples: in Lake Lilla Svansjön about 500 ind. $\cdot l^{-1}$ in July 1980 and 1981, in Lake Ramsjön about 1,500 ind. $\cdot l^{-1}$ in July 1979. In Lake Änketjärn, *Cosmarium* spp. were only counted in numbers of 3—5 ind. $\cdot l^{-1}$. *C. contractum* v. *ellipsoideum* and *C. subtumidum*, however, showed an increased number of individuals at the end of the period, about 50—100 ind. $\cdot l^{-1}$ in single samples.

Euastrum was represented by six taxa in Lake Änketjärn; in other lakes about three. As a rule, 4—6 ind. $\cdot l^{-1}$ were recorded with *E. ansatum*, *E. humerosum* and *E. pinnatum* more frequently occurring. *E. ansatum* showed an increase during phase III to 10—12 ind. $\cdot l^{-1}$.

Hyalotheca dissiliens was observed in seven lakes with the highest frequency during phase I, 5—6 filaments $\cdot l^{-1}$, later decreasing in number.

Micrasterias was most common in Lake Änketjärn — 11 taxa — in comparison with about 3 in all other lakes: 2—3 ind. $\cdot l^{-1}$ were observed and only *M. oscitans*, *M. radiosa* and *M. truncata* v. *quadrata* were more common. The two last mentioned species were counted in 6—8 ind. $\cdot l^{-1}$ during phase III (Lake Oxsjön and Lake Skållesjön). *M. oscitans*, however, decreased in number in lakes Grundasjön and Gärdsjön.

Penium polymorphum reached maximum development in May—June with 10—30 ind. $\cdot l^{-1}$. During phase III about 60 ind. $\cdot l^{-1}$ occurred in lakes Oxsjön and Ramsjön.

Spondylosium planum, 3—5 filaments $\cdot l^{-1}$, disappeared from all lakes after phase I (pH 5—6) but survived in Lake Änketjärn during the whole period.

Staurastrum spp. were represented by 22 taxa in Lake Änketjärn, 12 taxa in Lake Ramsjön and by 5 taxa (mean value) in other lakes. Maximum development was observed in August with 10—15 ind. $\cdot l^{-1}$. The following taxa were more common, 20—40 ind. $\cdot l^{-1}$: *S. anatinum* v. *vestitum*, *S. artiscon*, *S. hystrix*, *S. longipes*, *S. pelagicum* and *S. planctonicum*. *S. longipes* disappeared during phase II except in Lake Änketjärn and *S. planctonicum* decreased considerably in number at the end of the period, while *S. pelagicum* became more common during phase III.

Staurodesmus spp. showed maximum frequencies at pH 5.5—6.0. *S. triangularis* and, especially, *S. incus* increased markedly during phase III; *S. incus* was counted in 400—600 ind. $\cdot l^{-1}$ in some lakes (Stora Hålevatten, Lilla Svansjön and Oxsjön) compared with about 60 ind. $\cdot l^{-1}$ in Lake Änketjärn.

Xanthidium spp. were observed in single specimens in few lakes with slightly increasing numbers during phase

III. Exceptions were *X. antilopaeum* v. *hebridarum* and *X. armatum* occurring in most lakes and counted in 4—6 ind. $\cdot$ l^{-1} during phase III.

Euglenophyceae
Colacium vesiculosum was only observed during phase III with 15—40 ind. $\cdot$ l^{-1} in three lakes. In Lake Fixsjön it was attached to *Cyclops strenuus strenuus*.

Cryptophyceae
Cryptomonas spp. (mainly *C. ovata* and *C. erosa*) occurred in very variable numbers and reached, e.g., 2,000—3,000 cells $\cdot$ l^{-1} in lakes Grundasjön, Gärdsjön and Änketjärn in late summer.

Rhodomonas sp. was observed in about 2,000 cells $\cdot$ l^{-1} in Lake Ramsjön, August—September 1968 and 1969.

Chloromonadophyceae
Gonyostomum semen was only found in Lake Tolaredsjön where it was very abundant in 1968 and 1969: the plankton net was rapidly filled and the meshes became clogged by the mucilaginous trichocysts. The lake is regulated and in years with low water level *Gonyostomum* increases considerably in number before decreasing again at normal or high water levels.

Hyphomycetes
Cephalosporium sp., *Tricladium angulatum* INGOLD and *Centrospora filiformis* (GREATHEAD) PETERSEN were found in shallow water of lakes Oxsjön and Lilla Svansjön 1974—1976, growing on decaying leaves and branches.

4. Discussion

4.1. General aspects

In general, the phytoplankton population in non-acidified oligotrophic lakes is characterized by a predominance of chrysophyceans, chlorophyceans and diatoms, while cyanophyceans and dinoflagellates play a minor role. At increasing acidification the species composition changes in synchrony with a decreasing number of species of each algal group. Mostly there is a change of predominance in favour of dinoflagellates and chrysophyceans (ALMER et al. 1978; YAN 1979; HENDREY et al. 1980).

As a result of comprehensive synoptic studies knowledge has been gained about phytoplankton population changes at different phases of acidification. ALMER et al. (1974) give results from 115 west-Swedish lakes. A marked reduction in species number occurs between pH 5 and 6; in lakes of pH 5 only about ten species were recorded. Similar results are reported by HÖRNSTRÖM (1979) from an investigation of 60 west-Swedish lakes in August 1976. HÖRNSTRÖM & EKSTRÖM (1983) studied the effects of pH, nutrients and aluminium on phytoplankton and mentioned that in addition to about 20 phytoplankton taxa normally observed in acid, clear lakes with high metal concentration (Al), single individuals of about 40 taxa may occur in special nutrient-rich conditions.

HENDREY (1980) reports more than 40 taxa from a Canadian lake, pH 4.7—5.2, with predominance of chrysophytes. YAN & STOKES (1978) found 75 taxa in Carlyle Lake, pH 4.5—6.0, where dinoflagellates and cryptophyceans occurred most frequently. In these examples, the total number of taxa is based on samplings once a fortnight during the growing period.

From Sweden, three further investigations may be referred to: Lake Stora Neden (county of Halland, Sweden), pH ca. 5.0, was studied every month during the period April—November (WILLÉN 1983). The list of species comprises 80 taxa with chrysophyceans, dinoflagellates, and chlorophyceans predominating. In Lake Stora Tresticklan (county of Älvsborg, Sweden), pH 4.0—4.6, a number of 33 taxa were recorded from four sampling occasions (WILLÉN 1977). LYDÉN & GRAN (1985) describe the phytoplankton community in Lake Gårdsjön during three years before liming including measurements of primary production and biomass. The phytoplankton was composed mainly of chrysophyceans and dinophyceans.

It may be mentioned that the total number of taxa given in the studies cited comprises both nanoplankton and net plankton species. Therefore, it is not possible to directly compare these results with those from the present investigation which mainly deals with net plankton.

Reviews of the ecological effects of acidification on primary producers are given, among others, by GEELEN & LEUVEN (1986) and STOKES (1986). They discuss the causes for the reduced number of planktonic algae. The acidification is associated with a variety of physical and chemical changes like decrease in phosphorus concentration as a result of increased aluminium concentrations, and decrease in silicon concentration related to a decrease of pH. Changes in the carbon cycle may influence both species composition and photosynthesis. Decreasing pH values result in an obvious increase of metals in lake waters (DICKSON 1980): indications of direct toxic effects of Al on phytoplankton are given by HÖRNSTRÖM et al. (1984). Grazing is another mechanism controlling the phytoplankton composition and biomass since pH also affects the species composition and size distribution of zooplankton.

4.2. Changes in the number of taxa

The phytoplankton composition of the lakes investigated is presented in Appendix 1 where the species have been divided into groups, observed during the whole period, during phase I and/or phase III, respectively. Altogether 191 taxa of phytoplankton (euplankton) were observed in the lakes. Most of them were recorded during the whole investigation. It should be observed, however, that the total of 191 taxa only comprises plankton collected by a net with a mesh size of 63 µm. Consequently, the greatest part of the nanoplankton also occurring in the lakes is not included.

Among the 191 taxa mentioned, no less than 101 belong to Conjugatophyceae, while Chlorophyceae are represented by 26 taxa, Bacillariophyceae, Chrysophyceae, and Cyanophyceae by 16, 16 and 15, respectively, and Dinophyceae by 10 taxa.

The number of euplanktic taxa per lake ranged from 31 in Lake Tagsjön to 71 in Lake Ramsjön (Lake Änketjärn excluded). The values refer to phase I; during the following phases minor changes were recorded.

The number of taxa per class in the Lygners Vider lakes during acidification phases II and III has been compared to that of phase I (Table 3). Cyanophyceae decreased in number during phase II followed by unchanged number during phase III, while Bacillariophyceae remained at the same level. Species belonging to Dinophyceae were favoured by acidification, while Chlorophyceae and Conjugatophyceae showed a small decrease during phase II, followed by a "recovery".

In Lake Änketjärn the composition of the phytoplankton communities was quite different (MORLING et al. 1985). Cyanophyceae, Euglenophycae, and Crypto-

Table 3. Changes in number of species in the Lygners Vider lakes (except Lake Änketjärn) during phase II (1970—1977) and phase III (1978—1983) compared with the number of species during "normal" conditions, phase I (1966—1969). The table is based on about 260, 520 and 325 analyses, respectively, during phases I—III

	Phase I Number of species	Phase II	Phase III
		Changes in number	
Cyanophyceae	13	−5	−5
Chrysophyceae	14	−1	+2
Bacillariophyceae	16	±0	±0
Dinophyceae	7	±0	+2
Chlorophyceae	19	−2	+2
Conjugatophyceae	72	−3	+5

phyceae were represented by the same number of species from phase I to phase III, while an increase was noted for Chrysophyceae (2 species), Bacillariophyceae (2 species), Dinophyceae (1 species), and Chlorophyceae (15 species). The total number of euplanktic taxa in this lake was 171 which means a hundred more than in the nearby lakes.

The Table 4 is given to illustrate the seasonal distribution of different taxa in Lake Oxsjön during the three phases (mean values).

Table 4. Seasonal distribution of different taxa in Lake Oxsjön during three phases (mean values)

	April	May	June	July	Aug.	Sept.	Oct.
Phase I	18	22	26	27	27	24	21
Phase II	19	20	20	30	29	27	19
Phase III	21	22	23	29	32	28	17

Common to all phases are minimum values in April, May, and October and maximum values in July and August.

Concerning the total number of taxa in the different lakes considerable variations occurred:

	Number of taxa		
	Phase I	Phase II	Phase III
Lygners Vider lakes, mean value	44	39	43
Lake Änketjärn	113	111	133

Among the Lygners Vider lakes, Lake Ramsjön — apart from Lake Änketjärn — was richest in species with 71, 68, and 75, respectively, while the lowest number of species was recorded from lakes Tagsjön and Grundasjön. The development, however, was similar in all lakes: no marked differences in species number were recorded during the period 1966—1983. A common feature was a somewhat reduced number during phase II. The original values were reached again during phase III. Some changes in species composition, however, occurred during the investigation period.

4.3. Changes in species composition

The development of the different algal groups and taxa is described above and illustrated in Figs. 4—9. Appendix 1 also gives information on the occurrence of the various taxa during the three phases.

It is clear that most phytoplankton organisms were recorded from the whole investigated period. However, some species only occurred during phase I and were not found — with few exceptions — during the following phases:

Cyanophyceae: *Chroococcus limneticus, Coelosphaerium kuetzingianum, Gomphosphaeria lacustris, Microcystis aeruginosa, Anabaena flos-aquae, Anabaena spiroides, Aphanizomenon flos-aquae, Oscillatoria lacustris;*

Chrysophyceae: *Dinobryon bavaricum, Dinobryon divergens* (in some lakes), *Stichogloea doederleinii;*

Bacillariophyceae: *Cyclotella bodanica, Stauroneis anceps, Stenopterobia intermedia*;

Dinophyceae: *Ceratium cornutum*;

Chlorophyceae: *Volvox aureus, Crucigenia tetrapedia, Selenastrum bibraianum, Tetraedron minimum.*

A few "new" taxa appeared during phase III:

Cyanophyceae: *Anabaena spiroides* (found again in Lake Änketjärn);

Chrysophyceae: *Dinobryon pediforme, Stichogloea doederleinii* (in Lake Tolaredsjön);

Bacillariophyceae: *Melosira distans*;

Dinophyceae: *Peridinium cinctum v. tuberosum f. cristatum, Peridinium inconspicuum, (Peridinium pusillum), Ceratium cornutum* (reappeared in Lake Änketjärn);

Chlorophyceae: *Pseudosphaerocystis lacustris, Elakatothrix gelatinosa*;

Euglenophyceae: *Colacium vesiculosum.*

As mentioned above, taxa belonging to Cyanophyceae play a subordinate role in lakes with a pH below 5.8. The genus *Merismopedia* is one exception. Genera belonging to Hormogonales disappeared after phase I, and the cyanophyceans decreased in total number and biomass after that phase. In lakes of Florida predominated by cyanophyceans, this class was replaced by chlorophyceans at increasing acidification (Crisman et al. 1980).

The chrysophyceans also diminished in number after phase I. Two *Dinobryon* species disappeared, while *D. cylindricum* v. *alpinum* and *D. pediforme* were more tolerant. Hörnström & Ekström (1983) mention them in their list of taxa "normally occurring in highly acid lakes in spite of extreme nutrient depletion". *Isthmochloron trispinatum* (Fig. 5) also belongs to the acid-tolerant group of taxa just mentioned (Hörnström & Ekström 1983). Nygaard (1977) describes the species as a vernal alga occurring at pH 4.95—5.15. Rosén (1981) mentions that the species is rather frequent in extremely oligotrophic lakes and that it is one of few taxa tolerating pH values below 5. — *Isthmochloron* appeared in spring in the lakes investigated, 10—13 °C, at pH about 5.

Bacillariophyceae did not show any marked changes during the three phases and the class plays a minor role in strongly acid waters (Almer et al. 1974, etc.). *Tabellaria binalis* (Fig. 6) is reported from a few Swedish lakes. Quennerstedt (1949, 1955) found it in three lakes in northern Sweden, characterized by very low conductivity ($\mu S_{18°}$ 5—10) and pH values of 5.8—6.4. Almer et al. (1974) report *T. binalis* from extremely acid lakes. In the Lygners Vider lakes, where the species existed, the conductivity ranged between $\mu S_{20°}$ 65 and 80; maximum development was observed at pH 4.5—4.9.

In most investigations of acidified lakes, a marked increase of the class Dinophyceae is reported, especially from a quantitative point of view (Almer et al. 1978; Yan & Stokes 1978; Hendrey et al. 1980). In the Lygners Vider lakes the development was remarkably uniform with a smaller increase in species number during phase III. While *Peridinium bipes*, *P. cinctum*, and *P. willei* develop within a wide pH range, pH 4—8, *P. palustre*, here sparsely occurring in five lakes, is strongly linked to very acid waters. pH may not be higher than 5 and maximum development is reached at pH values of 4 and lower (HUBER-PESTALOZZI 1950). *Ceratium hirundinella* belongs to the group of species not appearing at pH lower than 5.8, according to HÖRNSTRÖM & EKSTRÖM (1983). In this investigation, however, the species was observed within the pH range of 5.8—4.7 in single samples.

Chlorophyceae were represented by about 25 taxa, most of them occurring sporadically and with few specimens. Some of them belonging to the genera *Volvox*, *Crucigenia*, *Selenastrum*, and *Tetraedron* disappeared after phase I. *Botryococcus braunii* was observed in all lakes, a species frequently occurring in strongly acid, nutrient-rich lakes (HÖRNSTRÖM & EKSTRÖM 1983). Several reports mention a considerable development of some chlorophycean genera in acid lakes, e.g., *Oocystis* (ALMER et al. 1974; SCHEIDER et al. 1975; HÖRNSTRÖM & EKSTRÖM 1983; WILLÉN 1983; HAVENS III & DECOSTA 1987).

The genus *Mougeotia* (Zygnematales) was found in all lakes investigated, often occurring in enormous quantities in the pelagial. A rich development in acid waters is reported by several authors. ALMER et al. (1974, 1978) describe *M. scalaris* as frequently free-floating in several acid lakes well adapted to these conditions. In experiments where sulphuric acid is added to natural lakes, HENDREY (1976) found an increased growth of attached algae, e.g., *Mougeotia* spp.; similar experiences are reported by YAN & STOKES (1978) and HAVENS III & DECOSTA (1987).

The order Desmidiales was represented by more than half of all taxa recorded in this study. Without doubt they constituted a dominant part of net plankton; their biomass, however, was of minor importance: their individual size often is considerable but the number of

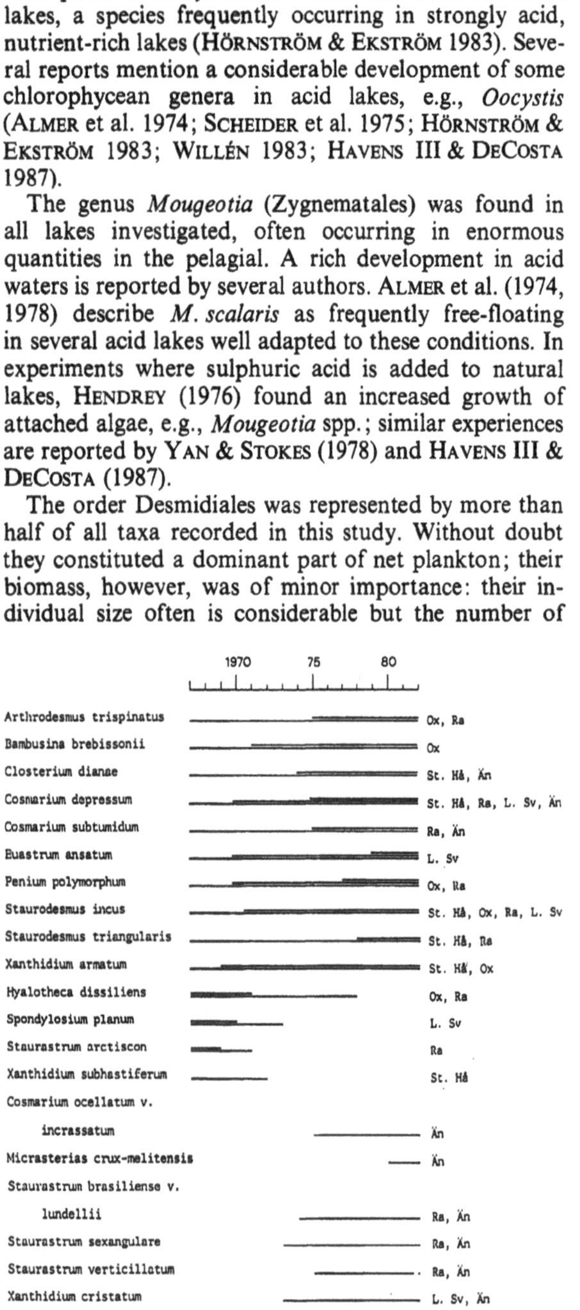

Fig. 9. Occurrence of some desmids 1967—1982 in lakes Lilla Svansjön (L. Sv.), Oxsjön (Ox), Ramsjön (Ra), Stora Hålevatten (St. Hå), and Änketjärn (Än). The frequency is indicated by 1, 2, or 3 lines

organisms per litre was very small. The large number of desmids included in Appendix 1 and Appendix 2 is largely explained by the fact that the lakes are small and shallow, thus containing numerous taxa normally living on the bottoms and in the littoral zone.

Most desmids are recorded from the lakes during the whole period. There are, however, examples of taxa only occurring during phase I and phase III, respectively. Some of them — from lakes Lilla Svansjön, Oxsjön, Ramsjön, Stora Hålevatten, and Änketjärn — are listed in Fig. 9, where the frequency is also marked by one, two or three lines.

Another species characteristic of acid lakes may be mentioned, viz. *Gonyostomum semen* (Chloromonadophyceae). In the area it was only observed in Lake Toläredsjön with a water colour of 10—20 mg Pt $\cdot l^{-1}$, pH 6.9 and alkalinity 0. At maximum development of *Gonyostomum* it was observed that the abundance of other plankton organisms decreased. The ecology of *Gonyostomum* is discussed by SÖRENSEN (1954), HÖRNSTRÖM et al. (1973), DICKSON (1975), ROSÉN (1981 b) and CRONBERG et al. (1988). In the 89 west-Swedish lakes where DICKSON found the species, the water colour had a mean value of 78 mg Pt $\cdot l^{-1}$. ROSÉN reports *Gonyostomum* from 161 Swedish lakes and states that the species is very pH-tolerant, requiring humus and occurring in lakes with a median water colour value of 60 mg Pt $\cdot l^{-1}$.

In summary, it can be concluded that acidification did not markedly change the number and composition of phytoplankton, which to some degree diverges from earlier results presented in the literature.

4.4. Biomass changes

Since no complete quantitative plankton calculations were performed, only the chlorophyll-**a** values give some information about the total phytoplankton biomass. As a rule, chlorophyll values lower than 2 mg $\cdot m^{-3}$ were recorded from the lakes (Table 2) without any marked changes between the different phases. No clear tendency is found as to the correlation between increasing acidification and phytoplankton biomass (chlorophyll concentration).

Similar chlorophyll values are found in other investigations, e.g., 0.55—1.6 mg $\cdot m^{-3}$ in Ontario lakes (SCHEIDER et al. 1975), 0.15—1.2 mg $\cdot m^{-3}$ in Norwegian lakes (RADDUM et al. 1980), 0.3—1.3 mg $\cdot m^{-3}$ in the west-Swedish lake Stora Neden (WILLÉN 1983), etc.

During experimental acidification, HAVENS III & DECOSTA (1987) observed that the total algal biomass, measured as chlorophyll **a**, increased markedly with acidification, especially at pH 4.5. This increase was mainly caused by an immense development of *Mougeotia viridis*. The occurrence of *Mougeotia* spp. in the Lygners Vider lakes is discussed above; these species were most abundant in the lakes Lilla and Stora Svansjön (1971 and 1973) and Stora Hålevatten (1973). The chlorophyll values from these lakes were differing: Lake Lilla Svansjön showed values of 10.2 and 6.6 mg $\cdot m^{-3}$ as a "normal" value for August.

Most acid lakes are extremely oligotrophic as a result of aluminium precipitation of phosphorus and humic compounds in the lake (HÖRNSTRÖM et al. 1984) and in the drainage area (JANSSON et al. 1986). However, the acid phosphatase activity is an important factor in the regeneration of orthophosphate in this type of lakes.

In the acid and oligotrophic Lake Gårdsjön the optimum potential phosphatase activity was found at pH 5. Small chrysophycean taxa were the major producers of acid phosphatases (OLSSON 1983).

The levels of phosphate-phosphorus in the Lygners Vider lakes were very low, 3—7 $\mu g \cdot l^{-1}$, and only Lake Änketjärn had values of 10—18 $\mu g \cdot l^{-1}$ (MORLING 1981). Due to the limited number of phosphate analyses it is not possible to discuss temporal development, correlations to chlorophyll data, etc.

The mean values of nitrate-nitrogen in all lakes during the various phases were:

	Phase I	Phase II	Phase III
NO_3^-	130	280	260

These values are also low but show a clearly increasing tendency during the last phases. As to the aluminium concentration, analyses were only performed during phase III. One value from Lake Oxsjön is given in Table 1, 230—480 $\mu g \cdot l^{-1}$; other values are in accordance with those given by HÖRNSTRÖM et al. (1984).

The chemical and biological data available indicate that all Lygners Vider lakes are oligotrophic, with one exception, Lake Änketjärn.

4.5. Lake Änketjärn

While all other lakes in the Lygners Vider area react in similar manner, Lake Änketjärn differs considerably from them in both chemical and biological respects (MORLING 1981; MORLING et al. 1985). During the period 1967—1977 the sulphate concentrations of the lake increased and parallel to that the pH decreased from about 7 down to a level of 5.5, a reduction which was much less pronounced than in the nearby lakes (Fig. 4). During the period 1977—1983 the sulphate concentrations decreased (MORLING 1984a; Fig. 1) and the pH values increased from 5.5 to 6.3. The water still contains small concentrations of bicarbonate (alkalinity 0.027 mequ $\cdot l^{-1}$, August 1986).

The macrophytic vegetation of the lake is more luxuriant than in the other lakes, and especially *Juncus bulbosus* L. has developed considerably after 1974, now dominating large areas of the soft sediments (WETZEL et al. 1984).

During the period 1967—1970 the composition and seasonal succession of phytoplankton showed only small variations. The changes in water quality during 1971—1974, when pH dropped to about 5.5, resulted in yearly variations in the periodicity of several taxa, e.g., *Merismopedia punctata*, *Stichogloea doederleinii* and *Quadrigula closterioides*. At the end of this period the cyanophyceans showed a considerable reduction as well as the populations of *Mallomonas caudata*, *Dinobryon bavaricum*, and *Peridinium cinctum*. Some species appeared as "new" for the lake: *Tabellaria binalis*, *Peridinium pusillum*, and *Binuclearia tatrana*.

During the following period, 1978—1982, when the pH values increased from about 5.5 to about 6.3, the populations of several species increased with a factor of 1.5 to 2, e.g., *Merismopedia punctata*, *Stichogloea doederleinii*, *Dinobryon bavaricum*, *Mallomonas caudata*, and *Quadrigula closterioides*. *Anabaena spiroides* reappeared (Fig. 4) as well as *Ceratium cornutum* (Fig. 8). During this period, thirteen "new" desmid species were recorded in the lake. — It has been mentioned above that the total number of taxa observed in the lake is 171, which means about 100 more than in the other lakes.

The deviating state of Lake Änketjärn may be explained by biological and/or geochemical processes. Through results from water analyses of samples from bored wells in the lake surroundings and by examining the geochemical conditions, it was concluded that the differences in alkalinity and pH between this lake and the other acidified lakes in the area can mainly be explained by geomorphological differences in the drainage basins (MORLING et al. 1985). The plankton composition of Lake Änketjärn is moderately changed, and the lake may be regarded as a reference lake for comparison with the other more acidified waters. Lake Änketjärn is also taken as an example of a possible reversibility of the acidification processes in lakes (FORSBERG & MORLING 1987).

4.6. Liming effects on phytoplankton

Liming of the Lygners Vider lakes started in 1979 and was performed until 1986 (Lake Oxsjön). Only Lake Änketjärn was saved as a reference lake. Very few plankton samples were collected during and after the liming period and the material is not large enough to draw any detailed or general conclusions about changes in phytoplankton composition.

One of the lakes investigated, Lake Stora Svansjön, however, is included in a study by HÖRNSTRÖM & EKSTRÖM (1986). Their methods do not correspond exactly to those used in the present investigation. They found 9—19 taxa at each sampling in July/August 1973—1982, with dinoflagellates and chrysomonads dominating. After liming in 1983 the total number of species increased to more than 40 with a predominance of *Dinobryon cylindricum* v. *palustre*, *Chrysochromulina parva*, and *Synedra* cf. *nana*. The phytoplankton biomass, dominated by chrysophyceans, was also larger than before liming.

In some cases the lime addition to a lake causes a rather unstable mixed phytoplankton population during the first years at the same time as some earlier "permanent" species are replaced (T. WILLÉN 1980).

According to the general results by HÖRNSTRÖM & EKSTRÖM (1986) obtained from summer plankton analyses, the number of phytoplankton species increases after liming. There is also an increase in the primary production, trophic lake indices and species diversity. LARSSON (1988) obtained the same results in his study of liming effects on phytoplankton, its biomass and productivity in Lake Gårdsjön. The phytoplankton community, however, was not stabilized three years after liming.

5. Acknowledgements

This investigation was supported in part by a grant from the Joint English-Scandinavian Surface Water Acidification Programme, directed in Sweden by the Swedish Royal Academy of Science.

6. References

ALMER, B., DICKSON, W., EKSTRÖM, C., HÖRNSTRÖM, E., & MILLER, U. (1974): Effects of acidification on Swedish lakes. Ambio 3, 30—36.

— (1978): Sulfur pollution and the aquatic ecosystem. In: NRIAGU, J. O. (ed.), Sulfur in the environment. II. Ecological impacts. New York, pp. 271—313.

CRISMAN, T. L., SCHULZE, R. L., BREZONIK, P. L., & BLOOM, S. A. (1980): Acid precipitation: the biotic response in Florida lakes. In: DRABLØS, D., & TOLLAN, A. (eds.). Proc., Int. Conf. Ecol. Impact Acid Precip., Norway 1980, SNSF project, pp. 296—297.

CRONBERG, G., LINDMARK, G., & BJÖRK, S. (1988): Mass development of the flagellate *Gonyostomum semen* (Raphidophyta) in Swedish forest lakes — an effect of acidification? Hydrobiologia **161**, 217—236.

DICKSON, W. (1975): The acidification of Swedish lakes. Rep. Inst. Freshw. Res. Drottningholm **54**, 8—20.

— (1980): Properties of acidified waters. In: DRABLØS, D., & TOLLAN, A. (eds.) Ecological impact of acid precipitation. SNSF project, Oslo—Ås, pp. 75—83.

FORSBERG, C., & MORLING, G. (1987): Examples of changes in water chemistry during lake acidi- and "deacidification". Verh. Int. Ver. Limnol. **23**, 193—199.

FOTT, B. (1971): Algenkunde. 2 Aufl. Stuttgart.

GEELEN, J. F. M., & LEUVEN, R. S. E. W. (1986): Impact of acidification on phytoplankton and zooplankton communities. Experientia **42**, 486—494.

HAVENS III, K. E., & DECOSTA, J. (1987): Freshwater plankton community succession during experimental acidification. Arch. Hydrobiol. **111**, 37—65.

HENDREY, G. R. (1976): Effects of low pH on the growth of periphytic algae in artificial stream channels. Internal report IR 25/76. SNSF, Oslo. 50pp.

— (1980): Effects of acidity on primary productivity in lakes: phytoplankton. In: SHRINER, D., RICHMOND, G. R., & LINDBERG, S. E. (eds.). Environmental impacts and health effects. Ann Arbor, Michigan.

— YAN, N. D., & BAUMGARTNER, K. J. (1980): Responses of freshwater plants and invertebrates to acidification. In: Restoration of lakes and inland waters. U.S. Environ. Prot. Agency; Office of Water Regul. and Stand., Washington D.C. EPA 440/5-81-010, pp. 68—74.

HÖRNSTRÖM, E. (1979): Kalknings- och försurningseffekter på växtplankton i tre västkustsjöar. Statens naturvårdsverk. Rapport PM **1220**, 1—69 (in Swedish; English summary).

— & EKSTRÖM, C. (1983): pH-, närings- och aluminiumeffekter på plankton i västkustsjöar. Statens naturvårdsverk. Rapport PM **1704**, 1—124 (in Swedish; English summary).

— (1986): Acidification and liming effects on phyto- and zooplankton in some Swedish West Coast lakes. Statens naturvårdsverk. Rapport **1864**, 1—108.

— & DURAINI, M. O. (1984): Effects of pH and different levels of aluminium on lake plankton in the Swedish west coast area. Rep. Inst. Freshw. Res., Drottningholm **61**, 115—127.

HUBER-PESTALOZZI, G. (1950): Das Phytoplankton des Süsswassers. 3. Cryptophyceen, Chloromonadinen, Peridineen. Die Binnengewässer **16** (3). Stuttgart.

JANSSON, M., PERSSON, G., & BROBERG, O. (1986): Phosphorus in acidified lakes: the example of Lake Gårdsjön, Sweden. Hydrobiologia **139**, 81—96.

LARSSON, L. S. (1988): Liming effects on phytoplankton in Lake Gårdsjön 1982—85. In: DICKSON, W. (ed.), Liming of Lake Gårdsjön. An acidified lake in SW Sweden. Nat. Swed. Environm. Prot. Board, Report **3426**, 245—280.

LYDÉN, A., & GRAHN, O. (1985): Phytoplankton species composition, biomass and production in Lake Gårdsjön — an acidified clearwater lake in SW Sweden. Ecol. Bull. (Stockholm) **37**, 195—202.

Monitor (1985): Sura och försurade vatten. Statens naturvårdsverk, 180 pp. (in Swedish).

MORLING, G. (1979): *Ceratium carolinianum* and the related *Ceratium carolinianum* var. *elongatum* and *Ceratium cornutum*. A study of distribution and morphology. Nova Hedwigia **31**, 937—968.

— (1981): Effects of acidification on some lakes in Western Sweden. Vatten **37**, 25—38.

— (1984a): Acidification and plankton in West-Swedish lakes 1966—1983. Ph. D. thesis, Uppsala Univ., Sweden.

— (1984b): Försurningens utveckling och planktonsammansättningen i några västsvenska sjöar. Manuscript, Limnol. Inst., Uppsala (in Swedish).

— & PEJLER, B. (1990): Acidification and zooplankton development in some west-Swedish lakes. *Limnologica* (Berlin) **20**, 307—318.

— FORSBERG, C., & WETZEL, R. G. (1985): Lake Änketjärn — a non-acidified lake in an acidified region. Oikos **44**, 324—330.

NYGAARD, G. (1977): New and interesting plankton algae — with a contribution on their ecology. K. danske Vidensk. Selsk. Biol. Skr. **21**, 1—107.

OLSSON, H. (1983): Origin and production of phosphatases in the acid Lake Gårdsjön. Hydrobiologia **101**, 49—58.

QUENNERSTEDT, N. (1949): Om diatoméerna *Actinella punctata* LEWIS och *Tabellaria binalis* (EHR.) GRUN. i svenska vatten. Svensk bot. Tidskr. **43**, 82—97 (in Swedish, English summary).

— (1955): Diatoméerna i Långans sjövegetation. Acta Phytogeogr. Suec. **36**, 1—208 (in Swedish, English summary).

RADDUM, G. G., HOBAEK, A., LØMSLAND, E. R., & JOHNSEN, T. M. (1980): Phytoplankton and zooplankton in acidified lakes in South Norway. In: DRABLØS, D., & TOLLAN, A. (eds.). Proc., Int. Conf. Ecol. Impact Acid Precip., Norway 1980. SNSF project, pp. 332—333.

ROSÉN, G. (1981a): Tusen sjöar. Växtplanktons miljökrav. (Phytoplankton ecology—results from the "Thousand Lakes Investigation"). Statens naturvårdsverk, Rapporter, 119pp. (in Swedish, English summary).

— (1981b): Phytoplankton indicators and their relations to certain chemical and physical factors. *Limnologica* (Berlin) **13**, 262—290.

ROUND, F. E. (1981): The ecology of algae. Cambridge.

SCHEIDER, W., ADAMSKI, J., & PAYLOR, M. (1975): Reclamation of acidified lakes near Sudbury, Ontario. Ontario Minist. Environ., Rexdale, 129pp.

SÖRENSEN, I. (1954): *Gonyostomum semen* (EHRENB.) DIESING — en vattenorganism av teoretiskt och praktiskt intresse. Svensk faun. Revy **1954** (2), 1—6 (in Swedish).

STOKES, P. M. (1986): Ecological effects of acidification on primary producers in aquatic systems. Water, Air, and Soil Pollut. **30**, 421—438.

TOLSTOY, A. (1969): Klorofyll. In: WILLÉN, T. (ed.). Metodik vid biologiska sjöundersökningar. Naturvårdsverkets limnol. unders. **30**, 25—28 (in Swedish).

UTERMÖHL, H. (1958): Zur Vervollkommnung der quantitativen Phytoplanktonmethodik. Mitt. Int. Ver. Limnol. **9**, 1—38.

WETZEL, R., BRAMMER, E., & FORSBERG, C. (1984): Photosynthesis of submerged macrophytes in an acidified lake. I. Carbon fluxes and recycling of CO_2 in *Juncus bulbosus* L. Aquat. Bot. **19**, 329—342.

WILLÉN, E. (ed.). (1977): Limnologiska undersökningar i Stora Tresticklan, Dals Ed, Älvsborgs län 1975. Naturvårdsverkets limnol. unders. Information **18**, 1—46 (in Swedish).

— (ed.). (1983): Limnologiska undersökningar i Stora Neden 1982, en försurad sjö i Hallands län. Statens naturvårdsverk. Rapport PM **1735**, 1—55 (in Swedish).

WILLÉN, T. (1980): Phytoplankton from Swedish lakes. III. Lake Hundsjön and other kettle lakes of Central Sweden. Arch. Hydrobiol. **89**, 135—159.

YAN, N. D. (1979): Phytoplankton community of an acidified, heavy metal-contaminated lake near Sudbury, Ontario: 1973—1977. Water, Air, and Soil. Pollut. **11**, 43—55.

— & STOKES, P. (1978): Phytoplankton of an acidic lake, and its responses to experimental alterations of pH. Environ. Conserv. **5**, 93—100.

Appendix 1

List of phytoplankton from the Lygners Vider lakes, 1966/1967—1982/1983.

Abbreviations of the lake names: Fi = Fixsjön; Gi = Gingsjön; Gr = Grundasjön; Gä = Gärdsjön; St. Hå = Stora Hålevatten; Kr = Kroksjön; Ox = Oxsjön; Ra = Ramsjön; Sk = Skållesjön; L. Sv = Lilla Svansjön; St. Sv = Stora Svansjön; Ta = Tagsjön; To = Toláredsjön; Än = Änketjärn. — * = euplanktonic species, according to ROUND (1981); × = observed during the whole investigation period; ⊕ = observed during phase I (disappearing during phase II); ○ = observed during phase III.

	Lygners Vider lakes													
	Fi	Gi	Gr	Gä	St. Hå	Kr	Ox	Ra	Sk	L. Sv	St. Sv	Ta	To	Än
Cyanophyceae														
Chroococcales														
Aphanocapsa siderosphaera NAUM.*														
Chroococcus dispersus (KEISSL.) LEMM.*								×		×				×
Chroococcus limneticus LEMM.*	⊕						⊕			⊕	⊕		×	×
Chroococcus turgidus (KÜTZ.) NÄG.*								×		×				×
Coelosphaerium kuetzingianum NÄG.*				⊕				⊕	⊕				×	×
Gomphosphaeria lacustris CHOD.*					⊕									×
Merismopedia elegans A. BR.*					×		×							×
Merismopedia punctata MEYEN*	×	×	×			×	×	×		×	×	×		×
Merismopedia tenuissima LEMM.*	×		×	×	×		×	×	×	×	×	×	×	×
Microcystis aeruginosa (KÜTZ.) KÜTZ.*						⊕	⊕							
Synechococcus linearis (SCHMIDLE & LAUT.) KOM.*							×	×	×	×	×	×		
Hormogonales														
Anabaena flos-aquae BRÉB.*										⊕	⊕		⊕	
Anabaena spiroides KLEB.*	⊕							⊕						⊕ ○
Aphanizomenon flos-aquae (L.) RALFS*								⊕		⊕		⊕		
Oscillatoria lacustris (KLEB.) GEITL.*														⊕
Chrysophyceae														
Chrysococcus major LACKEY*								×					×	×
Mallomonas caudata IVANOFF*		×		×										×
Mallomonas tonsurata TEIL.*								×						×
Rhipidodendron huxleyi KENT							×	×						×
Synura sphagnicola (KORSCH.). KORSCH.*				×										×
Synura uvella EHR.*	×													
Chrysosphaerella longispina LAUT.*		×						×						×
Dinobryon bavaricum IMH.		⊕		⊕				×			⊕			×
Dinobryon cylindricum IMH.	×	×	×	×			×	×	×			×	×	×
Dinobryon cylindricum v. *alpinum* (IMH.) BACHM.*	×	×		×	×	×	×	×	×	×	×			×
Dinobryon cylindricum v. *palustre* LEMM.*			×		×	×	×	×						×
Dinobryon divergens IMH.*	⊕		×	×	×	×	⊕	×	⊕	×	⊕		⊕	×
Dinobryon pediforme (LEMM.) STEINECKE*					○		○							
Dinobryon sertularia EHR.*	×			×	×	×	×		×	×	×	×	×	×
Epipyxis tabellariae (LEMM.) G. M. SMITH*														○
Salpingoeca frequentissima (ZACH.) LEMM.*		×												
Stichogloea doederleinii (SCHMIDLE) WILLE*							⊕	⊕				⊕	○	×
Xanthophyceae														
Isthmochloron trispinatum (W. & G. S. WEST) SKUJA*					○	○	×			○	○			
Tetraedriella regularis (KÜTZ.) FOTT*					×									
Bacillariophyceae														
Centrales														
Cyclotella bodanica EULENST.*	×			×				⊕		⊕				
Cyclotella planctonica BRUNNTH.*							×		×	×				
Melosira ambigua (GRUN.) O. MÜLL.*	×		×					×						
Melosira distans (EHR.) KÜTZ.*														○
Melosira islandica ssp. *helvetica* O. MÜLL.*		×				×							×	×

	Lygners Vider lakes													
	Fi	Gi	Gr	Gä	St. Hå	Kr	Ox	Ra	Sk	L. Sv	St. Sv	Ta	To	Än
Melosira italica ssp. *subarctica* O. MÜLL.*				×						×				
Melosira italica v. *valida* (GRUN.) HUST.*					×		×			×	×	×	×	
Rhizosolenia longiseta ZACH.		×												
Bacillariophyceae														
Pennales														
Asterionella formosa HASS.*		×			×	×	×			×	×		×	×
Diatoma vulgare BORY*					×			×						
Diatoma vulgare v. *capitulatum* GRUN.*					×		×		×					
Diatoma vulgare v. *ehrenbergii* (KÜTZ.) GRUN.*					×									
Eunotia elegans ØSTR.						×	×			×	×			
Eunotia pectinalis (KÜTZ.) RAB.			×	×				×						×
Eunotia robusta RALFS					×		×							×
Eunotia sp.							×							
Fragilaria cyclopum (BRUTSCH.) LANGE-BERT.							×	×		×			×	×
Frustulia rhomboides (EHR.) DE TONI							×	×		×	×	×	×	
Frustulia saxonica (RAB.) DE TONI	×		×				×		×	×			×	×
Navicula subtilissima CL.	×													
Nitzschia kuetzingiana HILSE								×						×
Nitzschia sigmoidea (EHR.) W. SM.					×	×				×	×			
Pinnularia cardinalis (EHR.) W. SM.	×						×							
Pinnularia dactylus EHR.										×	×			
Pinnularia legumen EHR.														×
Pinnularia nobilis EHR.			×				×							
Pinnularia viridis (NITZSCH) EHR.	×						×							
Stauroneis anceps EHR.								⊕						⊕
Stauroneis phoenicentron EHR.														×
Stenopterobia intermedia LEVIS		⊕	⊕			⊕				⊕	⊕			
Surirella elegans EHR.	×						×			×				
Surirella robusta EHR.			×	×										
Surirella tenera GREG.								×						
Tabellaria binalis (EHR.) GRUN.*					○		○	○		×	×			○
Tabellaria fenestrata (LYNGB.) KÜTZ.*	×	×	×	×	×	×	×	×	×	×	×	×	×	×
Tabellaria fenestrata v. *asterionelloides* GRUN.*							×			×	×			
Tabellaria flocculosa (ROTH) KÜTZ.*	×	×	×	×	×	×	×	×	×	×	×	×	×	×
Dinophyceae														
Peridinales														
Peridinium bipes STEIN*	×	×	×		×	×	×	×		×	×	×	×	
Peridinium cinctum (O. F. MÜLL.) EHR.*	×		×	×	×		×	×	×	×	×	×	×	×
Peridinium cinctum v. *tuberosum* f. *cristatum* (MEUNIER) LEF.*			○		○	○	○							○
Peridinium inconspicuum LEMM.*	○													
Peridinium palustre (LINDEM.) LEF.*	×	×			×	×	×							
Peridinium pusillum (PEN.) LEMM.*					○	○	×	○	○	○	○	○	○	○
Peridinium willei HUITF.-KAAS*	×	×	×	×	×	×	×	×	×	×	×	×	×	×
Ceratium carolinianum (BAIL.) JÖRG.*	×	×	×	×	×	×	×	×	×	×	×	×	×	×
Ceratium cornutum (EHR.) CLAP. & LACHM.*														⊕○
Ceratium hirundinella (O. F. MÜLL.) BERGH.*	×		×		×	×	×			×	×		×	×
Chlorophyceae														
Volvocales														
Eudorina elegans EHR.*					×			×		×	×			×
Gonium pectorale O. F. MÜLL.*			×								×			×
Pandorina morum (O. F. MÜLL.) BORY*		×		×						×	×			
Volvox aureus EHR.*		⊕		⊕										
Tetrasporales														
Asterococcus limneticus G. M. SMITH*														×
Asterococcus superbus (CIENK.) SCHERF.*		×			×		×					×	×	×
Pseudosphaerocystis lacustris (LEMM.) NOVÁK.*								○						○

	Lygners Vider lakes													
	Fi	Gi	Gr	Gä	St. Hå	Kr	Ox	Ra	Sk	L. Sv	St. Sv	Ta	To	Än
Chlorococcales														
Ankistrodesmus falcatus (CORDA) RALFS*														×
Ankistrodesmus stipitatus (CHOD.) KOM.-LEG.*				×										×
Botryococcus braunii KÜTZ.*	×	×	×	×	×	×	×	×	×	×	×	×	×	×
Coelastrum indicum TURN.*						×		×						×
Coelastrum microporum NÄG.*														×
Crucigenia quadrata MORREN*	×						×			×				
Crucigenia tetrapedia (KIRCHN.) W. & G. S. WEST*					⊕			×		⊕	⊕			×
Crucigeniella rectangularis (NÄG.) KOM.*										×	×			×
Dictyosphaerium pulchellum WOOD*		×	×	×										×
Elakatothrix gelatinosa WILLE*				○										○
Micractinium pusillum FRES.*														×
Oocystis lacustris CHOD.*						×				×	×			
Pediastrum boryanum (TURP.) MENEGH.*		×		×	×	×	×		×		×		×	×
Pediastrum duplex v. *rugulosum* RAC.*														×
Pediastrum limneticum THUNM.*							×	×						
Quadrigula closterioides (BOHL.) PRINTZ*				×				×	×					×
Quadrigula pfitzeri (SCHRÖDER) G. M. SMITH														×
Selenastrum bibraianum REINSCH.*		⊕					⊕							
Sphaerocystis schroeteri CHOD.*	×	×				×		×	×					×
Tetraëdron minimum (A. BR.) HANSG.		⊕		⊕				○						×
Willea irregularis (WILLE) SCHMIDLE*			×									×		
Ulotrichales														
Binuclearia tatrana WITTR.	×			×	×		×	×	×	×	×		×	○
Chlorhormidium rivulare (KÜTZ.) STARM.					×		⊕							
Microspora palustris WICHM.					×	×	⊕		×		×	×		
Microspora palustris v. *minor* WICHM.							×							
Ulothrix zonata (WEB. & MOHR) KÜTZ.						×	×			×				×
Bulbochaete setigera (ROTH.) AG.			×		×		×			×	×			×
Oedogonium sphaerandrium WITTR. & LUND										×	×			×
Oedogonium sp. *steril*						×				×			×	
Conjugatophyceae														
Zygnematales														
Mougeotia parvula HASS.			×		×			×	×	×				
Mougeotia recurva (HASS.) DE TONI					×	×								
Mougeotia sp. *steril*	×	×	×	×	×	×	×	×	×	×	×	×	×	×
Spirogyra sp. *steril*			×			×		×	×			×	×	
Zygnema stellinum (VAUCH.) AG.						×								
Zygnema sp. *steril*	×				×				×	×	×	×	×	×
Gonatozygales														
Gonatozygon aculeatum HAST			×		×				×				×	×
Gonatozygon sp.*							×			×				
Desmidiales														
Closterium dianae EHR.*		×			×	×			×	×	×		×	×
Closterium intermedium RALFS	×				×	×	×		×	×	×			×
Closterium kuetzingii BRÉB.*	×		×		×			×						×
Cosmarium contractum v. *ellipsoideum* (ELFV.) W. & G. S. WEST*			×		×		×		×	×	×	×		×
Cosmarium contractum v. *minutum* (DELP.) W. & G. S. WEST*	×					×				×	×		×	×
Cosmarium depressum (NÄG.) LUND.*	×	×	×		×	×	×	×		×	×	×	×	×
Euastrum ansatum RALFS				×	×	×				×	×	×		
Euastrum humerosum RALFS			×			×					×			
Euastrum humerosum v. *affine* (RALFS) WALLICH	×				×		×							
Euastrum pinnatum RALFS			×				×	×	×					×
Hyalotheca dissiliens (J. E. SMITH) BRÉB.				×			×	×		×	×		×	×
Micrasterias furcata RALFS*		×					×			×	×	×		×
Micrasterias oscitans RALFS	×		×	×	×	×	×	×		×	×	×	×	
Micrasterias radiosa RALFS	×			×	×	×	×		×					×
Micrasterias truncata v. *quadrata* BULNH.		×	×				×	×	×	×	×		×	×
Penium polymorphum (PERTY) PERTY*						×	×	×	×	×	×		×	

	Lygners Vider lakes													
	Fi	Gi	Gr	Gä	St. Hå	Kr	Ox	Ra	Sk	L. Sv	St. Sv	Ta	To	Än
Spondylosium planum (WOLLE) W. & G. S. WEST*	⊕	⊕							⊕	⊕	⊕	⊕	⊕	×
Staurastrum anatinum f. *vestitum* (RALFS) BROOK*	×		×	×		×	×		×			×		×
Staurastrum arachne RALFS*							×	×				×	×	×
Staurastrum artiscon (EHR.) LUND.*					×			×			×		×	×
Staurastrum longipes (NORDST.) TEIL.*	⊕	⊕		⊕		⊕			⊕			⊕		×
Staurastrum ophiura LUND.*					×	×				×	×	×		×
Staurastrum paradoxum MEYEN*	×	×					×			×		×		
Staurastrum pelagicum W. & G. S. WEST*			×		×	×			×	×	×			×
Staurodesmus cuspidatus (BRÉB.) TEIL.*				×	×	×		×	×	×	×			×
Staurodesmus extensus (BORGE) TEIL.*					×			×				×	×	
Staurodesmus incus (BRÉB.) TEIL.*	×	×	×		×	×	×	×	×	×	×		×	×
Staurodesmus quiriferus (W. & G. S. WEST) TEIL.*								×	×	×	×	×	×	
Staurodesmus triangularis (LAGERH.) TEIL.*	×			×	×	×		×						×
Xanthidium antilopaeum v. *hebridarum* W. & G. S. WEST*		×			×	×		×	×	×	×	×	×	×
Xanthidium armatum (BRÉB.) RAB.*			×	×	×		×		×	×	×	×	×	×
Xanthidium armatum v. *fissum* NORDST.*						×		×	×					×
FLAGELLATE														
Euglenophyceae														
Colacium vesiculosum EHR.*	O											O	O	
Trachelomonas volvocina EHR.*														×
Cryptophyceae														
Cryptomonas ovata EHR.*			×	×						×			×	×
Cryptomonas spp.*	×	×			×									
Rhodomonas sp.*								×					×	×
Chloromonadophyceae														
Gonyostomum semen (EHR.) DIES.*													×	

Appendix 2

Taxa belonging to Desmidiales recorded sporadically in one or a few lakes; details on their distribution are given by MORLING (1984b).

* = euplanktonic species, according to ROUND (1981).

**Arthrodesmus octocornis* (EHR.) RALFS, **A. trispinatus* W. & G. S. WEST.

**Bambusina brebissoni* KÜTZ.

**Closterium aciculare* T. WEST, **C. acutum* BRÉB., *C. angustatum* KÜTZ., **C. baillyanum* (BRÉB.) BRÉB., **C. decorum* BRÉB., *C. ehrenbergii* MENEGH., *C. gracile* BRÉB., **C. idiosporum* W. & G. S. WEST, **C. juncidum* RALFS, **C. parvulum* NÄG., **C. pseudolunula* BORGE, **C. ralfsii* BRÉB., **C. rostratum* EHR., **C. setaceum* EHR., **C. subscoticum* GUTW., *C. turgidum* EHR., **C. venus* KÜTZ.

Cosmarium abbreviatum RAC., *C. contractum* KIRCHN., **C. granulatum* W. & G. S. WEST, **C. margaritiferum* MENEGH., **C. ocellatum* v. *incrassatum* W. & G. S. WEST, **C. ovale* RALFS, **C. punctulatum* v. *subpunctulatum* (NORDST.) BØRG., **C. quinarium* v. *pictum* SKUJA, **C. subtumidum* NORDST., **C. subtumidum v. klebsii* (GUTW.) W. & G. S. WEST.

**Cylindrocystis brébissonii* MENEGH.

**Docidium bacculum* BRÉB., **D. undulatum* BAIL.

**Euastrum bidentatum* v. *speciosum* (BOLDT) SCHMIDLE, **E. crassum* (BRÉB.) KÜTZ., **E. elegans* (BRÉB.) KÜTZ., *E. oblongum* (GREV.) RALFS, **E. pectinatum* BRÉB., **E. sinuosum* LENORM., *E. verrucosum* EHR., *E. verrucosum* v. *alatum* WOLLE.

Hyalotheca mucosa (MERT.) EHR.

**Micrasterias americana* (EHR.) RALFS, **M. apiculata* (EHR.) MENEGH., **M. crux-melitensis* (EHR.) HASS., **M. denticulata* v. *angulosa* (HANTZSCH) W. & G. S. WEST, *M. jenneri* RALFS, *M. papillifera* BRÉB., *M. pinnatifida* (KÜTZ.) RALFS, *M. radiosa* v. *ornata* NORDST., *M. rotata* (GREV.) RALFS **M. thomasiana* v. *rotata* (NORDST.) GRÖNBL., *M. truncata* v. *semiradiata* (NÄG.) WOLLE.

Netrium digitus (BRÉB.) ITZ. & ROTHE, **N. interruptum* v. *digitiforme* RUZ.

Pleurotaenium ehrenbergii (BRÉB.) DE BARY.

**Staurastrum anatinum* COOKE & WILLS, **S. anatinum* f. *cortum* (G. M. SMITH) BROOK, **S. anatinum* f. *denticulatum* (G. M. SMITH) BROOK, **S. anatinum* f. *longibrachiatum* (W. & G. S. WEST) BROOK, **S. anatinum* v. *grande* W. & G. S. WEST, **S. brasiliense* v. *lundellii* W. & G. S. WEST, **S. bullardii* G. M. SMITH, **S. cerastes* LUND., **S. cingulum* (W & G. S. WEST) G. M. SMITH, **S. clevei* (WITTR.) ROY & BISS., **S. hystrix* RALFS, **S. longispinum* (BAIL.) ARCH., **S. lunatum* RALFS, **S. manfeldtii* DELP., **S. planctonicum* v. *bullosum* TEIL., **S. pseudopelagicum* W. & G. S. WEST, **S. rotula* NORDST., **S. sexangulare* (BULNH.) LUND., **S. sexcostatum* BRÉB., **S. teliferum* RALFS, **S. verticillatum* ARCH.

**Staurodesmus cuspidatus* v. *curvatus* (W. WEST) TEIL., **S. dejectus* (BRÉB.) TEIL., **S. glaber* v. *limnophilus* TEIL., **S. indentatus* (W. & G. S. WEST) TEIL., **S. pachyrynchus* (NORDST.) TEIL., * *S. triangularis* v. *limneticus* TEIL., **S. triangularis* v. *subparallelus* (G. M. SMITH) THOM.

Teilingia granulata (ROY & BISS.) BOURR.

Tetmemorus brebissonii (MENEGH.) RALFS, *T. granulatus* (BRÉB.) RALFS, *T. laevis* (KÜTZ.) RALFS.

**Xanthidium antilopaeum* v. *dimoazum* NORDST., **X. antilopaeum* v. *laeve* SCHMIDLE, **X. antilopaeum* v. *oligacanthum* SCHMIDLE, **X. controversum* W. & G. S. WEST, **X. cristatum* BRÉB., **X. smithii* ARCH., **X. subhastiferum* W. & G. S. WEST.

Adress of the authors:

GRETA MORLING
TORBJÖRN WILLÉN
Institute of Limnology
University of Uppsala
P. O. Box 557
S-751 22 Uppsala
Sweden

Limnologica (Berlin) **20** (2), 307—318, Juni 1990

Institute of Limnology, University of Uppsala, Uppsala, Sweden

Acidification and Zooplankton Development in Some West-Swedish Lakes 1966—1983

By GRETA MORLING & BIRGER PEJLER

With 16 Figures

Abstract

Zooplankton development was followed through two decades in some small west-Swedish lakes, starting with a relatively well-buffered, circumneutral situation, passing through a period with rather rapid decrease in pH and ending when new buffering systems had stabilized pH at a low level. Many species disappeared or decreased strongly in connection with the process of acidification, but some euryoecious forms, as well as some indicators of oligotrophy, survived with about the same frequency and abundance. All cladocerans were found to be more or less sensitive, while the copepods survived the changes. No specialists of acid environments were encountered, but some normally non-planktic forms invaded the pelagic zone during the course of acidification.

1. Introduction

The process of acidification in some small west-Swedish lakes (Gothenburg area) was followed by G. MORLING in a unique long-term investigation during the period 1966—1983. The changes of chemical conditions were treated by MORLING (1981), and those of phytoplankton by MORLING & WILLÉN (this volume). In the present paper the effects on zooplankton are reported.

2. Material and Methods

The sampling took place from a boat in the central part of the respective lakes. The plankton samples were mainly taken from the upper water layers. Qualitative determination was based on nets with 63 μm and 110 μm mesh size. These samples were studied alive, as well as after fixation with IIK solution or formalin. For quantitative analysis a Ruttner sampler was used, whereby a total volume of 20 l was collected, sieved through a 63 μm plankton net and then treated according to the method described by UTERMÖHL (1958).

Samples were taken continuously from 1966 onwards, once a month and sometimes every fortnight from April to October. During years with late formation of ice cover, sampling was also done during a period from November to January.

3. The Lakes

The lakes of the investigation are situated in an area SE of Gothenburg, called Lygners Vider (see MORLING & WILLÉN, this volume, Fig. 2). For a closer description of the lakes the reader is referred to MORLING (1981).

From the referred papers it is evident that three developmental phases can be distinguished during the period when the lakes were studied: phase 1, from 1966 to 1969, with a relatively original situation, well-buffered conditions, pH close to the neutral point, and no remarkable biological deviation; phase 2, 1970—1977, when alkalinity in most lakes reached zero and pH declined to around 5, with great variation during the course of the year, and when extensive biological changes occurred, including killing of fish; phase 3, when new buffering systems stabilize pH at a low level, new communities develop and the fishes often have entirely disappeared (see also MORLING & WILLÉN, this volume, Fig. 3 and

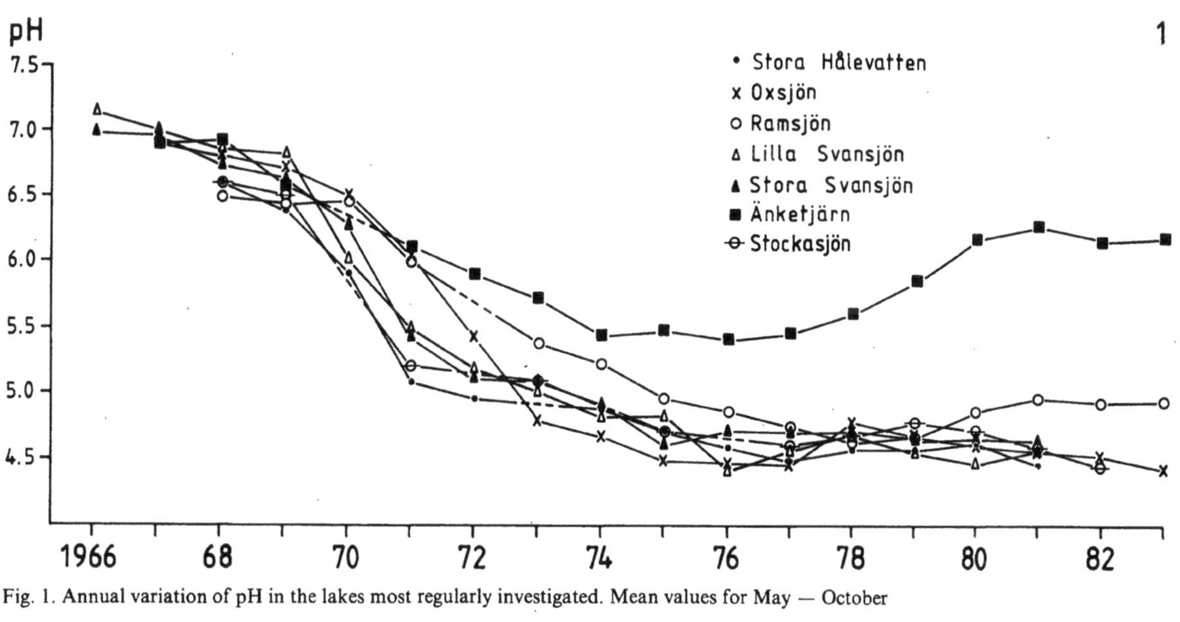

Fig. 1. Annual variation of pH in the lakes most regularly investigated. Mean values for May — October

p. 292). Although this trend was followed more or less typically in most of the lakes, L. Änketjärn, however, remained relatively unchanged, for reasons discussed by MORLING et al. (1985). The change in pH is shown in Fig. 1 for the seven lakes most intensively studied.

Some of the lakes have been treated with lime, starting in 1982, 1983 or 1984 (in L. Gingsjön already from 1979) but no samples after this treatment are reported in the present paper, which thus exclusively deals with the effect of acidification.

4. Results

All zooplankters encountered are recorded in Table 1. Some species which normally are restricted to benthic and periphytic environments apparently may invade the pelagic zone in special lakes like those treated in the present paper. This is the reason why, e.g., *Keratella serrulata* and *Polyarthra minor* are here designated as planktic.

Table 1. Zooplankters in the investigated lakes: Fixsjön = Fi; Gingsjön = Gi; Grundasjön = Gr; Gärdsjön = Gä; St. Hålevatten = St. Hå; Kroksjön = Kr; Oxsjön = Ox; Ramsjön = Ra; Skållesjön = Sk; L. Svansjön = L. Sv; St. Svansjön = St. Sv; Tagsjön = Ta; Tolaredsjön = To; Änketjärn = Än.

	Lygners Vider lakes													
	Fi	Gi	Gr	Gä	St. Hå	Kr	Ox	Ra	Sk	L. Sv	St. Sv	Ta	To	Än
Testacea														
Difflugia limnetica LEVANDER										×	×		×	
Difflugia pelagica PEJLER	×				×									
Heliozoa														
Acanthocystis perpusilla BOYE PETERS. & HANS.	×	×		×			×	×				×	×	×
Ciliata														
Cothurnia sp.					×					×	×			
Tintinnidium fluviatile STEIN										×				
Tintinnopsis lacustris ENTZ								×		×	×			
Rotatoria														
Ascomorpha ecaudis PERTY					×		×	×	×	×	×	×		×
Ascomorpha ovalis (BERGENDAL)							×		×					×
Asplanchna priodonta GOSSE		×			×		×				×		×	×
Brachionus angularis GOSSE								×		×	×			×
Collotheca mutabilis (HUDSON)							×						×	×
Conochilus unicornis ROUSSELET	×	×		×				×		×	×		×	×
Epiphanes pelagica (JENNINGS)							×							
Gastropus stylifer IMHOF			×				×		×					×
Kellicottia longispina (KELLICOTT)	×	×	×		×	×	×	×		×	×	×	×	×
Keratella cochlearis (GOSSE)	×	×	×	×	×	×	×	×	×	×	×	×	×	×
Keratella quadrata (MÜLL.)					×				×					
Keratella serrulata (EHRBG.)	×				×		×	×	×	×	×			×
Polyarthra euryptera (WIERZEJSKI)	×													
Polyarthra major (BURCKHARDT)														×
Polyarthra minor VOIGT				×	×	×	×	×		×			×	×
Polyarthra remata (SKORIKOV)	×		×	×	×	×	×	×		×	×	×	×	×
Polyarthra vulgaris CARLIN	×	×			×	×	×	×		×	×	×	×	×
Synchaeta grandis ZACHARIAS		×												
Trichocerca capucina (WIERZEJSKI & ZACHARIAS)			×	×	×	×	×	×	×	×	×	×		×
Trichocerca pusilla (LAUTERBORN)	×		×											
Cladocera														
Bosmina (Eubosmina) longispina LEYDIG	×	×	×	×	×	×	×	×	×	×	×	×	×	×
Bythotrephes longimanus LEYDIG					×	×			×	×	×			×
Ceriodaphnia quadrangula hamata SARS	×	×		×		×								×
Daphnia cucullata SARS	×							×						
Daphnia longispina MÜLL. s. str.	×	×						×						
Diaphanosoma brachyurum (LIÉVIN)	×	×	×	×	×	×	×	×	×	×	×	×	×	×
Holopedium gibberum ZADDACH	×	×	×	×	×	×	×	×	×	×	×	×		×
Leptodora kindti (FOCKE)					×			×	×					×
Polyphemus pediculus (L.)	×		×		×	×	×	×	×	×	×	×	×	×
Copepoda														
Eudiaptomus gracilis SARS	×	×	×	×	×	×	×	×	×	×	×	×	×	×
Eudiaptomus graciloides LILLJ.						×	×	×						
Cyclops strenuus strenuus (FISCHER) KOZMINSKI	×	×		×			×			×	×			×
Mesocyclops leuckarti (CLAUS)	×						×							×
Thermocyclops oithonoides (SARS)	×													

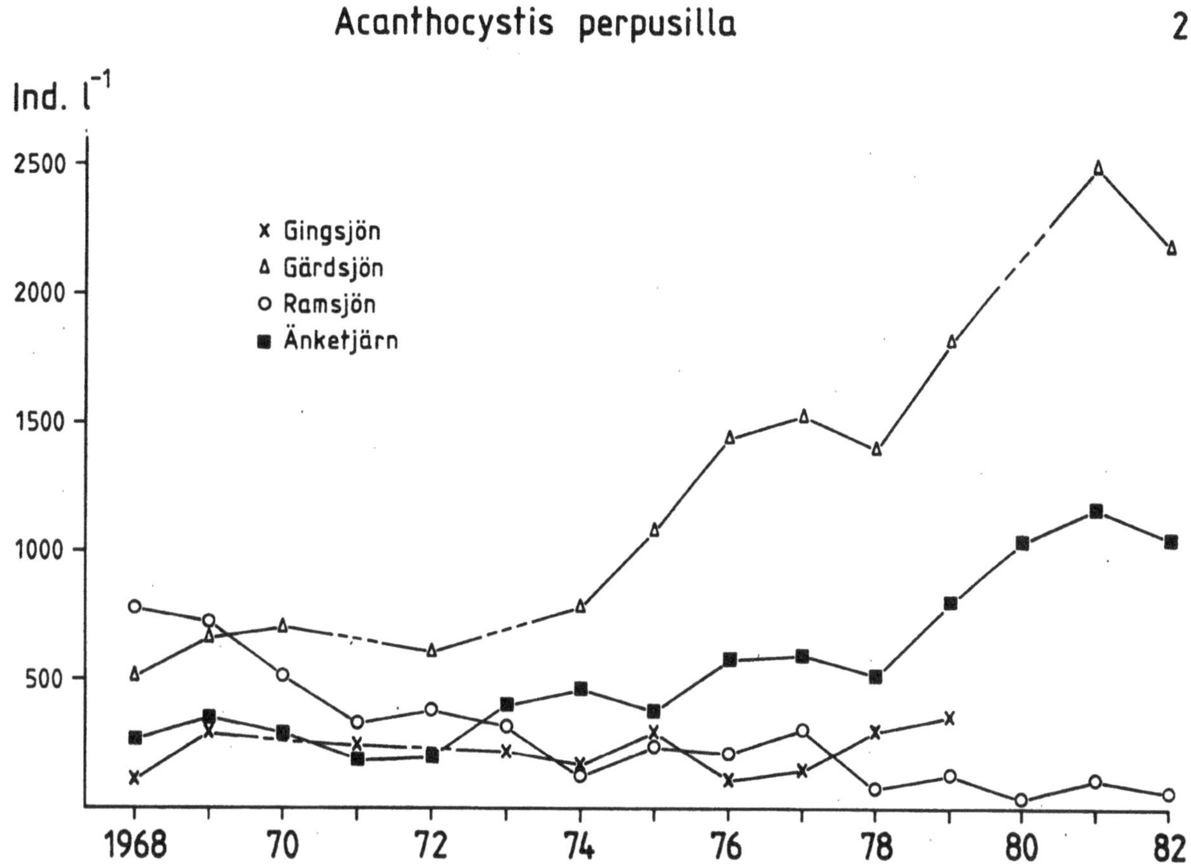

Figs. 2—16. Annual variation of import zooplankters in regularly studied lakes where they had a high abundance. Mean values for May—October (Figs. 5, 15 and 16), May—September (Figs. 3—4, 7—9, 11 and 14), June—October (Fig. 12), June—September (Figs. 2, 10 and 13) or July—September (Fig. 6)

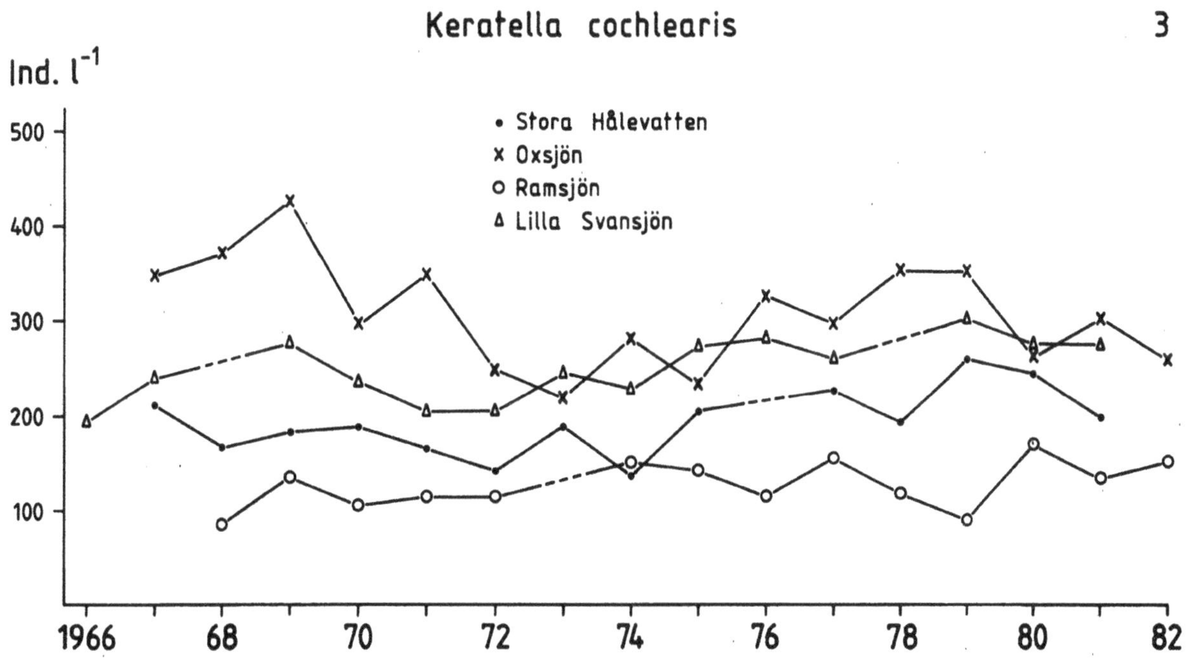

Keratella serrulata 4

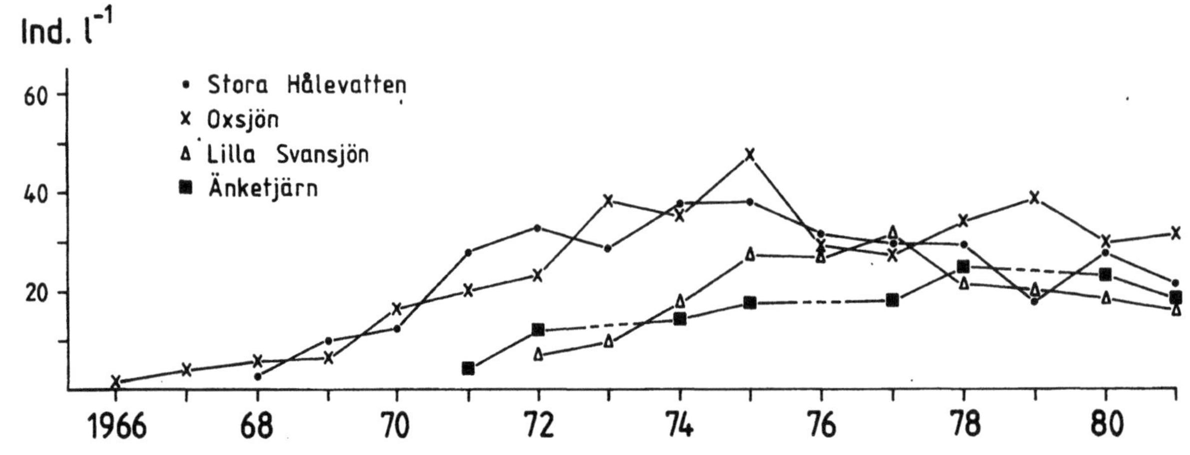

Kellicottia longispina 5

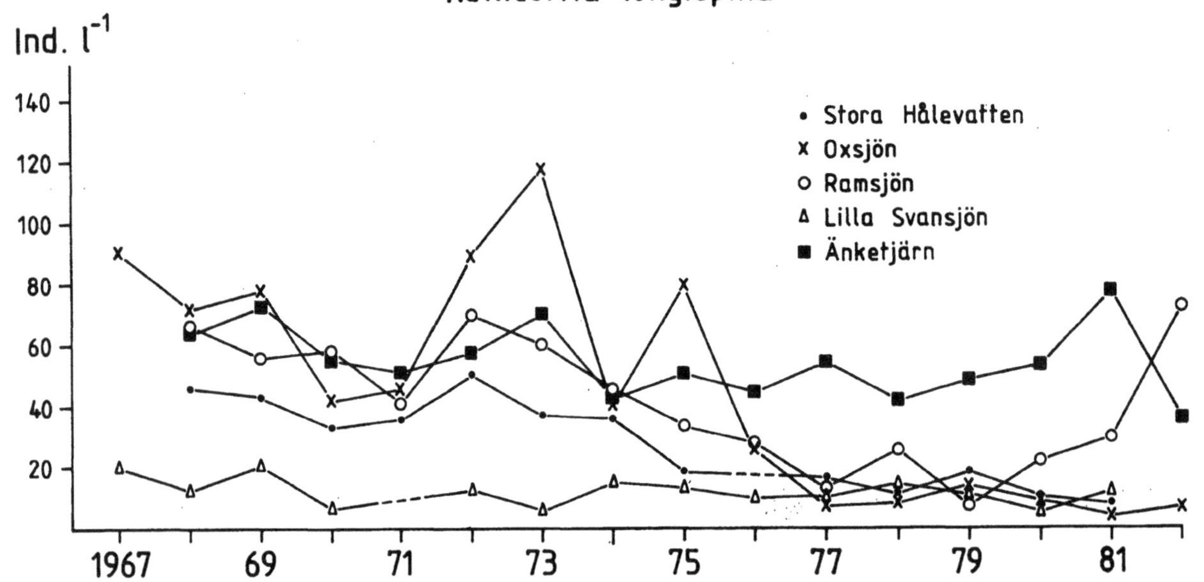

Asplanchna priodonta 6

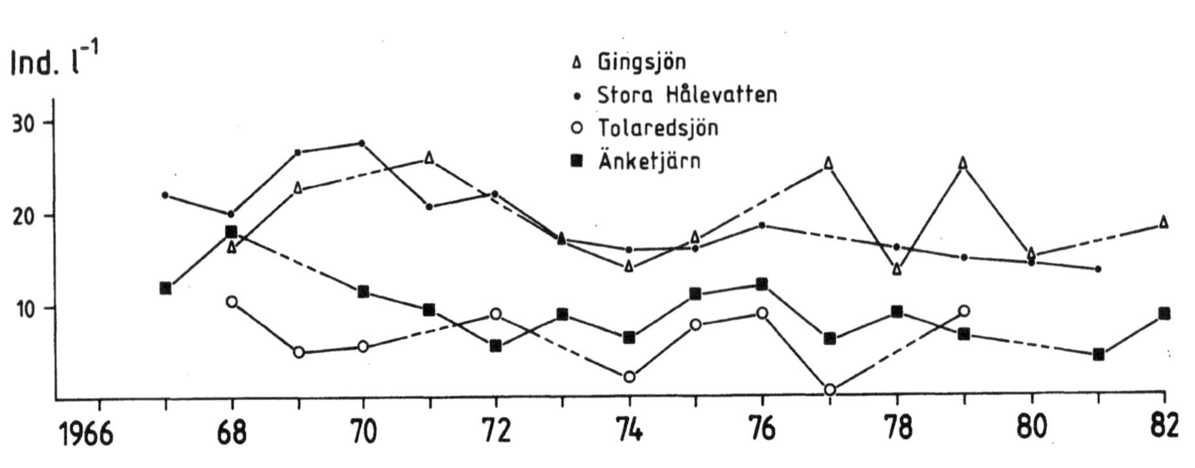

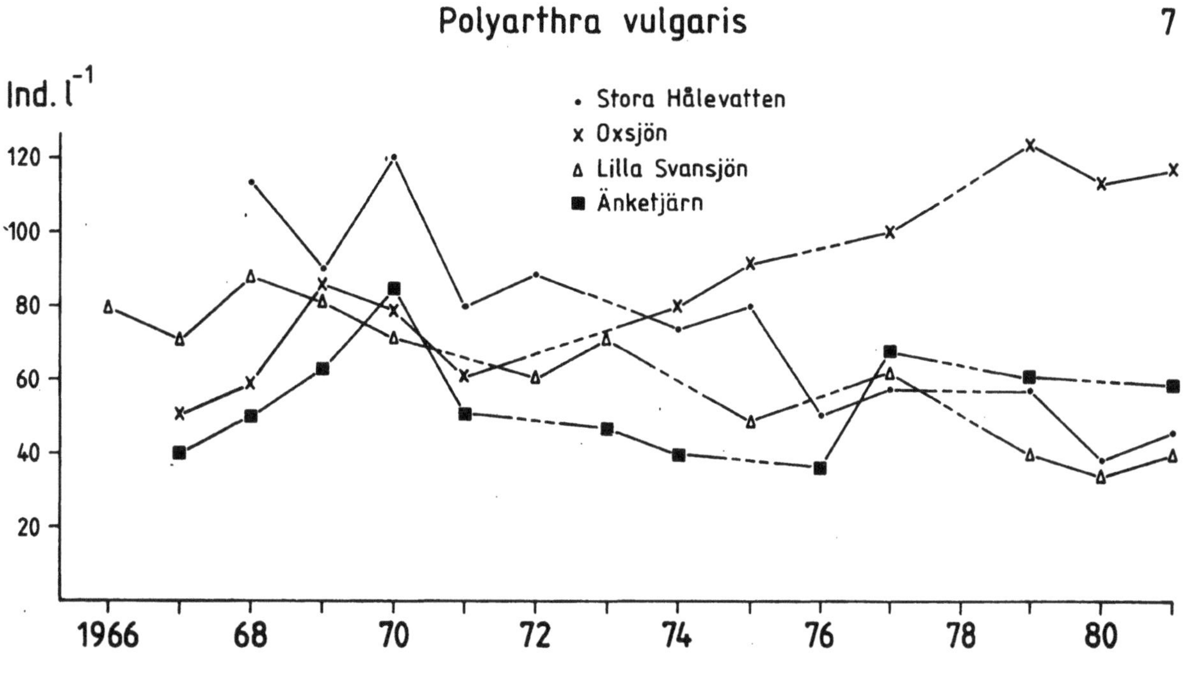
Polyarthra vulgaris
7
Ind. l^{-1}
• Stora Hålevatten
x Oxsjön
Δ Lilla Svansjön
■ Änketjärn
120
100
80
60
40
20
1966
68
70
72
74
76
78
80

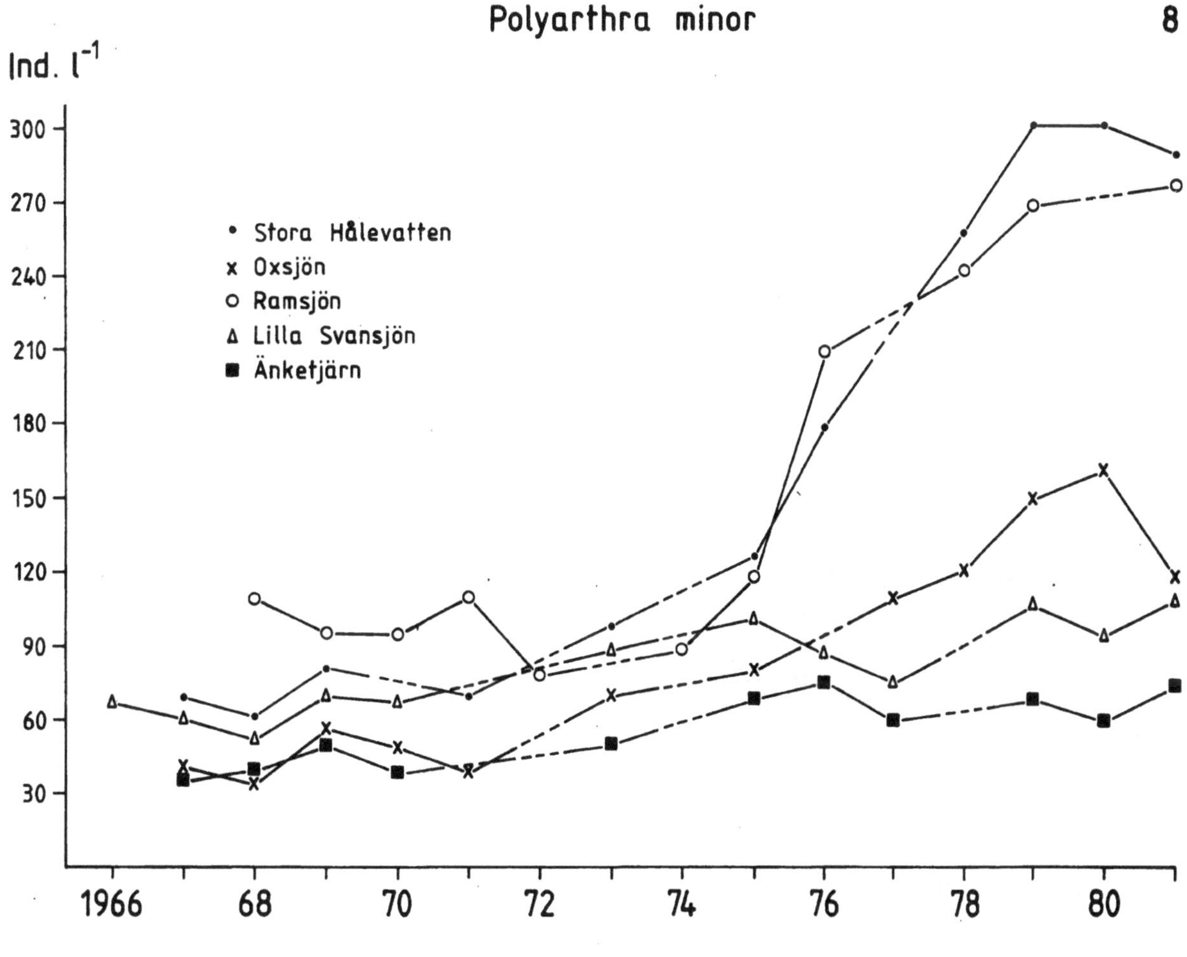
Polyarthra minor
8
Ind. l^{-1}
• Stora Hålevatten
x Oxsjön
o Ramsjön
Δ Lilla Svansjön
■ Änketjärn
300
270
240
210
180
150
120
90
60
30
1966
68
70
72
74
76
78
80

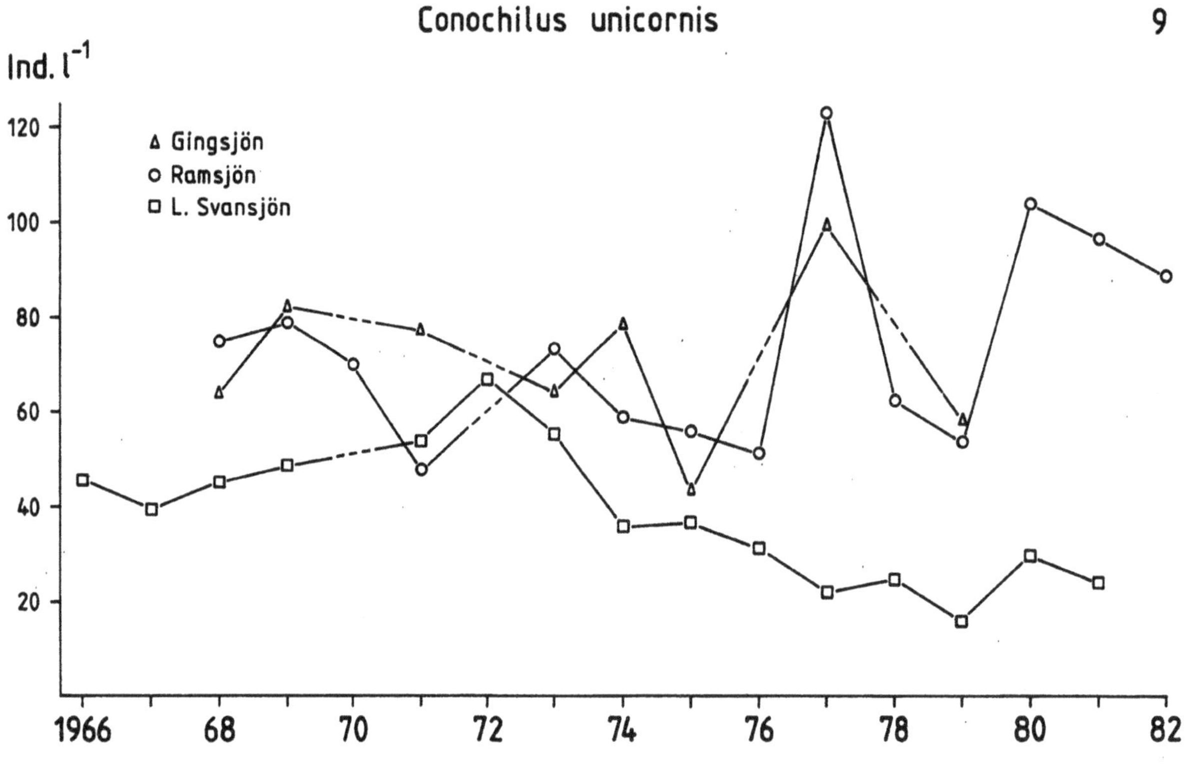
Conochilus unicornis
9
Ind. l⁻¹
Gingsjön
Ramsjön
L. Svansjön
120
100
80
60
40
20
1966
68
70
72
74
76
78
80
82

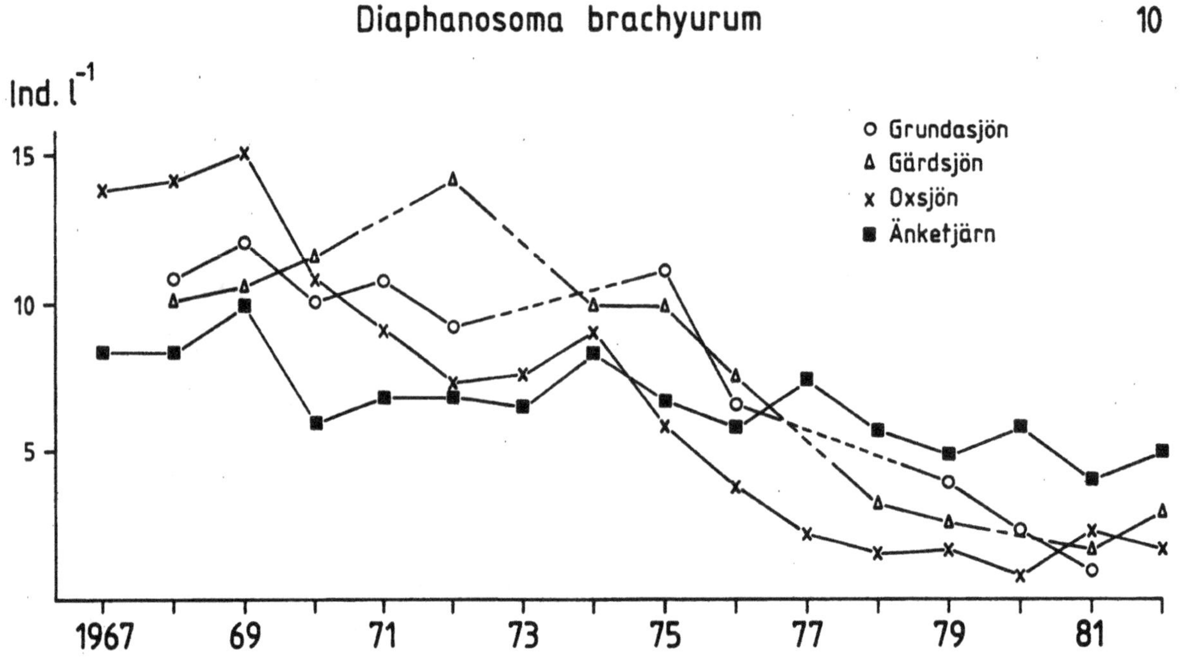
Diaphanosoma brachyurum
10
Ind. l⁻¹
Grundasjön
Gärdsjön
Oxsjön
Änketjärn
15
10
5
1967
69
71
73
75
77
79
81

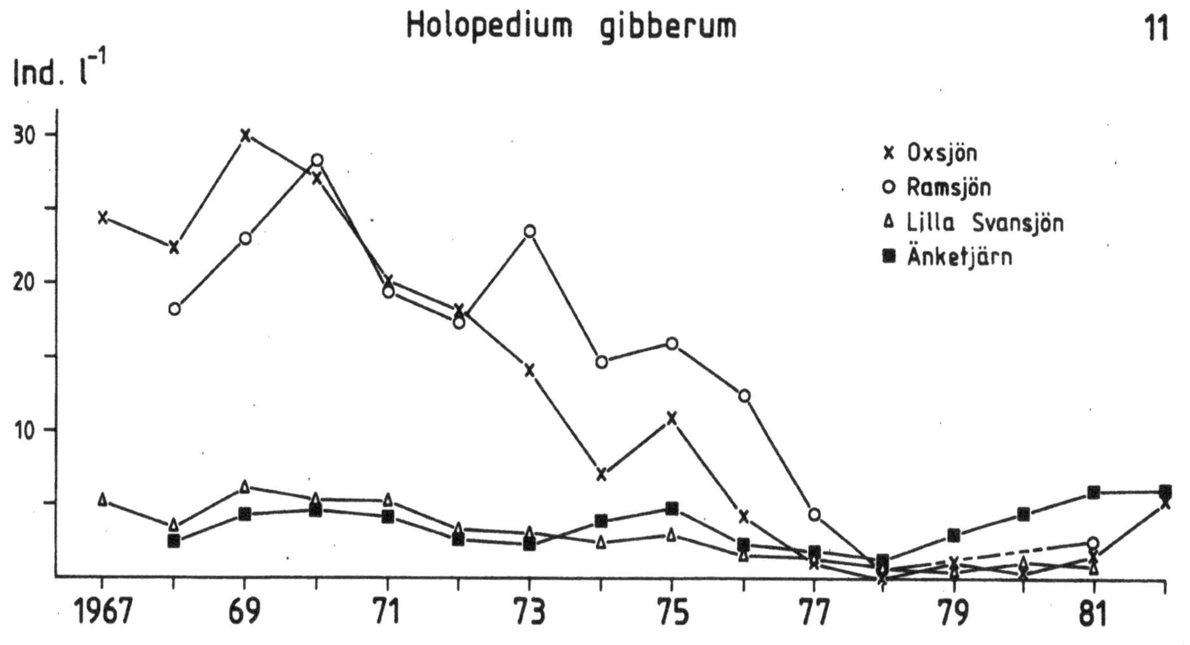
Holopedium gibberum
11
Ind. l⁻¹
30
20
10
x Oxsjön
o Ramsjön
Δ Lilla Svansjön
■ Änketjärn
1967
69
71
73
75
77
79
81

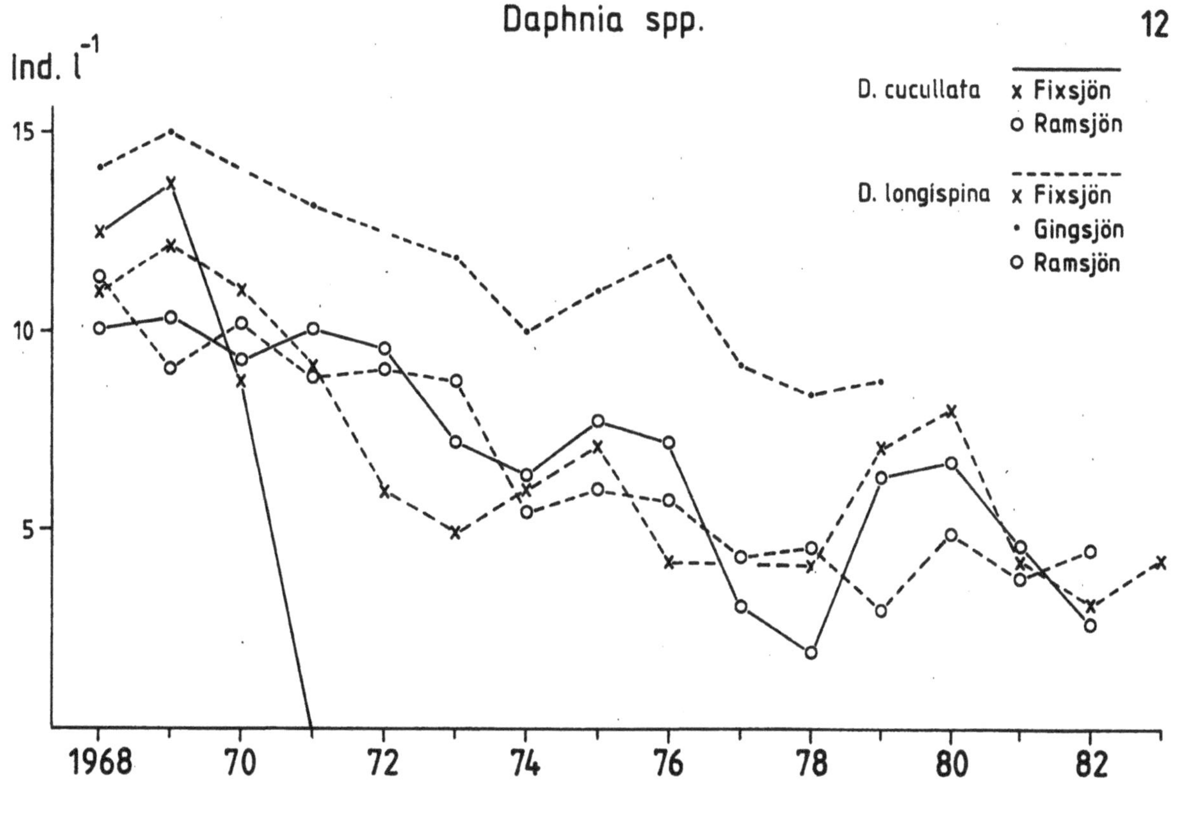
Daphnia spp.
12
Ind. l⁻¹
15
10
5
D. cucullata
x Fixsjön
o Ramsjön
D. longispina
x Fixsjön
• Gingsjön
o Ramsjön
1968
70
72
74
76
78
80
82

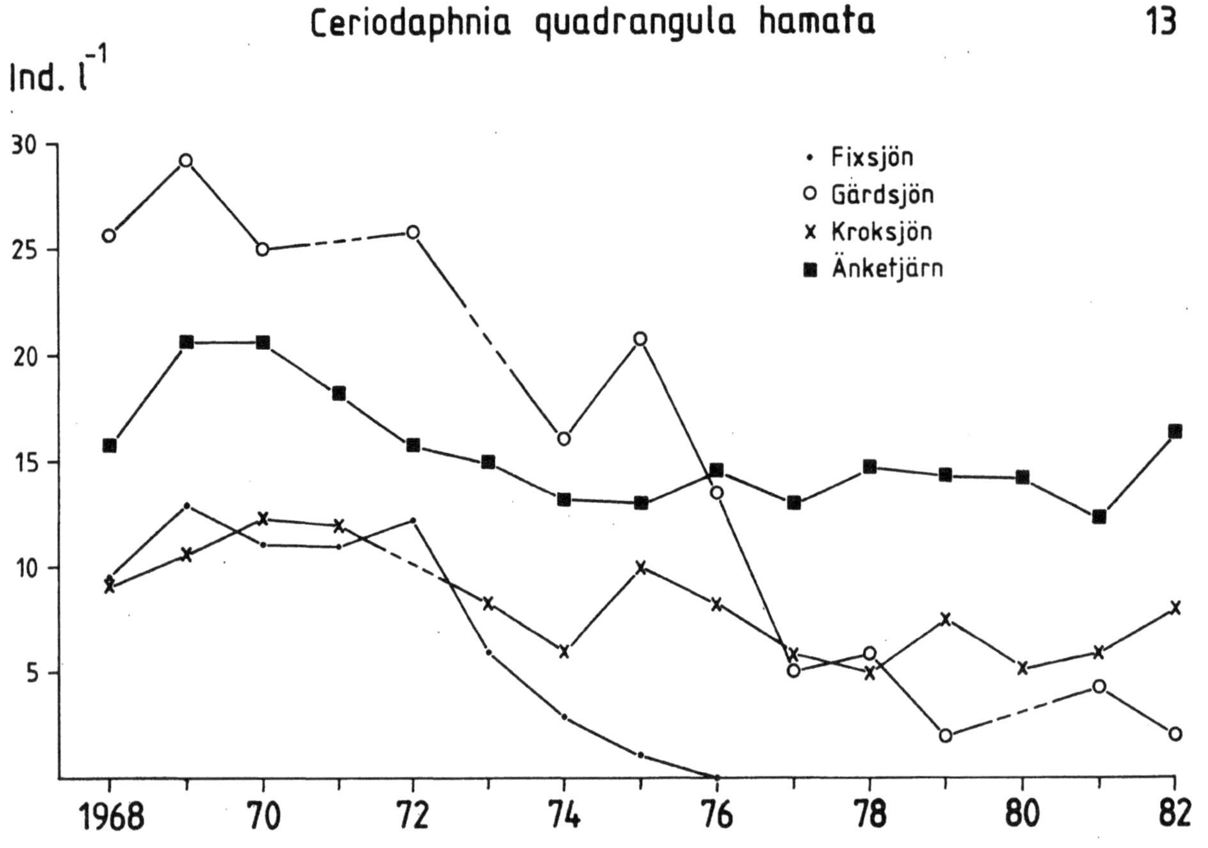
Ceriodaphnia quadrangula hamata
13
Ind. l⁻¹
30
25
20
15
10
5
Fixsjön
Gärdsjön
Kroksjön
Änketjärn
1968
70
72
74
76
78
80
82

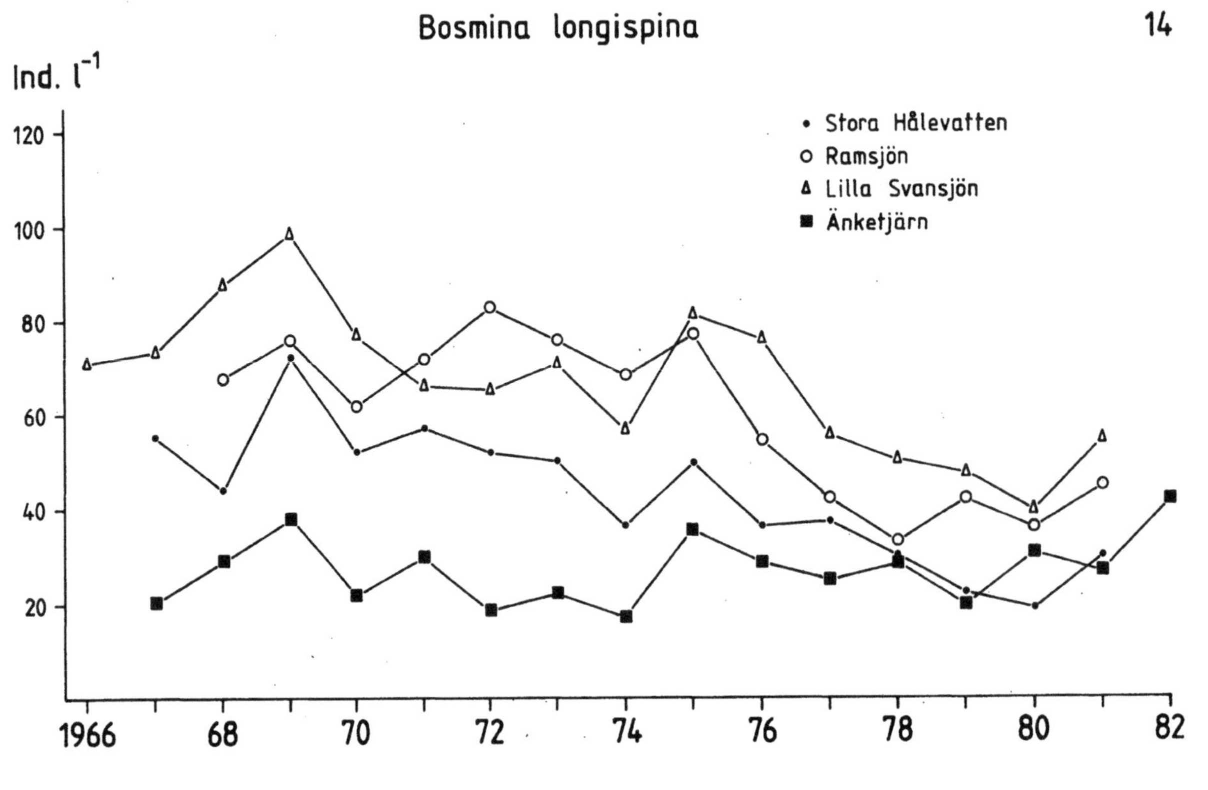
Bosmina longispina
14
Ind. l⁻¹
120
100
80
60
40
20
Stora Hålevatten
Ramsjön
Lilla Svansjön
Änketjärn
1966
68
70
72
74
76
78
80
82

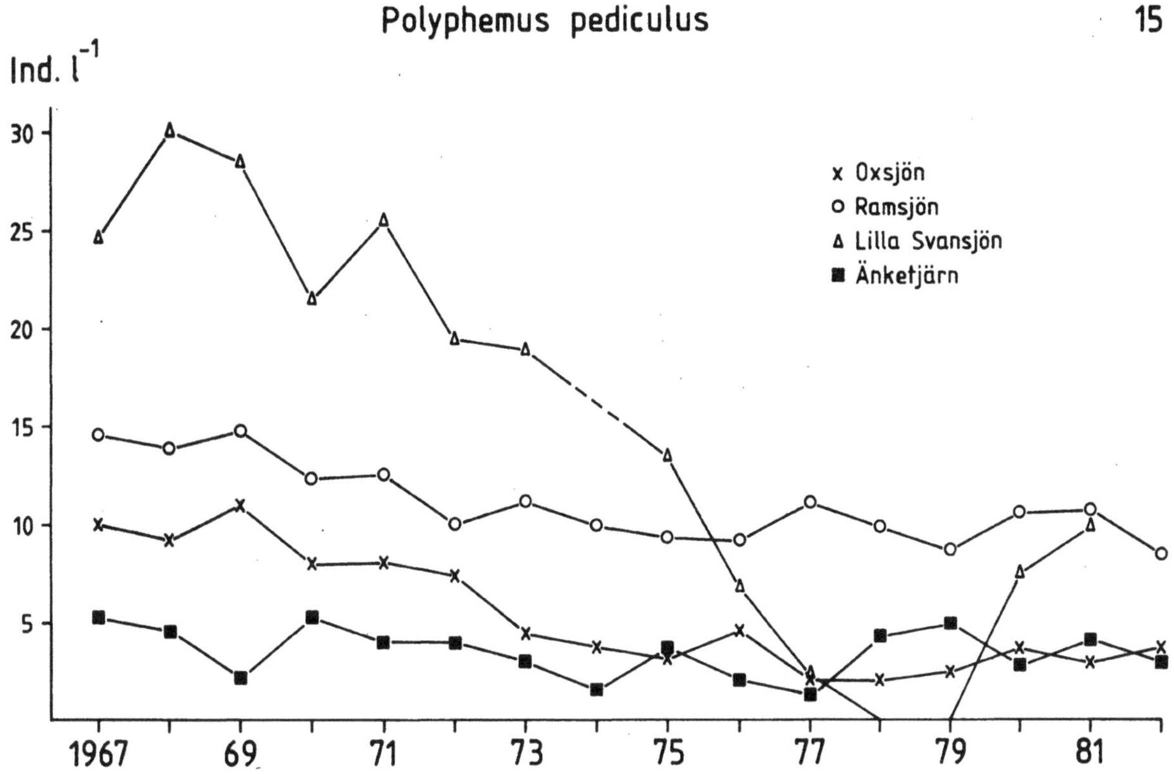

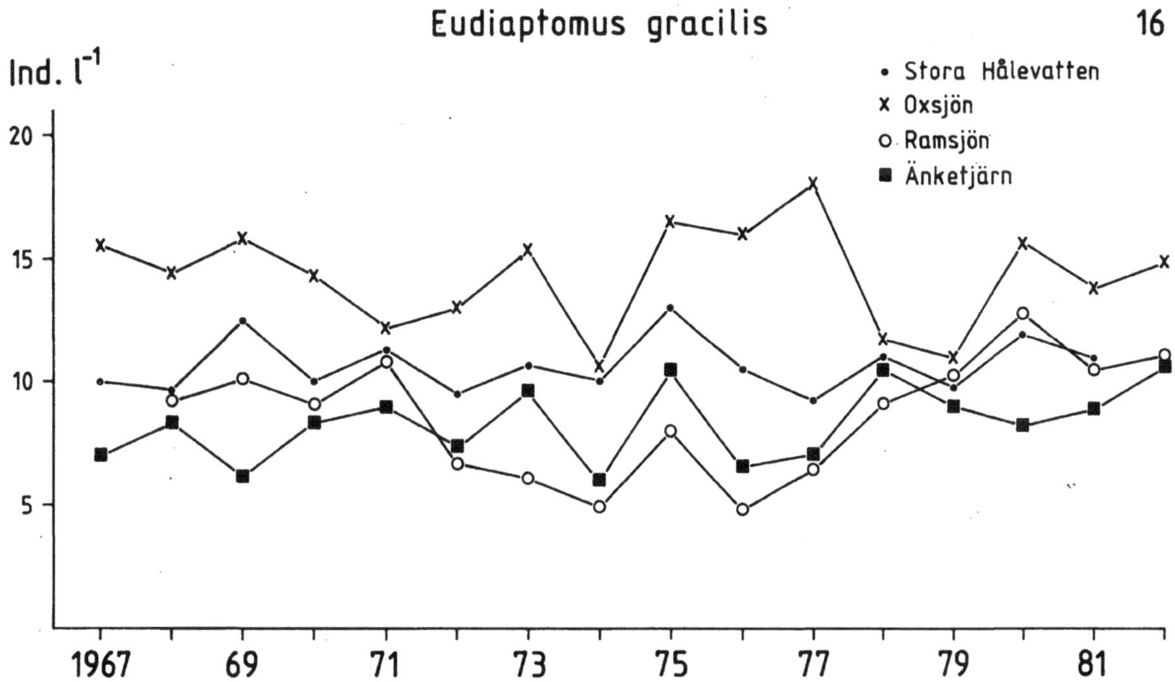

As indicated in the methods section, a very large quantitative material was collected, which is archived at the Limnological Institute, Uppsala. For international publication the material had to be compressed which is why only mean values for the individual years are reported in Figs. 2—16. In this connection, the data used were from the months when the respective species occurred in a fair number, i.e. from a longer period for perennial species than for the summer species (for details, see the text of the respective figures).

Such diagrams are only given for species which were exposed to remarkable population changes during the course of the investigation period, as well as for some dominating taxa which seem to be rather insensitive to the prevailing environmental variation. For each separate diagram those lakes were chosen where the species occurred in its highest abundance and where also continuous sampling was made. Furthermore, lakes with a well-known fish fauna have been given priority in this choice. The two lakes which have been investigated most thoroughly are L. Oxsjön (with strong decrease in pH) and L. Änketjärn (with comparably stable conditions; cf. above), and these lakes have been included in all cases when the respective species occurred in sufficient abundance.

Of course, the composition of the zooplankton differs between the different localities (for details see MORLING 1984 and, especially, the archived material). However, since the lakes all belong to about the same type and have been influenced in much the same way, some generalizations might be made:

1) Three of the species found belong to the indicators of oligotrophy mentioned by PEJLER (1965), viz. *Difflugia pelagica*, *Synchaeta grandis* and *Holopedium gibberum*. If also the frequency is considered, *Kellicottia longispina* and *Bosmina longispina* may be regarded as such indicators as well (see PEJLER 1965, 1983; LIEDER 1983, and NILSSEN & LARSSON 1980). Of these species, *D. pelagica* and *S. grandis* are relatively unimportant in the zooplankton although the former might be more important than stated because of its small size which allows it to pass easily through the meshes of the plankton net.
2) Only one of the eight eutrophy indicators according to PEJLER (1965) was found, viz. *Trichocerca pusilla*, in only two lakes and with comparatively low frequency. *Daphnia cucullata*, indicating alkaline conditions rather than high production, occurred in two lakes before the increasing acidification and afterwards only sporadically and in decreasing numbers in one of them.
3) The material contains only one of the three species which PEJLER (op. cit.) found only in humic environments, viz. *Keratella serrulata*.
4) According to previous literature, none of the species observed can be regarded as cold stenothermal, which ought to be connected with the fact that sampling was mainly done in the warmer seasons and in the epilimnion.
5) Euryoecious forms dominate in the investigated lakes, both as regards frequency and abundance. Especially the following should be mentioned: *Keratella cochlearis*, *Polyarthra vulgaris*, *P. remata*, *Conochilus unicornis*, *Eudiaptomus gracilis* and *Cyclops strenuus strenuus*. However, also three indicators of oligotrophy are very important, viz. *Kellicottia longispina*, *Bosmina longispina* and *Holopedium gibberum*.
6) Species which are generally regarded as semiplankters, or even non-planktic forms, often appear in the plankton samples in high frequency. Especially the following should be mentioned: *Keratella serrulata*, *Polyarthra minor* and *Polyphemus pediculus*.
7) Many species (those with abundances enabling this to be proved) show a descending curve during the course of increasing acidification. Such a development is evident for *Kellicottia longispina*, *Bosmina longispina*, *Ceriodaphnia quadrangula hamata*, *Daphnia cucullata*, *D. longispina*, *Diaphanosoma brachyurum*, *Holopedium gibberum* and *Polyphemus pediculus* (Figs. 5 and 10—14). In several cases it is possible to observe an ascent during phase 3, when conditions have improved. When reading the diagrams it should be borne in mind that Lake Änketjärn forms an exception by keeping fairly stable conditions.
8) The following common species occur with about the same abundance (or develop differently in different lakes) throughout the whole investigation period, thus being apparently relatively insensitive to pH variation: *Keratella cochlearis*, *Trichocerca capucina*, *Ascomorpha ecaudis*, *Asplanchna priodonta*, *Polyarthra remata*, *P. vulgaris*, *Conochilus unicornis*, *Eudiaptomus gracilis*, *E. graciloides*, *Cyclops strenuus strenuus*, *Mesocyclops leuckarti* and *Thermocyclops oithonoides* (Figs. 3, 6, 7, 9 and 16).
9) Only two of the zooplankters can clearly be demonstrated to be favoured by a decrease in pH (within the scope of the present investigation), viz. the rotifers *Keratella serrulata* and *Polyarthra minor* (Figs. 4 and 8). A protozoan, *Acanthocystis perpusilla*, increased remarkably in one of the acidified lakes, as well as in the "reference lake", L. Änketjärn (Fig. 2).

Besides the real zooplankters the following non-planktic species were observed (with abbreviations of the find localities within brackets):

Testacea: *Arcella angulata* (PLAYFAIR) DEFLANDRE (Sk), *A. artocrea* LEIDY (Gä, Kr, Ox), *A. costata* EHRBG. (Ox), *A. discoides* EHRBG. (L. Sv., St. Sv), *A. hemisphaerica* PERTY (Ox, L. Sv, St. Sv, Ta), *A. vulgaris* EHRBG. (Gr, Kr, Ox, Sk, L. Sv, To, Än).

Ciliata: *Vorticella* sp. (Gä, Ox, Ra, L. Sv, St. Sv, To, Än).

Rotatoria: *Cephalodella auriculata* (MÜLL.) (L. Sv, St. Sv, Ta), *Epiphanes brachionus* EHRBG. (Fi), *Euchlanis dilatata* EHRBG. (St. Hå, Ox, Sk, L. Sv, St. Sv), *Lecane flexilis* (GOSSE) (Ox, Sk), *L. lunaris* (EHRBG.) (Fi, Gr, Gä, St. Hå, Kr, Ox, Sk, L. Sv, St. Sv, Ta, To, Än), *L. stichaea* HARRING (Fi, Gä, St. Hå, Kr, Ox, L. Sv, St. Sv, Ta, To, Än), *Testudinella parva* (TERNETZ) (Ox, L. Sv), *Trichocerca gracilis* (TESSIN) (Ox), *T. longiseta* (SCHRANK) (Än), *Trichotria tetractis* (EHRBG.) (Gä, Ox, L. Sv).

Cladocera: *Acroperus harpae* BAIRD (Gr, St. Hå, Ox, Sk, L. Sv), *Alona affinis* (LEYDIG) (L. Sv), *A. guttata* SARS (Sk), *Alonella nana* (BAIRD) (Fi, Gi, Gr, St. Hå, Kr, Ox, Ra, Sk, L. Sv, St. Sv, Ta, To, Än), *Eurycercus lamellatus* (MÜLL.) (Ox, Ra, Sk, L. Sv, St. Sv, Ta), *Pleuroxus striatus* SCHOEDLER (Fi, Gr, Ox, L. Sv, St. Sv, Ta, To), *Scapholeberis mucronata* (MÜLL.) (Ox, Än).

Copepoda: *Megacyclops viridis* (JURINE) (Fi, Gi, Ra, Än).

In this connection also an epibiontic species should be mentioned, viz. the ciliate *Cothurnia* sp. It was especially common on filamentous algae of the genera *Mougeotia* and *Batrachospermum*, which invaded the pelagic zone in connection with acidification (see MORLING & WILLÉN, this volume). As a consequence, *Cothurnia* became more common also in the plankton samples (see MORLING 1984, Fig. 14).

5. Discussion

On the whole, the zooplankton fauna in the acidified lakes does not differ radically from that in non-acidified lakes, e. g. the "reference lake" Änketjärn. There are no species exclusively found in these lakes and missing in other localities of central Sweden. Obviously, the acidified environments are too young for a development of new species to have taken place. A comparison might be made with the Baltic, where the low number of species and the paucity of endemics is ascribed to its young age, seen from a geological point of view. However, lakes which have been acid for a long period of time often contain at least one specific zooplankter, viz. *Brachionus urceolaris sericus* (ROUSSELET) (sometimes regarded as a species of its own). This is found in such widely different parts of the world as Japan

(UÉNO 1958), Australia (KOSTE & SHIEL 1983), northern Canada (HAVAS & HUTCHINSON 1983), Egypt (HAUER 1963) and the Soviet Union (KUTIKOVA 1970). The find localities closest to those of the present study ought to be a pond ("Tonteich") in northern Germany (OHLE 1936, 1981) and the "acidotrophic" lakes Sladan and Blåmisusjön in northern Sweden (VALLIN 1953). There are two conceivable reasons why this rotifer has not found its way to the lakes now discussed: 1) It still has not had an opportunity to become dispersed here. 2) The west-Swedish lakes are not extreme enough for the species to settle down. (It ought to be mentioned that the previous find localities of *B. urceolaris sericus* mostly have a pH below 4, down to 2.8).

Thus, the zooplankton fauna in the lakes now discussed consists of ordinary species, predominantly euryoecious ones with a large morphological and physiological variation. This seems to be the case also in the acidified lakes investigated by ALMER et al. (1974), HÖRNSTRÖM et al. (1984) and HOBAEK & RADDUM (1980). Because the lakes have a low primary production it is natural that indicators of oligotrophy are found more often than indicators of eutrophy. However, all these indicators, including those of oligotrophy, decrease in importance during the course of acidification (see above p. 316, generalization no. 7), while some euryoecious forms are more indifferent (generalization no. 8 above). Perhaps this ought not to be too surprising considering that indicators should on the whole be more specialized and therefore sensitive to sudden changes.

Laboratory studies show that *Daphnia* species are unusually susceptible to diverse chemical influences (see, e. g., BESCH 1977; ADEMA 1978 and references there), and they are also among the first to decrease in number during acidification (the present material; ALMER et al. 1974; NILSSEN 1984, etc.) as well as after regulation of the water level for hydroelectric purposes (LINDSTRÖM 1973 and references). Similarly, *Holopedium gibberum* declined in a drastic way when a road was built close to Lake Fixsjön (the present investigation), whereby masses of minerogenous material entered the lake, causing perturbation. Furthermore, *Holopedium* is probably sensitive to mass development of the planktic alga *Gonyostomum semen*, as it is missing in the samples from L. Tolaredsjön, the only lake in the present investigation where such bloom occurs. Similar experiences were made in northern Germany by W. OHLE (pers. comm.).

As is evident from p. 316 generalization no. 7, all the seven more common cladocerans decrease in abundance during the course of acidification. The same seems to be true for the more sporadically occurring *Bythotrephes longimanus* and *Leptodora kindti*. On the other hand, all the five planktic copepods survive in about the same abundance (gen. 8 above). Similar results were obtained in the Scandinavian investigations cited above, where lakes with different degrees of acidification were compared. The causes of this condition are certainly of a complex nature but some facts ought to be pointed out: According to NILSSEN et al. (1984) copepods show, in contrast to daphnids, "good sodium balance in acidic waters primarily because their internal fluids are more effectively insulated from the ambient water". Competitive and predatory interactions are also considered to be important. In this connection, it is interesting to state that several investigations point at cladocerans (especially *Daphnia*) being on the whole more vulnerable to fish predation than copepods (see, e. g., LINDSTRÖM 1955, KLEMETSEN 1973, NILSSON 1978, LESSMARK 1983). They could be supposed to be still more so vis-à-vis invertebrate predators, and this is also what NYBERG (1984) found in some acidified Swedish lakes.

Of course, it would be interesting to know more about the relations to the other trophic levels in the lakes now studied. However, the small phytoplankters, being most important to the herbivorous zooplankters, are not included by MORLING & WILLÉN (this volume), because they passed through the meshes of the plankton net. Likewise, only sporadic observations were made on the fish fauna: In L. Änketjärn (the "reference lake") both roach, perch and pike have survived during the entire investigation and may be found in all stages. In the other lakes only adult perch and pike were caught, if any fish, with one exception: in L. Oxsjön pike fry were observed in 1980. Probably fish predation plays a minor role in all lakes except L. Änketjärn. In acid lakes invertebrate predators often take over the role as dominating zooplanktivores (see, e. g., STENSON et al. 1978; HENRIKSON & OSCARSON 1984; NILSSEN et al. 1984; NYBERG 1984). However, these animals were not investigated in the present context.

As is evident from p. 316, gen. 9, only two species seem to be favoured by the acidification. Both of them normally appear as periphytic or benthic (among *Sphagnum* etc.; see KOSTE 1978 and references), but in the lakes now studied they live as true plankters in rather high abundance. Possibly they have invaded the pelagic zone as a result of diminished competition and/or predation. The same species were also reported to be important in a similar lake by HENRIKSON et al. (1985). It is of interest in this connection that the lorica of *Keratella serrulata* is much thicker than that of the euplanktic brachionid rotifers (for details, see HENDELBERG et al. 1979), a trait which it shares with the acidobiontic *Brachionus urceolaris sericus*. This might be interpreted as being protection against the harsh chemical environment or against invertebrate predators.

Acknowledgements

This investigation was supported in part by a grant from the Joint English-Scandinavian Surface Water Acidification Programme, directed in Sweden by the Swedish Royal Academy of Science.

References

ADEMA, D. M. M. (1978): *Daphnia magna* as a test animal in acute and chronic toxicity tests. Hydrobiologia **59**, 125—134.

ALMER, B., DICKSON, W., EKSTRÖM, C., HÖRNSTRÖM, E., & MILLER, U. (1974): Effects of acidification on Swedish lakes. Ambio **3**, 30—36.

BESCH, W. (1977): Studien zum Gewässerschutz. 2. Bioteste in der limnischen Toxikologie. Karlsruhe.

HAUER, J. (1963): Zur Kenntnis der Rädertiere (Rotatoria) von Ägypten. Arch. Hydrobiol. **59**, 162—195.

HAVAS, M., & HUTCHINSON, Th. C. (1983): The Smoking Hills: natural acidification of an aquatic ecosystem. Nature (Lond.) **301**, 23—27.

HENDELBERG, M., MORLING, G., & PEJLER, B. (1979): The ultrastructure of the lorica of the rotifer *Keratella serrulata* (EHRBG.). Zoon **7**, 49—54.

HENRIKSON, L., & OSCARSON, H. G. (1984): Lime influence on macro-invertebrate zooplankton predators. Rep. Inst. Freshw. Res., Drottningholm **61**, 93—103.

— NYMAN, H. G., OSCARSON, H. G., & STENSON, J. A. E. (1985): Changes in the zooplankton community after lime treatment of an acidified lake. Verh. Int. Ver. Limnol. **22**, 3008—3013.

HOBAEK, A., & RADDUM, G. G. (1980): Zooplankton communities in acidified lakes in South Norway. SNSF-project IR 75/80. Oslo-Ås.

HÖRNSTRÖM, E., EKSTRÖM, C., & DURAINI, O. (1984): Effects of pH and different levels of aluminium on lake plankton in the Swedish west coast area. Rep. Inst. Freshw. Res., Drottningholm **61**, 115—127.

KLEMETSEN, A. (1973): Pelagic, plankton-eating perch. Astarte **6**, 27—33.

KOSTE, W. (1978): Rotatoria. Die Rädertiere Mitteleuropas. Ein Bestimmungswerk. Begr. von MAX VOIGT. Überordnung Monogononta. Vol. 1—2. Berlin, Stuttgart.

— & SHIEL, R. J. (1983): Morphology, systematics and ecology of new monogonont Rotifera (Rotatoria) from the Alligator Rivers region, Northern Territory. Trans. R. Soc. S. Aust. **107**, 109—121.

KUTIKOVA, L. A. (1970): The rotifer fauna of USSR. Fauna USSR, **104**. Leningrad (in Russian).

LESSMARK, O. (1983): Competition between perch (*Perca fluviatilis*) and roach (*Rutilus rutilus*) in south Swedish lakes. Diss., Inst. Limnol., Lund.

LIEDER, U. (1983): Revision of the genus *Bosmina* BAIRD, 1845 (Crustacea, Cladocera). Int. Rev. Hydrobiol. **68**, 121—139.

LINDSTRÖM, T. (1955): On the relation fish size — food size. Rep. Inst. Freshw. Res., Drottningholm **36**, 133—147.

— (1973): Life in a lake reservoir: Fewer options, decreased production. Ambio **2**, 145—153.

MORLING, G. (1981): Effects of acidification on some lakes in western Sweden. Vatten **37**, 25—38.

— (1984): Försurningens utveckling och planktonsammansättningen i några västsvenska sjöar. Manuscript, Limnol. Inst., Uppsala (in Swedish).

— FORSBERG, C., & WETZEL, R. G. (1985): Lake Änketjärn, a non-acidified lake in an acidified region. Oikos **44**, 324—330.

— & WILLÉN, T. (1990): Acidification and phytoplankton development in some west-Swedish lakes. *Limnologica* (Berlin) **20** (2), 291—306

NILSSEN, J. P. (1984): An ecological jig-saw puzzle: reconstructing aquatic biogeography and pH in an acidified region. Rep. Inst. Freshw. Res., Drottningholm **61**, 138—147.

— & LARSSON, P. (1980): The systematical position of the most common fennoscandian *Bosmina* (*Eubosmina*). Z. Zool. Syst. Evol. Forsch. **18**, 62—68.

— ØSTDAHL, T., & POTTS, W. T. W. (1984): Species replacements in acidified lakes: physiology, predation or competition? Rep. Inst. Freshw. Res., Drottningholm **61**, 148—153.

NILSSON, N.-A. (1978): The role of size-biased predation in competition and interactive segregation in fish. In: GERKING, S. D. (ed.), Ecology of freshwater fish production, pp. 305—325. Oxford.

NYBERG, P. (1984): Impact of *Chaoborus* predation on planktonic crustacean communities in some acidified and limed forest lakes in Sweden. Rep. Inst. Freshw. Res., Drottningholm **61**, 154—166.

OHLE, W. (1936): Der schwefelsaure Tonteich bei Reinbek. Monographie eines idiotrophen Weihers. Arch. Hydrobiol. **30**, 604—662.

— (1981): Photosynthesis and chemistry of an extremely acidic bathing pond in Germany. Verh. Int. Ver. Limnol. **21**, 1172—1177.

PEJLER, B. (1965): Regional-ecological studies of Swedish freshwater zooplankton. Zool. Bidr. Uppsala **36**, 407—515.

— (1983): Zooplanktic indicators of trophy and their food. Hydrobiologia **101**, 111—114.

STENSON, J. A. E., BOHLIN, T., HENRIKSON, L., NILSSON, B. I., NYMAN, H. G., OSCARSON, H. G., & LARSSON, P. (1978): Effects of fish removal from a small lake. Verh. Int. Verh. Limnol. **20**, 794—801.

UÉNO, M. (1958): The disharmonious lakes of Japan. Ibid. **13**, 217—226.

UTERMÖHL, H. (1958): Zur Vervollkommnung der quantitativen Phytoplankton-Methodik. Mitt. Int. Ver. Limnol. **9**, 1—38.

VALLIN, S. (1953): Zwei azidotrophe Seen im Küstengebiet von Nordschweden. Rep. Inst. Freshw. Res., Drottningholm **34**, 167—189.

Address of the authors:

GRETA MORLING
BIRGER PEJLER
Institute of Limnology
University of Uppsala
P.O.B. 557
S-751 22 Uppsala
Sweden

Limnologica (Berlin) **20** (2), 319—327, Juni 1990

Institute of Freshwater Research, Drottningholm, Sweden

Population Regulation of the Crayfish *Astacus astacus* L. after Liming an Oligotrophic, Low-alkaline Forest Lake

By MAGNUS APPELBERG

With 8 Figures

Key words: Crayfish; *Astacus astacus*; Population regulation; Lime treatment; Acidification.

Abstract

The regulation of a sparse crayfish population (*Astacus astacus* L.) was studied for 1 year before and 4 years after liming. Two hypotheses for population regulation were suggested; 1) the population was mainly restricted by abiotic factors such as low pH and low Ca^{2+} concentration, and/or 2) the population was food-limited. Liming resulted in an increase in pH, alkalinity, and Ca^{2+} concentration as well as an increase in the total littoral biomass of benthic fauna. Although crayfish fecundity and growth rates were high after liming the crayfish abundance did not increase. As no change in crayfish abundance was noted after liming, despite ameliorated water quality and improved food availability, both hypotheses were rejected. Instead, the results support an alternative hypothesis suggesting that the crayfish population possibly was restricted by predation by perch on YOY (Young-of-year) crayfish.

Introduction

Despite high sensitivity to acid water, the crayfish *Astacus astacus* L. is also regulated by other factors in low-alkaline forest lakes (SVÄRDSON 1974; NYBERG et al. 1986; APPELBERG 1987). In a comparative study of 15 crayfish lakes it was found that the development of *A. astacus* populations after liming acid lakes was dependent on both water quality and biotic factors (NYBERG et al. 1986).

Liming freshwaters causes major chemical changes such as increased buffering capacity, increased pH and Ca^{2+}, and changes in speciation of soluble metals (see ERIKSSON et al. 1983, and FRASER et al. 1985 for review). At a pre-neutralization pH level of 5.5, the impact of liming on biota is often small, although both fish and benthic fauna species vulnerable to acid water (e. g. snails and roach) have been found to increase in abundance (ERIKSSON et al. 1983; NYBERG et al. 1986).

In this study I wanted to test two possible hypotheses for regulation of a sparse population of *A. astacus* in connection with lime addition to a low-alkaline, oligotrophic, forest lake.

The first hypothesis suggests that abiotic factors such as low pH and low Ca^{2+} concentration are the main cause of the low abundance of crayfish. As earlier studies show, *A. astacus* is affected by acid water even at moderately low pH levels (5.5—6.0), especially during early stages of its life history (SVÄRDSON 1974; APPELBERG 1983, 1984, 1985, 1987). Low pH has also been found to cause sub-lethal effects such as disturbance of ion-regulation and acid-base status (MALLEY 1980; MORGAN & MCMAHON 1982; APPELBERG 1985). Additionally, CAPELLI (1975) reported lower survival of crayfish in lakes with low calcium concentrations compared to the survival in hard water lakes. After liming the crayfish population should be expected to increase in abundance due to an increased survival of young individuals and reduced sub-lethal stress.

The second hypothesis proposes that the crayfish population is mainly regulated by food shortage. In oligotrophic waters, crayfish are suggested to be food-limited and dependent on lake productivity (MOMOT 1984; FRANCE 1985). According to WHITE (1978), relative food shortage will also be reflected in the growth rate of the young individuals. Thus, the increase in abundance of detritivores after liming, which has been attributed to reduced pH and aluminium toxicity as well as to an increased microbial decomposition rate (HULTBERG & ANDERSSON 1982; RADDUM et al. 1984), should be expected to increase crayfish food availability and population growth.

Material and Methods

Site description

I chose to study two lakes, situated in central Sweden, 60 m above sea level, at 17°20′ east and 59°80′ north. Each is separately connected to the Trosaån River. The lake that was treated with lime, Lake Holmsjön, has a larger watershed and less maximum depth than the lake used for control for changes in water chemistry and benthic fauna, Lake Västra Stensjön. Both the lime-treated L. Holmsjön and the control L. Västra Stensjön are clear-water lakes with a low level of nutrients. The bedrock of the watersheds mainly consists of red and grey gneiss, the loose deposits of boulder and clay and a small proportion of bog-peat. Basic physical data are given in Table 1.

The abundance of macrophytes was low in both lakes, except for a rich occurrence of *Myriophyllum alterniflorum* in shallower parts of L. Holmsjön. *Phragmites communis, Lysimachia thyrsiflora, Menyanthes trifoliata, Lobelia dortmanna, Equisetum fluviatile* and *Carex* sp. occurred in both lakes. In L. Holmsjön *Nymphaea alba* was also noted, whereas *Potamogeton natans* was more common in L. Västra Stensjön.

According to local fishermen both lakes were inhabited by rich populations of the crayfish *A. astacus* until the late 1950th. However, at the start of the study the number of crayfish was low in both lakes.

Table 1. Physical data for L. Holmsjön (limed lake) and L. Västra Stensjön (control lake)

	L. Holmsjön	L. V. Stensjön
Precipitation area (ha)	45.8	22.7
Lake area (ha)	7.5	6.9
Maximum depth (m)	6.3	11.3
Lake area as percentage of precipitation area	18	29
Volume	209,000	242,000
Retention time (years)	2.1	5.1

Lime treatment

Lime treatment of L. Holmsjön was conducted on 5 March, 1980. The amount of lime needed ($CaCO_3$, in the 0 to 1 mm diameter fraction) was calculated, after titration to lake water, to produce an alkalinity of 0.2 meq/l and compensate for acid input over 5 years. 12 metric tons were distributed on ice throughout the littoral zone.

Water chemistry

In both lakes, samples from epi- and hypolimnion were taken at the deepest points using a Ruttner sampler. In April 1979, 1980, 1982, and 1983 littoral samples were taken at a depth of 1 m in L. Holmsjön. From 1979 to 1983, pH, alkalinity, conductivity, water colour, Ca^{2+}, Mg^{2+}, Na^+, K^+, Cl^-, and SO_4^{2-} were analysed monthly. Oxygen concentration was measured at the end of each stagnation period (i.e., August and March). Nutrients, in terms of total- and MR-phosphorus, NO_2, NO_3, and Kjeldahl-N, were measured during circulation periods (May and October). Methods of analysis were as described by AHLGREN & AHLGREN (1976). Atomic absorption spectrophotometer (Unicam SP 1900) was used for all metal analyses.

A linear model was set up for chemical parameters in the epilimnion of L. Holmsjön and L. Västra Stensjön during the pre-neutralization period (1979) to eliminate the influence of non-predicted, large-scale changes in water chemistry after liming. These correlations were used to predict water chemistry in L. Holmsjön in a "non-limed" form from 1980 to 1983. Only parameters with $r^2 > 0.4$ and $p < 0.05$ were used for this purpose.

Benthic fauna

Benthic fauna was sampled during winter in L. Holmsjön in 1979, 1980, 1981, and 1983, whereas samples in L. Västra Stensjön were taken from 1978 to 1981. In both lakes, two replicate samples were taken on soft bottoms along a transect at 2 m intervals from the shoreline to the deepest part of the lake with an Ekman grab (0.25 m^2 area). Six samples from the crayfish habitat (depth 0—1.5 m) on a hard bottom were taken in July 1983 in L. Holmsjön with a dredge-sieve normally used for sampling juvenile crayfish (see below). Each sample was sieved through a 0.6 mm mesh and preserved in 10% formaldehyde. Animals were identified to species, counted, and weighed to the nearest µg. Biomass was estimated within a depth of 0 to 4 m in the littoral zones as g wet weight/m^2.

Fish

To sample fishes, I used two to four gill-nets in June/July and in September each year. Gill-nets, 36 m long and 1.5 m deep, consisted of 12 different mesh sizes that were randomly distributed. All catches were identified to species, measured (to the nearest mm), and weighed (to the nearest g). Stomach analyses of perch (*Perca fluviatilis*), roach (*Rutilus rutilus*) and ruffe (*Gymnocephalus cernua*) were performed in July 1983.

Crayfish

In L. Holmsjön a monthly sampling programme was conducted from May to October in 1979 to 1983 in the area (4,000 m^2) shown in Fig. 1. Four different sampling methods were used with the primary, and most efficient method being catching crayfish by hand during scuba diving.

Fishing with cylindrical traps with two entrances (mesh size 15 mm), baited with either roach or perch was occasionally used to catch individuals with a carapace length (CL) of over 30 mm. Traps were placed at 5 m intervals along the shoreline at depths of 0.5 to 3 m. Normally, they were set at nightfall and retrieved at the latest at 09^{00} the following morning. Electric fishing was used to sample crayfish larger than 10 mm CL in May and September/October, when the risk of chela loss was small (WESTMAN et al. 1977). A dredge-sieve, constructed for sampling juvenile crayfish, was used for estimating the abundance or YOY (Young-of-year) crayfish by total count on sample plots (ODELSTRÖM 1983). 5 to 10 sampling areas of 0.25 to 0.5 m^2 were sampled on each occasion.

For each crayfish sex, maturity, total and carapace lengths (CL), wet weight, stage in the moulting cycle (according to STEVENSON 1975), and injuries were recorded. All crayfish larger than 10 mm CL were individually marked with a cauterization technique modified after ABRAHAMSSON (1965). After handling, each crayfish was returned to the location where it was caught. Crayfish stomach con-

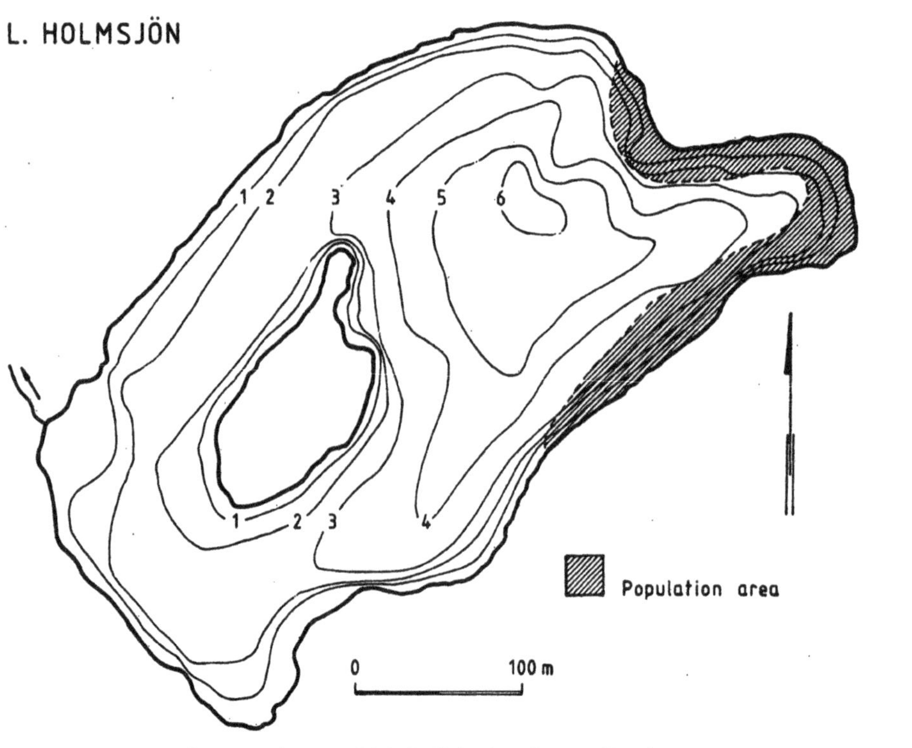

Fig. 1. Area (4.000 m^2) inhabited by crayfish in L. Holmsjön. Contour lines in m

tents were regularly sampled by flushing the stomachs with a double-acting syringe.

Data from recaptured individuals were used to estimate crayfish growth. Percentage length increase at each moult and the time between subsequent moults were used to describe growth characteristics in the population, as described by MAUCHLINE (1977). The number of moults was mainly discerned by the appearance of the marks. However, in some cases, when there were more than two moults, the number was extrapolated from data on crayfish which had only moulted once. The length increment during each moult was used to calculate the percentage moult increment (PCMI) for each individual, whereas the length of the intermoult period was estimated by dividing the number of days between mark and recapture by the number of moults in each of the separate 5 mm length classes. Individuals with a CL over 45 mm were assumed to moult only once a year (ABRAHAMSSON 1972).

Growth of juvenile crayfish during the post-liming period was estimated from size-frequency distribution pooled for the years 1980 to 1983. Size-classes were readily distinguishable during the first two growth seasons, and by assuming a negligible difference in growth between the four years after liming, mean individual growth in the first two years of the life history during the post-liming period was estimated.

Adjusted Petersen estimates were used to calculate the population size of crayfish older than two summers (24 mm CL) (SEBER 1982). Two to three different sampling methods were used during each sampling period, and sampling was performed both in daylight and in the dark over 2 to 5 days. Because the population was sparse during the whole experimental period, sexes were combined. To avoid bias due to sex ratio differences (MALLEY & REYNOLDS 1979), several sampling methods were used on each sampling occasion. However, due to their reduced activity, egg-bearing females were underrepresented in samples from early summer. To reduce the effects of the gradual loss of marks due to repeated moulting, only crayfish which were recaptured within four sampling occasions following marking were used for population estimates. When adjacent samples were close in time, they were in some cases treated as one sample so as to increase the precision.

Statistical methods

Differences between water chemistry parameters measured and predicted "non-limed" were tested using Student's t-test. A non-parametric ANOVA (Kruskal-Wallis) was used to test differences between benthic fauna estimates as well as between crayfish food items. Where nothing else is mentioned, parametric ANOVA (general linear model) was used for all other analyses of fish and crayfish parameters. The Statistical Analyses System was used in all cases (SAS INST. INC. 1982).

Results

Water chemistry

No marked changes in nutrients were recorded during the years 1979 to 1983 in any of the two lakes (Table 2). Following circulation periods in spring and autumn stable thermal stratification developed in L. Västra Stensjön, whereas L. Holmsjön was only weakly stratified at the end of each stagnation period. Oxygen saturation never fell below 50% at any time at any depth.

Table 2. Nutrients in L. Holmsjön and L. Västra Stensjön. Samples from circulation periods (May and October) in the years 1979 to 1983. Mean values in μg/l, $n = 6$

	Total-P	MR-P	NO_2—N	NO_3—N	Kjeldahl-N
L. Holmsjön					
Mean	8.6	2.8	2.7	21.4	437
S. D.	1.3	1.3	1.2	10.8	93
L. Västra Stensjön					
Mean	11.4	5.8	3.7	12.2	388
S. D.	3.5	2.9	2.7	9.0	187

Table 3. Linear regression for some water chemistry parameters in L. Holmsjön and L. Västra Stensjön during the pre-liming year (1979)

	N	r^2	Intercept	Slope	Probability
pH	17	0.41	1.97	0.67	$p < 0.05$
Alkalinity	17	0.44	0.02	0.51	$p < 0.01$
Conductivity	10	0.07	19.7	0.49	NS
Ca^{2+}	11	0.74	−0.11	1.41	$p < 0.001$
Mg^{2+}	11	0.94	0.00	1.07	$p < 0.001$
Na^+	11	0.33	−0.01	1.16	NS
K^+	11	0.54	0.00	0.57	$p < 0.05$

Table 4. Mean values of water chemistry parameters in L. Holmsjön after lime treatment (years 1980 to 1983), as measured and as calculated in a "non-limed" form. All values in meq/l except for pH. Differences according to Student's t-test

	N	Calculated "non-limed"		Actual limed		Probability
		Mean	S. E.	Mean	S. E.	
pH	37	6.32	0.30	6.59	0.38	$p < 0.01$
Alkalinity	37	0.075	0.012	0.139	0.031	$p < 0.001$
Ca^{2+}	32	0.163	0.042	0.211	0.036	$p < 0.001$
Mg^{2+}	32	0.097	0.020	0.100	0.025	NS
Na^+	32	0.107	0.022	0.102	0.019	NS
K^+	32	0.011	0.003	0.014	0.005	$p < 0.01$

The most pronounced change after liming in L. Holmsjön was noted for alkalinity, which increased from 0.07 meq/l to 0.14 meq/l. Ca^{2+} increased by about 30% and other significant changes were increases in pH, conductivity, Mg^{2+} and K^+. Differences between the annual mean values measured and those simulated for "non-limed" conditions revealed that alkalinity, pH, Ca^{2+} and K^+ concentrations increased significantly for the limed lake compared to the calculated "non-limed" situation (Tables 3 and 4). On the other hand, Mg^{2+} and Na^+ concentrations did not show any significant differences (Table 4, Fig. 2). Littorally measured pH values in spring in L. Holmsjön were 0.5 units below the corresponding pelagic measurements (Kruskal-Wallis, $p < 0.05$, Fig. 4).

Benthic fauna

The total biomass of the benthic soft bottom fauna in L. Holmsjön increased from 1.8 g/m² in the pre-neutralization year to 6.2, 4.8, and 4.4 g/m² in 1980, 1981, and 1983, respectively (Fig. 3). The biomass values in the same depth interval in L. Västra Stensjön did not differ significantly between years and were 1.0, 1.1, 2.2, and 1.7 g/m² during the years 1978, 1979, 1980, and 1983, respectively. Before liming (1979) total biomass did not differ significantly between the two lakes (1.8 g/m² in L. Holmsjön and 1.1 g/m² in L. Västra Stensjön, $p < 0.05$, $n = 6$). However, the biomass in L. Holmsjön was significantly higher in the 2 years after liming ($p < 0.05$, $n = 12$).

Asellus aquaticus dominated the littoral fauna of L. Holmsjön in each of the years. Different species of Chironomidae were found to be abundant in 1979, and 1 year after liming (1980) the numbers of Pisidae, *Sialis lutaria*, Tubificidae, and trichopterans were non-significantly higher. During 1981 and 1983 some large trichop-

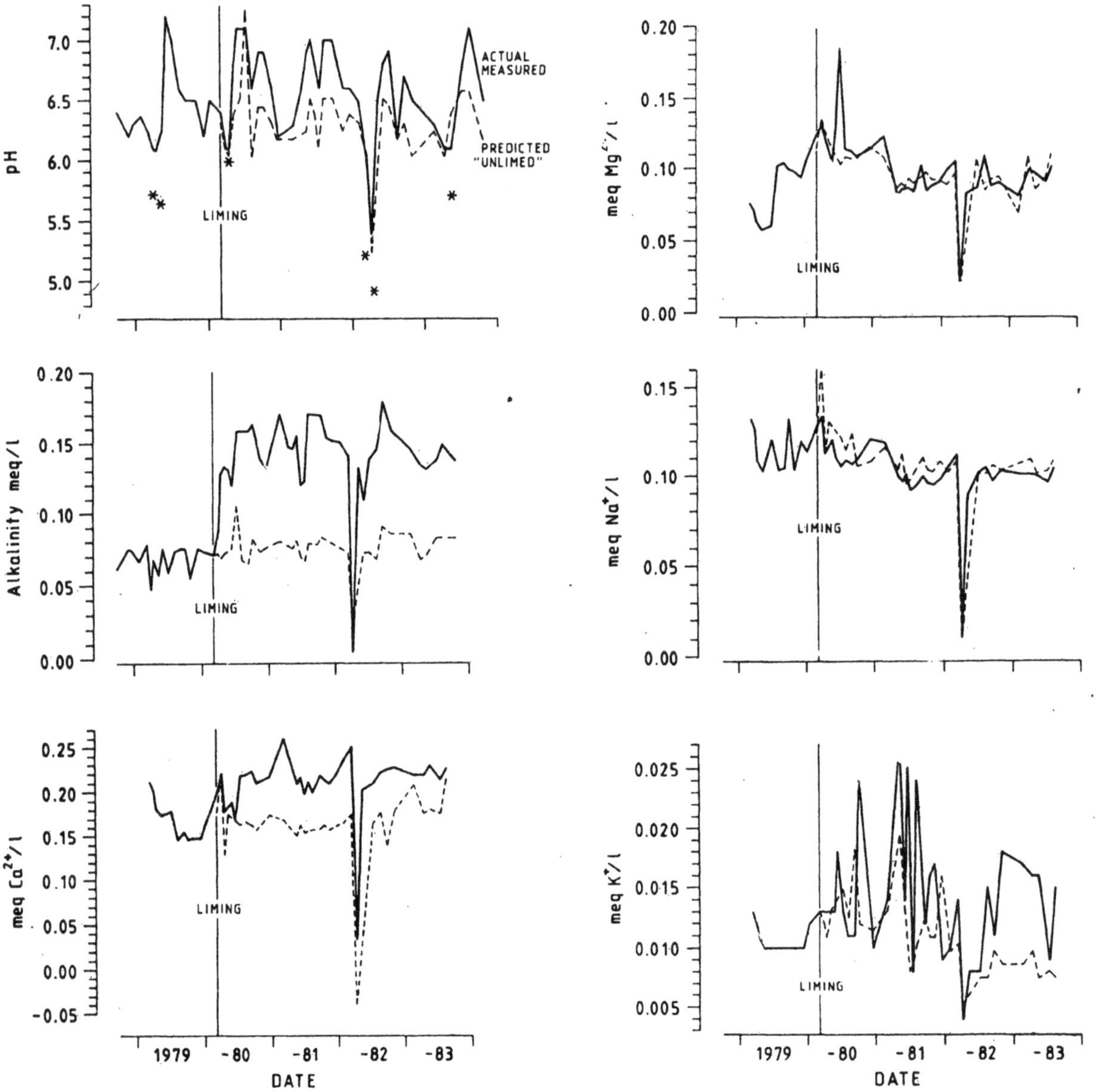

Fig. 2. Some water chemistry parameters in L. Holmsjön (epilimnion) before and after liming (solid line). Dotted line shows calculated values for the lake as "non-limed" over the same period. Asterisks indicate littoral samples at a depth of 1.0 m

terans were also abundant. In L. Västra Stensjön no significant changes in total biomass were observed. However, *Asellus aquaticus* slowly decreased in biomass, whereas *Sialis lutaria* and Pisidae increased from 1979 to 1980.

Mean total biomass on hard bottoms in L. Holmsjön in July, 1983, was about 5.7 g/m^2 in the 0.2 to 1.5 depth zone (Fig. 4). Highest biomass was recorded at the depth of 0.2 m, and after a dramatic decline at 0.5 m ($p < 0.05$, $n = 6$), biomass increased slightly with increasing depth (not significantly). Species composition in the hard bottom samples was similar to that of the soft bottom from the same year. *Asellus aquaticus* dominated, but molluscs, such as *Lymnaea peregra* and *Pisidium* sp., as well as various species of Chironomidae and Trichoptera, were represented. The small micro-crustaceans *Sida crystallina* and *Latona setifera* were also present.

Fish

European perch (*Perca fluviatilis*), roach (*Rutilus rutilus*), ruffe (*Gymnocephalus cernua*), rudd (*Scardinius erythrophtalmus*), and pike (*Esox lucius*) constituted the fish fauna of L. Holmsjön. Between 1979 and 1983 the catch of roach per unit effort (CPUE) decreased from 10.0 ± 5.7 (mean ± SD, $n = 4$) to 1.3 ± 0.9 ($p < 0.05$), whereas mean length increased from 24.3 ± 2.3 ($n = 35$) to 27.4 ± 1.6 cm ($n = 19$) in 1982. In 1983 three small individuals were caught, reducing mean length to 16.0 ± 10.8 cm ($n = 5$, Fig. 5).

CPUE of perch fluctuated and no significant change was observed during 1979 to 1983 ($p > 0.05$, $n = 14$). Mean CPUE in 1983 was 33.8 ± 5.2 ($n = 4$) and mean length 12.0 ± 3.2 cm ($n = 146$). Perch was visually observed to be abundant in the crayfish habitat during scuba-diving at all times of day. Ruffe was regularly caught during electro-fishing for crayfish but was not analysed quantitatively. Pikes were caught occasionally in gill-nets, but were not observed in the crayfish habitat during crayfish sampling.

Stomach analyses on fish caught in gill-nets in July, 1983, in L. Holmsjön revealed that perch in the size group 10 to 15 cm in length ($n = 110$) fed on zooplank-

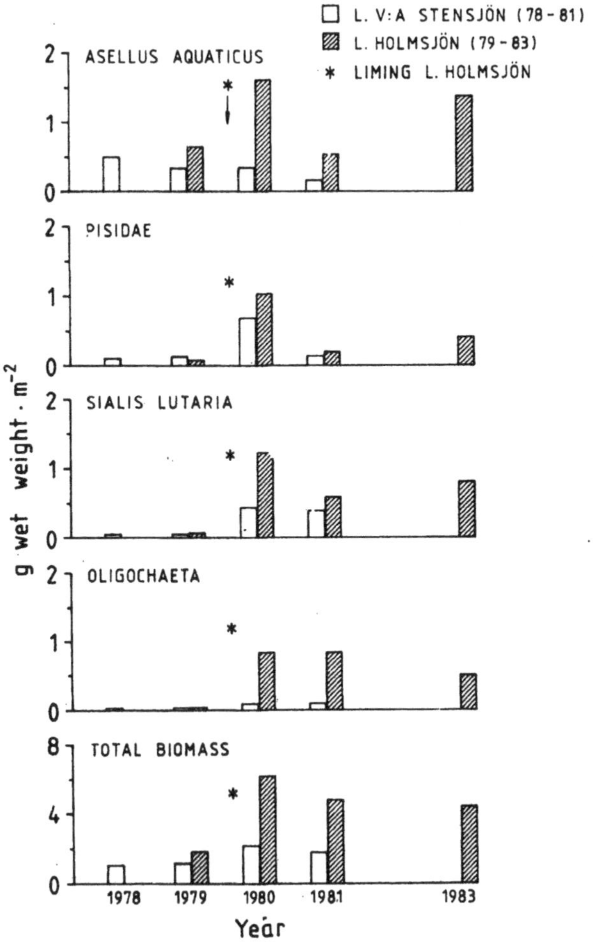

Fig. 3. Mean biomass of the benthic fauna sampled on soft bottoms (Ekman grab, $0.25\,m^2$, area, $n = 6$) at depth 0 to 4 m in L. Holmsjön in winter 1979, 1980, 1981 and 1983. The same for L. Västra Stensjön, years 1978 to 1981

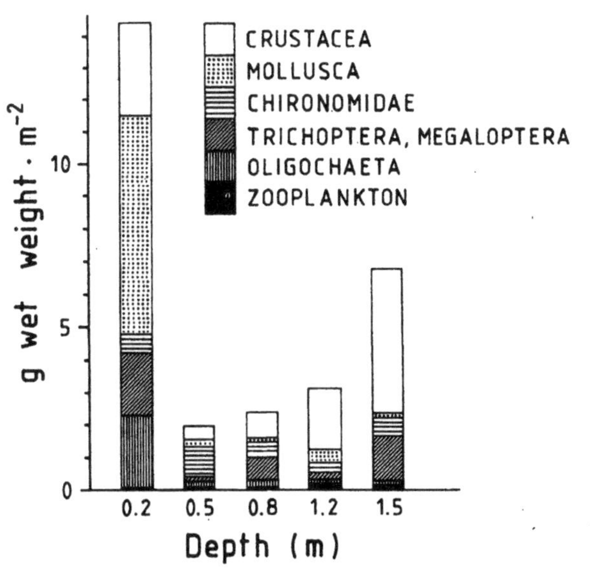

Fig. 4. Mean biomass of benthic fauna on hard bottoms within the crayfish habitat in L. Holmsjön. Two replicate samples were taken at each depth in July 1983

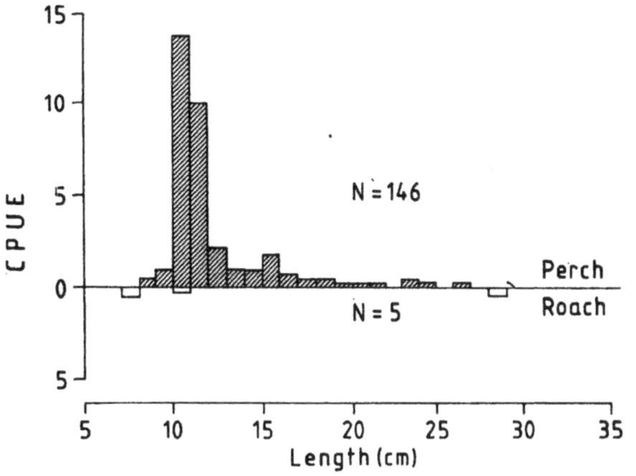

Fig. 5. Mean catch per unit effort (CPUE) of perch and roach in L. Holmsjön in July 1983 ($n = 4$)

ton, benthic micro-crustaceans, Ephemeroptera, Chironomidae and *Pisidium* sp. *Asellus aquaticus* was only found in perch less than 10 cm in length. A shift towards larger organisms, such as fish, was noted with increasing size of perch. However, crayfish were not found in any of the stomachs.

In roach ($n = 2$), benthic micro-crustaceans, zooplankton, Chironomidae and other benthic animals were found. The ruffe's food ($n = 13$) was dominated by Chironomidae and *Asellus aquaticus.*

Crayfish habitat

The crayfish population in L. Holmsjön was concentrated to hard-bottom habitats at depths of 0 to 3 m in one part of the lake (Fig. 1). The substrate in the crayfish habitat was dominated by rocks and pebbles on sand, covered with allochthonous material such as birch and alder leaves and pine and spruce needles. The latter material also predominated in the uppermost part of the soft littoral bottoms.

Where large substrate particles were lacking, adult crayfish were commonly found to inhabit burrows underneath the shoreline. According to capture-recapture data and visual observations, no other habitat segregation or migrational patterns were observed within the littoral zone during the growth seasons. Although soft bottoms were searched using scuba-diving on each sampling occasion, crayfish were only occasionally observed on soft bottoms below the littoral zone.

Crayfish food choice

Aquatic macrophytes and/or detritus dominated the stomach content of all sizes of crayfish (Fig. 6). The proportion of detritus decreased from 41.4% in YOY to 7% in adult crayfish ($p < 0.05$). Also the proportion of benthic micro-crustaceans decreased with increasing size of the crayfish. The share of aquatic macrophytes, on the other hand, increased with increasing size of crayfish ($p < 0.001$), and constituted 82% of the stomach content of adult crayfish. Benthic fauna, primarily *Asellus aquaticus*, Ephemeroptera, and Chironomidae, constituted about 20% of the stomach content of crayfish ranging between 10 mm and 50 mm CL, whereas their percentage was smaller in YOY and larger in larger individuals ($p < 0.05$).

When data from different years were pooled, some seasonal change in the food ingested during the growth season was noted. Macrophytes and detritus were evenly distributed over time ($p > 0.05$), whereas the occur-

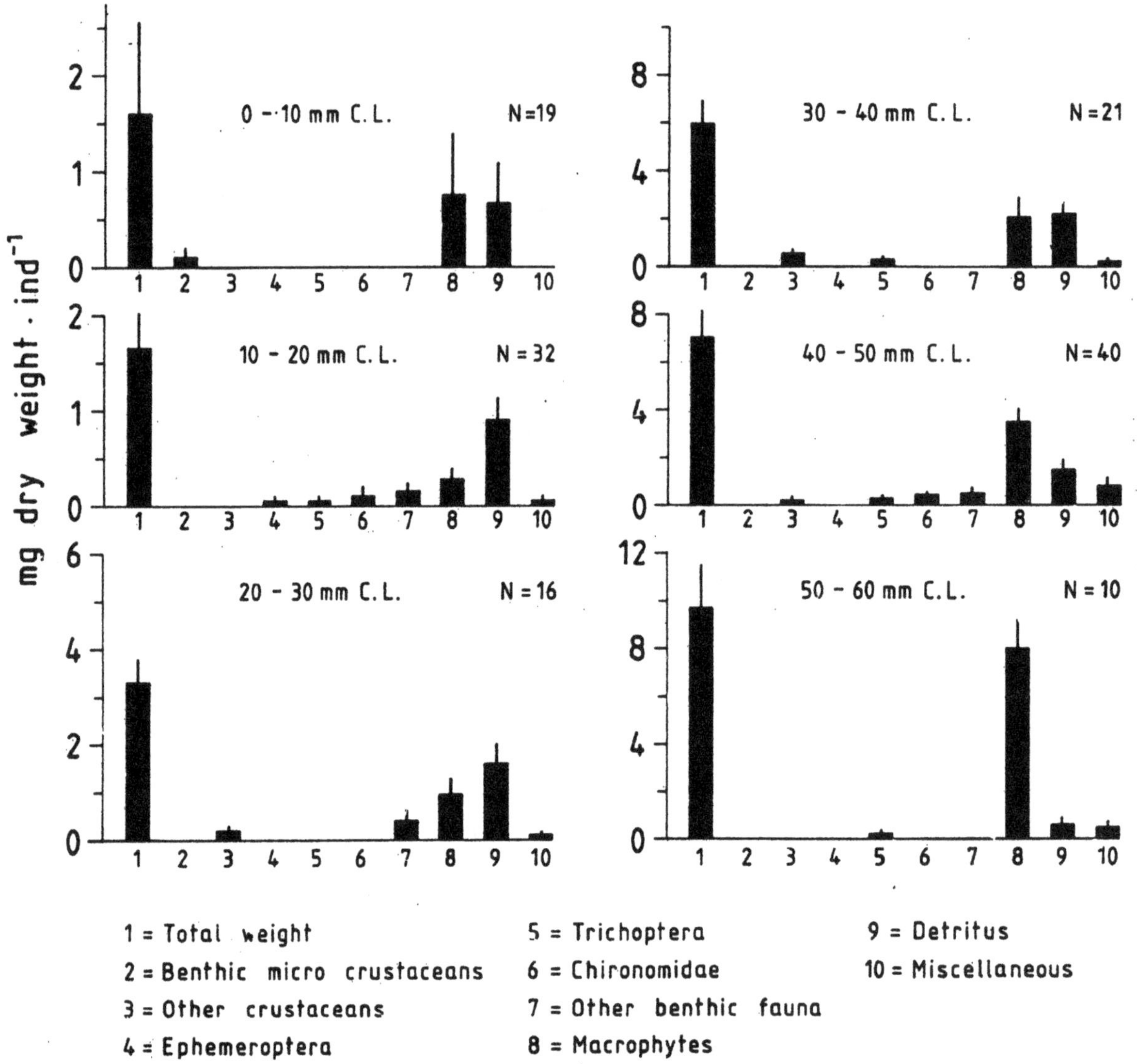

Fig. 6. Pooled values for food composition in different size classes of crayfish during the period 1980 to 1983 in L. Holmsjön. Expressed as mean dry weight of stomach content for an average individual of each size class

rence of benthic micro-crustaceans was highest in June ($p < 0.05$). The variation of other benthic organisms was not significantly related to season ($p > 0.05$). Remains of *A. astacus* were found in the stomachs in June and July. However, this may be attributed to ingestion of shed exoskeletons after moulting.

Crayfish growth

Due to a low number of recaptured animals in 1979, differences in the percentage carapace moulting increment (PCMI) could not be compared. When PCMI was compared for each of the years after liming (1980 to 1983), no significant differences could be observed (ANOVA, least-square-means). PCMI for mature females decreased more rapidly during individual development than for that of mature males, whereas immature crayfish showed no significant difference between sexes (ANOVA, $p < 0.05$, Table 5).

Table 5. Growth constant (percentage carapace moulting increment) for the L. Holmsjön crayfish population after liming. Estimated according to Mauchline (1977)

Sex	Maturity	Intercept	Slope	S. E. (slope)	Probability
Male	Non-mature	1.483	−0.114	0.031	p < 0.001
Male	Mature	1.638	−0.141	0.065	p < 0.05
Female	Non-mature	1.436	−0.114	0.025	p < 0.001
Female	Mature	1.647	−0.163	0.049	p < 0.01

The logarithm of the intermoult time was linearly correlated to individual length, and the relation between intermoult time and carapace length classes may be described as

$$\mathrm{LOG(Days)} = 0.849 + 0.327 \times (\mathrm{CL})\ (p < 0.001,\ r^2 = 0.92,\ n = 12)$$

for crayfish with a CL of less than 45 mm in L. Holmsjön. There was no significant difference in intermoult time between years, nor was there any difference between the sexes ($p < 0.05$, ANOVA). Mean individual length increment for the two youngest age-groups after liming was calculated from the pooled size-frequency distributions for 1980 to 1983 (Fig. 7).

Crayfish reproduction

The percentage of mature females larger than 40 mm CL fluctuated between 80 and 100% in autumn in 1979 to 1983. No egg-bearing females were caught in 1979. After

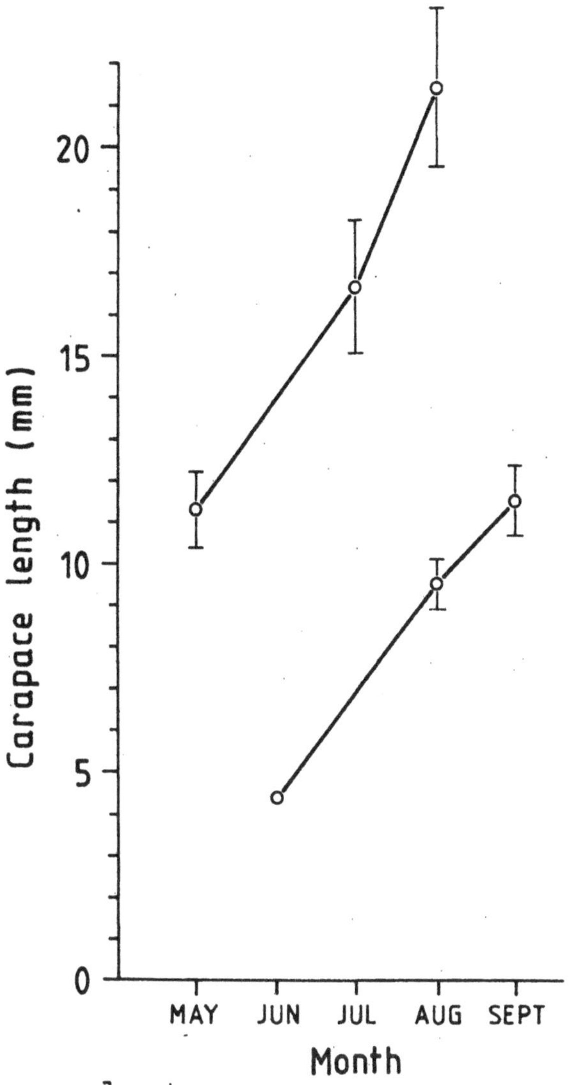

Fig. 7. Mean carapace length of the young-of-the-year crayfish (lower curve) and yearling crayfish (upper curve) in L. Holmsjön (X ± SD). Data for carapace length are pooled from the years 1980 and 1983

liming no visible damage of eggs was noted. In May, in the years after liming (1980—1983), the number of pleopod-eggs per female averaged 129 ± 15.1 for the mean size 44.5 ± 4.3 mm (mean ± SD, $n = 10$) and was not significantly correlated to female size ($p < 0.12$, $r^2 = 0.27$). Hatching success was found to be normal when four egg-bearing females were collected and placed in aquaria in spring 1983.

Crayfish population size

Despite great efforts to obtain adequate sample sizes, each population sample usually contained no more than 30 to 80 individuals. The number of marked individuals averaged 5 to 20% of the population throughout the study.

Population size fluctuated between 300 to 1000 individuals (0.08 to 0.25 ind./m^2) in 1979 to 1983 and there was no significant correlation to time ($p > 0.05$, $n = 16$; Fig. 8). In autumn the year before liming (1979), the total population of crayfish larger than 24 mm CL (older than two summers) was 515 ± 175 (mean ± SD) individuals. The following year population size decreased from 805 ± 262 in early June, to 302 ± 97 in September. The second summer after liming (1981), the estimated population size increased from 506 ± 137 individuals in May to 826 ± 356 at the end of the season. In 1982, the difference within the growth season was similar to that of 1981; 595 ± 204 individuals in June and 805 ± 261 in September. The population estimate in the middle of May 1983 was 485 ± 126, whereafter it decreased to 315 ± 94 in July.

The proportion of juvenile crayfish in the samples (YOY excluded) averaged 31.6 ± 9.4% (mean ± SD, $n = 14$). A slight, non-significant reduction within the

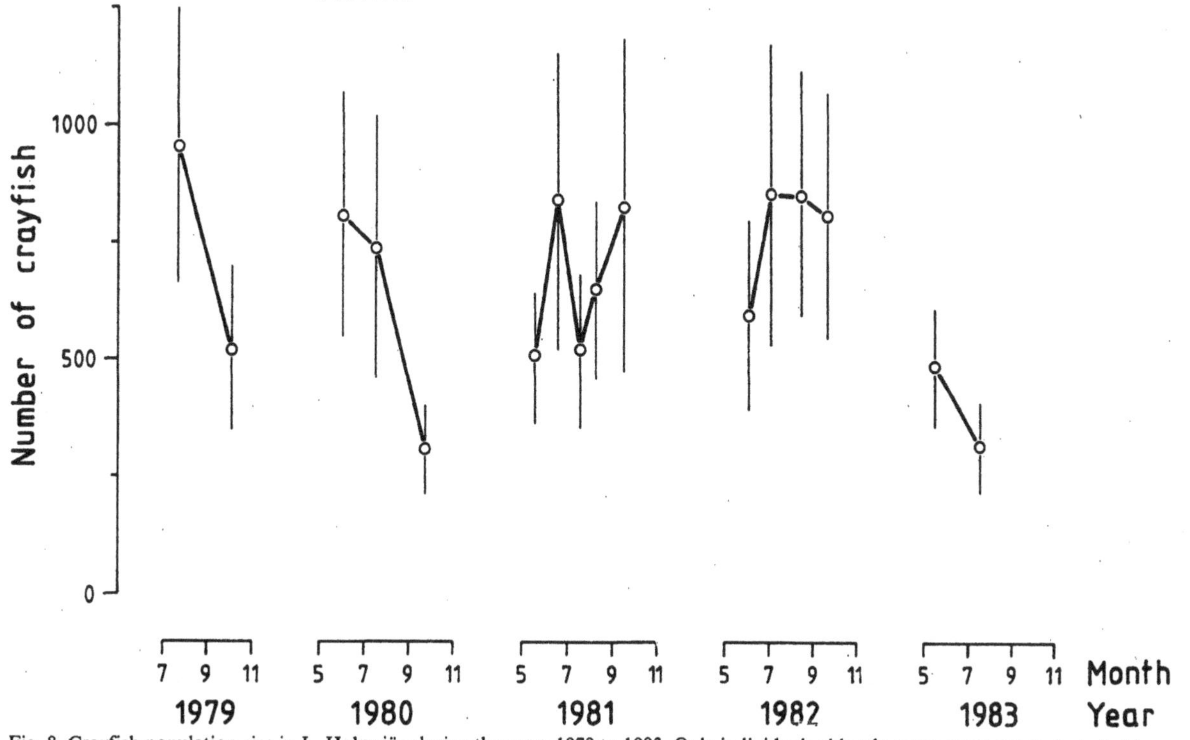

Fig. 8. Crayfish population size in L. Holmsjön during the years 1979 to 1983. Only individuals older than two summers are included (mean ± S. D.)

growth seasons was noted from 36% in May to 22% in September ($p < 0.21$, $r^2 = 0.42$), and no significant differences between the years after liming (1980 to 1983) were recorded ($p > 0.05$, ANOVA).

Due to low abundance and contagious distribution, dredge-sieve sampling failed to give reliable abundance estimates for YOY crayfish. However, when samples for the years 1980 to 1983 ($n = 136$) were pooled, the abundance of YOY was estimated at 0.39 ± 1.28 individuals/m^2 in July and August. As the population area was about 4.000 m^2, this value corresponds to a total of about 1500 YOY in the first few months after hatching.

Discussion

Despite the fact that liming of L. Holmsjön caused both abiotic and biotic responses to the ecosystem, no significant change in crayfish population size was recorded over time.

If the crayfish population was mainly restricted by abiotic factors, as suggested by the first hypothesis, then the increases in pH, alkalinity and Ca^{2+} after lime addition would have been expected to favour crayfish reproduction success and increase the abundance of young individuals. However, although there was no indication of crayfish reproductive failure after liming the abundance of YOY crayfish was still low.

Low pH in the littoral zone during spring may have affected the egg-bearing females which inhabited the uppermost part of that zone. However, the eggs of *A. astacus* are in an undeveloped and relatively pH-tolerant stage during early spring, and a short period of low pH does not necessarily damage them (APPELBERG 1983, 1984).

After liming, buffering capacity, pH and Ca^{2+} concentrations exhibited values above those which have earlier been shown to cause decreased survival and sublethal stress to crayfish (CAPELLI 1975; MALLEY 1980; MORGAN & MCMAHON 1982; MCMAHON & MORGAN 1983; APPELBERG 1985).

The biotic responses to liming were primarily observed as an increase in the biomass of benthic fauna. Similar changes have been reported from several studies, and are mainly attributed to the reduced influence of low pH and toxic agents after liming (HULTBERG & ANDERSSON 1982; ERIKSSON et al. 1983; RADDUM et al. 1984; FRASER et al. 1985). The occurrence of young roach in 1983 may have been the result of increased hatching success after liming. This species is vulnerable to acidification, and reproductive failures have been found at pH levels between 5.5 and 6.0 (MILBRINK & JOHANSSON 1975).

I found no support for the second hypothesis, which proposed that the crayfish population was mainly food-limited. Firstly, the rich occurrence of allochthonous detritus in L. Holmsjön, which also was reflected in the crayfish food choice, would be of benefit for crayfish production both before and after liming (JONES & MOMOT 1981). Secondly, as the crayfish food situation possibly was improved by the increase in biomass of the benthic fauna after liming, the abundance of crayfish should have increased.

The contribution of animal food items is of great importance to crayfish due to the higher efficiency of assimilation (MOSHIRI & GOLDMAN 1969; SÖDERBÄCK et al. 1988). As young individuals were found to consume a higher proportion of animal food compared to older ones, food limitations would primarily influence the growth of the first year-classes of crayfish (WHITE 1978). When comparing the annual mean growth rate of the two youngest year-classes of *A. astacus* in L. Holmsjön with that of other comparable populations, the former does not seem to be reduced. In September YOY crayfish were between 11 and 12 mm CL and yearlings about 22 mm in L. Holmsjön, whereas the length of YOY and yearlings ranged between 8 to 11 mm and 19 to 22 mm, respectively, in four other oligotrophic lakes (APPELBERG & ODELSTRÖM 1986). Also, growth of crayfish older than two summers measured as length increment per moult, was large compared to that of other populations (ABRAHAMSSON 1972; APPELBERG & ODELSTRÖM 1986).

Additionally, if there was a shortage of food after liming, more energy should be spent on growth instead of reproduction. However, the proportion of mature females larger than 40 mm CL in the autumn was high compared to the situation in other similar lakes (APPELBERG & ODELSTRÖM, unpubl. data). The mean number of pleopod eggs in spring was about 40% higher than was earlier noted for *A. astacus* females of the same size by ABRAHAMSSON (1966) and about 50% higher than the number of pleopod eggs recorded in four other, low-alkaline, forest lakes (APPELBERG & ODELSTRÖM, unpubl. data).

The results from earlier studies indicate that perch is the main conceivable predator on the L. Holmsjön crayfish population (DEHLI 1981; MORALES & APPELBERG 1984; APPELBERG & ODELSTRÖM 1988). Despite the lack of direct evidence from the single occasion when food analysis of perch was carried out, other observations indicate that predation may be of importance in regulating the crayfish population. Predation by fish on crayfish has been shown to be strongly size selective (STEIN 1977). YOY crayfish are in general found to be most vulnerable to fish predation, although size of the predator, substrate particle size, and stage in the moulting cycle also influence selectivity (MOMOT 1967; STEIN 1977). According to the size distribution of crayfish used as food for perch in other lakes (DEHLI 1981; MORALES & APPELBERG 1984), predation by perch less than 15 cm in length would primarily be restricted to YOY crayfish in L. Holmsjön. Recent experimental data also emphasize the importance of perch as a predator on YOY crayfish affecting both the crayfish abundance and behaviour (APPELBERG & ODELSTRÖM 1988).

The relative abundance of perch in L. Holmsjön was high compared to the situation in other limed lakes. In a gill-netting program including 136 lakes with perch, DEGERMAN & NYBERG (1987) report 11.1 as the median CPUE for perch, and only in less than 5% of the lakes perch was more abundant than in L. Holmsjön. Thus, several factors indicate that predation may be a plausible control mechanism for the crayfish population in L. Holmsjön: 1) the abundance of perch in L. Holmsjön was high, 2) perch is known as a predator on *A. astacus* and found to prey on crayfish in similar lakes, 3) due to size-selective predation perch should prefer YOY crayfish as food, and 4) the abundance of YOY crayfish was low during the study period.

In conclusion, I found no support for neither of the two hypotheses suggested for crayfish regulation in L. Holmsjön. As no change in crayfish population size was noted after liming, despite increased pH, alkalinity and Ca^{2+} concentration, as well as an increase in the biomass of benthic fauna, both hypotheses have to be rejected. Instead, the results support an alternative hypothesis suggesting that the crayfish population was restricted by predation by perch on YOY crayfish.

Acknowledgements

I would like to express my gratitude to BIRGER PEJLER and STELLAN HAMRIN for constructive criticism. Especially I thank ROY STEIN for comments on an earlier draft of this paper. TOMMY ODELSTRÖM, KJELL TABERMAN and JOHANN HAMMAR are thanked for help during scuba-diving. Benthic fauna samples were analysed by ANDERS HASSELROT. BJÖRN ANDERSSON is thanked for preparing illustrations. Financial support was provided by a grant from the Swedish National Board of Fisheries.

References

ABRAHAMSSON, S. S. A. (1965): A method of marking crayfish *Astacus astacus* L. in population studies. Oikos **16**, 228—231.

— (1966): Dynamics of an isolated population of the crayfish *Astacus astacus* L. Oikos **17**, 96—107.

— (1972): Fecundity and growth of some populations of *Astacus astacus* L. in Sweden. Rep., Inst. Freshw. Res., Drottningholm **52**, 23—27.

AHLGREN, I., & AHLGREN G. (1976): Vattenkemiska analysmetoder sammanställda för undervisningen i limnologi (Methods for water chemical analyses compiled for the education in limnology) Inst. of Limnology, Univ. of Uppsala (in Swedish).

APPELBERG, M. (1983): Response of acid stress upon the oxygen uptake in eggs of the crayfish *Astacus astacus* L. In: GOLDMAN, C. R. (ed.), Freshwater Crayfish V, pp. 59—70. AVI Publishing, Company, Westport, Conn.

— (1984): Early development of the crayfish *Astacus astacus* L. in acid water. Rep., Inst. Freshw. Res., Drottningholm **61**, 48—59.

— (1985): Changes in haemolymph ion concentration of *Astacus astacus* L. and *Pacifastacus leniusculus* (DANA) after exposure to low pH and aluminium. Hydrobiologia **121**, 19—25.

— (1987): Some factors regulating the crayfish *Astacus astacus* L. in acid and neutralized waters. Annls. Soc. r. zool. Belg. **117**, 167—179.

— & ODELSTRÖM, T. (1986): Habitat distribution, growth and abundance of the crayfish *Astacus astacus* L. in the littoral zone of four neutralized lakes. (Manuscript).

— (1988): Interaction between European perch (*Perca fluviatilis*) and juvenile *Pacifastacus leniusculus* (DANA) in a pond experiment. In: GOELDLIN, P. (ed.), Freshwater Crayfish VII, Lausanne, Switzerland. (In press).

CAPELLI, G. M. (1975): Distribution, life history, and ecology of crayfish in northern Wisconsin, with emphasis on *Orconectes propinquus* (GIRARD). Ph. D. Thesis, University of Wisconsin-Madison, USA, 215p.

DEGERMAN, E., & NYBERG, P. (1987): Fiskfaunans sammansättning och täthet i försurade och kalkade sjöar — en arbetsrapport (English summary: The composition and abundance of the fish fauna in acidified and limed lakes in Sweden). Information, Inst. Freshw. Res., Drottningholm 7, 1987, 71pp.

DEHLI, E. (1981): Perch and freshwater crayfish. Fauna **34**, 64—67 (English summary).

ERIKSSON, F., HÖRNSTRÖM, E., MOSSBERG, P., & NYBERG, P. (1983): Ecological effects of lime treatment of acidified lakes and rivers. Hydrobiologia **101**, 145—163.

FRANCE, R. L. (1985): Relationship of crayfish (*Orconectes virilis*) growth to population abundance and system productivity in small oligotrophic lakes in the experimental lakes area, Northwestern Ontario. Can. J. Fish. Aquat. Sci. **42**, 1096—1102.

FRASER, J. E., SAUNDERS, P. W. JR., OETTINGEN, S. L., BRITT, D. L., & SVERDRUP, H. U. (1985): Ecological effects of base addition to lake ecosystems. General Res. Corp., Intern. Sci. and Tech., Inc. 1985, 66p.

HULTBERG, H., & ANDERSSON, I. (1982): Liming of acidified lakes: Induced long-term changes. Wat., Air and Soil Poll. **18**, 311—331.

JONES, P. D., & MOMOT, W. T. (1981): Crayfish productivity, allochthony and basin morphometry. Can. J. Fish. Aquat. Sci. **38**, 175—183.

MALLEY, D. F. (1980): Decreased survival and calcium uptake by the crayfish *Orconectes virilis* in low pH. Can. J. Fish. Aquat. Sci. **37**, 364—372.

— & REYNOLDS, J. B. (1979): Sampling strategies and life history of non-insectan freshwater invertebrates. J. Fish. Res. Bd. Can. **36**, 311—318.

MAUCHLINE, J. (1977): Growth of shrimps, crabs and lobsters — an assessment. J. Cons., Int. Expl. Mer. **37**, 162—169.

MACMAHON, B. R., & MORGAN, D. D. (1983): Acid toxicity and physiological responses to sub-lethal acid exposures in crayfish. In: GOLDMAN, C. R. (ed.), Freshwater Crayfish V, pp. 71—85. AVI Publishing Company, Westport, Conn.

MILBRINK, G., & JOHANSSON, N. (1975): Some effects of acidification on roe of roach, *Rutilus rutilus* L. and perch, *Perca fluviatilis* L. — with special reference to the Åvaå Lake System in eastern Sweden. Rep., Inst. Freshw. Res., Drottningholm **54**, 52—62.

MOMOT, W. T. (1967): Effects of brook trout predation on a crayfish population. Trans. Amer. Fish. Soc. **107**, 202—209.

— (1984): Crayfish production: a reflection of community energetics. J. Crust. Biol. **4**, 35—54.

MORALES, J., & APPELBERG, M. (1984): Studier av en signalkräftpopulation före och efter en sjökalkning (Studies of a crayfish population before and after liming a lake). Inst. of Limnology, Univ. of Uppsala, **B:13**, 1—20.

MORGAN, D. O., & MCMAHON, B. R. (1982): Acid tolerance and effects of sublethal acid exposure on iono-regulation and acid-base status in two crayfish *Procambarus clarkii* and *Orconectes rusticus*. J. exp. Biol. **97**, 241—252.

MOSHIRI, G. A., & GOLDMAN, C. R. (1969): Estimation of assimilation efficiency in the crayfish *Pacifastacus leniusculus* (DANA) (Crustacea: Decapoda). Arch. Hydrobiol. **66**, 298—306.

NYBERG, P., APPELBERG, M., & DEGERMAN, E. (1986): Effects of liming on crayfish and fish in Sweden. Wat., Air and Soil Poll. **31**, 669—687.

ODELSTRÖM, T. (1983): A portable hydraulic diver-operated dredgesieve for sampling juvenile crayfish. Description and experience. In: GOLDMAN, C. R. (ed.), Freshwater Crayfish V, pp. 270—274. AVI Publishing Company, Westport, Conn.

RADDUM, G. C., HAGENLUND, G., & HALVORSEN, A. (1984): Effects of lime treatment on the benthos of Lake Söndre Boksjö. Rep., Inst. Freshw. Res., Drottningholm **61**, 167—177.

SAS INSTITUTE INC., (1982): SAS User's Huide: Statistics, 1982 Edition. Cary N. C. SAS Institute Inc., 1982.

SEBER, R. A. Jr. (1982): The estimation of animal abundance and related parameters. London.

STEIN, R. A. Jr. (1977): Selective predation, optimal foraging, and the predator-prey interaction between fish and crayfish. Ecology **58**, 1237—1253.

STEVENSON, J. R. (1975): The molting cycle in the crayfish: Recognizing the molting stages, effects of ecdyson, and changes during the cycle. In: AVAULT, J. W. Jr. (ed.), Freshwater Crayfish II, pp. 255—270. Louisiana, USA.

SVÄRDSON, G. (1974): Översikt över laboratoriets verksamhet med plan för år 1974 (Overview over the activities of the laboratory with plans for year 1974). Information, Inst. Freshw. Res., Drottningholm **1**, 1974.

SÖDERBÄCK, B., APPELBERG, M., ODELSTRÖM, T., & LINDQVIST, U. (1988): Food consumption of the crayfish *Astacus astacus* L. in the laboratory. In: GOELDLIN, P. (ed.), Freshwater Crayfish VII, Lausanne, Switzerland. (In press).

WESTMAN, K., SUMARI, O., & PURSIAINEN, M. (1977): Electric fishing in sampling crayfish. In: LAURRENT, P.-J. (ed.), Freshwater Crayfish IV, pp. 251—256. Thonon-les-Bains, France.

WHITE, T. C. R. (1978): The importance of a relative shortage of food in animal ecology. Oecologia **33**, 71—86.

Address of the author:

MAGNUS APPELBERG
Institute of Freshwater Research
S-170 11 Drottningholm
Sweden

Limnologica (Berlin) **20** (2), 329—338 Juni 1990

Institute of Limnology, University of Uppsala, Uppsala, Sweden

Effects of Arsenate Additions to Pelagic Mesocosms on Phytoplankton Biomass and Composition

By PETER BLOMQVIST & ULF HEYMAN

With 5 Figures

Key words: Arsenate additions; In-situ enclosures; Phytoplankton; Tychoplankton.

Abstract

Experiments with arsenate additions to four large in-situ enclosures were performed throughout the ice-free season in Lake Sörmogen, a mesotrophic brownwater lake in Central Sweden. Only a small amount, in most cases less than 10%, of total As was found in the particulate fraction in the corrals in which arsenate was added.

The majority of the particulate As was usually found in the 2—40 µm size fraction (the phytoplankton size fraction).

Phytoplankton responded to the arsenate additions with a significant decrease in total biomass, while species diversity remained unchanged. No obvious species- or group-selective effects of As could be traced among the 87 taxa included in the biomass estimates. Heavy non-motile taxa (i. e. Bacillariophyceae), however, seemed generally most affected and showed the lowest degree of recovery in periods when concentrations of As decreased.

Tychoplanktic species responded to initial As dosing by forming a peak in plankton shortly afterwards.

Introduction

As a trace element, arsenic is a common constituent of virtually all natural environments. Natural import of arsenic to aquatic environments occurs through weathering of the bedrock and soils in the lithosphere. However, in many lakes and estuaries, anthropogenic loadings of arsenic have for many years been the dominant source of arsenic in the water (FERGUSON & GAVIS 1972; SANDERS 1986).

In natural waters, arsenic is present in all its states of oxidation and in both inorganic and organic form. Inorganic As (V) compounds, mainly in the form of arsenate, usually dominate under aerobic conditions (ANDREAE & FROELICH 1984). In Swedish freshwaters, natural concentrations of As usually range between 0.1 and 0.5 µg/l (BORG 1983).

Arsenate is a chemical analogue of phosphate and is believed to affect the physiology of algae in two different ways: by acting as an uncoupler during the formation of organic phosphorus esters and by acting as a competitive inhibitor of phosphorus uptake to the cells (cf. the review by PLANAS & HEALEY 1978).

Studies of the effects of arsenic compounds on the organisms of the pelagic ecosystem of freshwaters as well as of the sea, have mainly been carried out on a laboratory scale using single test organisms. In these laboratory studies, arsenate has been found to depress the growth rate of most phytoplankton species tested. The levels of arsenate necessary to cause a depression of growth rates have, however, been found to vary considerably between species. CONWAY (1978) found that arsenate levels of up to 160 µg/l neither caused a reduction of growth rate nor affected the uptake of nutrients in continuous-culture experiments with *Asterionella formosa*. PLANAS & HEALEY (1978), testing five species of freshwater phytoplankton, found that 75 µg/l of arsenate depressed the growth rates of two of the species, a third species required 750 µg/l for the same degree of depression, while two species were unaffected up to concentrations of 7500 µg/l. However, the latter two species began to die at the higher concentrations when phosphorus was depleted in the cultures. BOTTINO et al. (1978) did not discover any effects of arsenate on two marine species of phytoplankton at concentrations of up to 1000 µg/l. SANDERS (1979) found that both the growth rate and the rate of carbon uptake of the marine alga *Skeletonema costatum* were depressed at arsenate concentrations of as little as 5 and 12.5 µg/l. CHRISTENSEN & ZIELSKI (1980) reported depression of the growth rate of *Chlamydomonas* at 75 µg As/l. SANDERS & VERMERSCH (1982) found a considerable variability in the reduction of growth rates among seven marine species of phytoplankton exposed to arsenate concentrations of up to 100 µg/l, although only one species was not affected at all. These highly variable results, even among species within the same phylogenetic group, indicate that the effects of arsenate on phytoplankton vary not only between species but also within species, depending on both the physiological status of the cells at the start of the experiment and the supply of nutrients in the culture medium (SANDERS 1979; BRUNSKILL et al. 1980; THURSBY & STEELE 1984). Thus, extrapolation of these test results to species growing under natural conditions is extremely dangerous.

Few experiments on the effects of arsenate on natural mixed populations of phytoplankton have been performed. The collected results from the laboratory tests cited above, indicate that arsenate may have a pronounced effect on the seasonal succession of phytoplankton species and thereby on all higher trophic levels of the pelagic environment. PLANAS & LAMARCHE (1983), using small in-situ enclosures, however, did not discover any covariation between changes in the species composition of algae and the presence of different forms of arsenic. By contrast, effects were demonstrated in large-scale outdoor experiments in tanks using natural mixed populations of marine phytoplankton (SANDERS & VERMERSCH 1982). Apart from a general lowering of the total number of algal cells in the tanks containing arsenic, the latter authors reported major effects on the succession of species. In the low arsenic tank (8.7 µg/l) the dominant species *Skeletonema costatum*, a centric diatom, was replaced by *Nitzschia seriata*, a pennate diatom, while in the high arsenic tank (24 µg/l) *Skeletonema* was replaced by a mixture of *Nitzschia* and microflagellates.

These results together with the finding made by SANDERS (1986) that low levels of arsenate can cause a shift in dominance within natural populations of phytoplankton in large outdoor cultures, and that such changes in food conditions can have dramatic effects on herbivore populations, indicate a need for further enlarged studies on the effects of arsenate on the pelagic ecosystem.

The study presented here is a part of a larger Swedish research programme (ESTHER, LANDNER et al. 1982) investigating various hazard assessment systems using arsenic as one test substance. Arsenic was chosen partly because it was found a suitable test-substance, but primarily because the level of arsenic pollution in Sweden is one of the highest in the world (BLANCK et al. 1989). The largest discharges are from ore smelters, glass industries and from the use of arsenic as a fungicide in wood preparation. The objective of the study is to investigate the effects, as well as the fate, of As in the pelagic ecosystem using large-scale in-situ enclosures.

The present paper reports on the response of phytoplankton to long-term sub-lethal exposure to arsenate. The experiments were performed in large limnocorrals placed in a mesotrophic, mesohumic lake in Central Sweden and at As concentrations of between 1.5 and 15 µg/l.

Materials and Methods

Limnocorral experiments with additions of As were performed in Lake Sörmogen, an elongated (lmax 4750m, bmax 1700m), fairly large (total area 3.7 km²), shallow (Zmax 10m), mesohumic and mesotrophic forest lake situated at 59°44′N, 15°26′E.

Four limnocorrals, arranged in two pairs and linked by a central working platform, were placed in the lake some 400 m from the eastern shore. The corrals had glassfibre-reinforced polyethylene walls and were 10 m in diameter and 4 m in depth, with approximate volumes of 300 cubic m. The enclosures were open to the sediment into which the walls were sunk by means of 15 mm thick iron rods sewn into the plastic. To ensure sufficient inocula of species from the lake, peristaltic pumps were used to pump water in and out of the corrals at such a rate that the theoretical turnover time of the water was 100 days.

The walls of the corrals were let down on May 23, 1986, about two weeks after ice breakup in the lake. However, the walls of corral 2 were lifted to the surface for adjustment, and let down again within the same day, one week later. Two weeks after the walls were let down the water entrapped within the corrals was dosed with arsenate, gradually over four days, to initial theoretical total arsenic concentrations of 3, 10 and 20 µg/l respectively. One corral, used as a control, was not dosed. After the initial dosage As was added through the flowthrough system in amounts yielding concentrations in the inflowing water equal to the initial theoretical concentrations in the corrals.

Sampling of the corrals, and of the surrounding lake water close to the corrals, was performed at roughly weekly intervals using a 2 m tube sampler with a volume of 2 l. From each of the corrals and from the lake, a composite sample was taken using four surface (0—2 m) and four bottom (2—4 m) samples. From each 16 l sample, subsamples for phytoplankton counting were taken and preserved with IIK supplemented with acetic acid. The remainder of the sample was filtered through a 40 µm screen and 1 l of the filtrate was saved for chemical analysis. The "zooplankton" (40 µm) fraction was transferred to a 100 ml bottle and bubbled with carbon dioxide, to anaesthetize the animals. In the laboratory, immediately after sampling, this fraction was split and a portion equivalent to 3/10 of the fraction was obtained for both As and P analysis.

Table 1. Concentrations of As in different particle size fractions. Mean values for the different time periods (see text)

Period and treatment	Arsenic µg/l				Arsenic per biomass of phytoplankton
	Total	>40 µm	2—40 µm	0.2—2 µm	
Period 1					
A	1.0	0.03	0.14	0.05	0.22
B	1.0	0.05	0.16	0.05	0.17
C	1.2	0.03	0.10	0.04	0.16
D	1.1	0.01	0.10	0.03	0.08
E	0.9	0.02	0.10	0.06	0.16
Period 2					
A	1.0	0.02	0.23	0.03	0.22
B	1.1	0.02	0.22	0.03	0.21
C	2.0	0.02	0.24	0.05	0.49
D	9.6	0.03	0.99	0.24	1.54
E	11.3	0.02	0.37	0.35	1.55
Period 3					
A	1.2	0.01	0.20	0.06	0.16
B	1.3	0.01	0.10	0.03	0.07
C	1.4	0.01	0.10	0.03	0.08
D	4.2	0.02	0.17	0.11	0.71
E	6.4	0.01	0.19	0.14	0.50
Period 4					
A	1.6	0.01	0.35	0.03	0.16
B	0.7	0.01	0.15	0.03	0.08
C	1.4	0.01	0.15	0.04	0.14
D	2.9	0.01	0.15	0.04	0.16
E	4.9	0.01	0.29	0.09	0.21
Period 5					
A	1.2	0.02	0.29	0.10	0.36
B	0.6	0.03	0.14	0.04	0.21
C	1.2	0.02	0.20	0.09	0.26
D	2.9	0.02	0.25	0.63	1.25
E	3.1	0.01	0.38	0.18	0.96

Treatments: A = lake reference, B = control corral, C = 3 µg As/l, D = 10 µg As/l and E = 20 µg As/l.

Part of the previously obtained filtrate was then filtered successively through 2 µm Nucleopore and 0.2 µm Gelman membrane filters. As a result of these procedures, the As/P results could be divided into three size fractions: 0.2—2 µm, 2—40 µm and >40 µm.

Total and particulate As was determined by hydride generation and flameless atomic absorption (BORG et al. 1985). The reduced solutions from this analysis were also used to determine phosphorus in the particulate fractions by means of a normal Murphy & Riley colour reaction.

After sedimentation of 10 ml of the IIK-preserved water, phytoplankton counting was performed in accordance with the Utermöhl method (UTERMÖHL 1958), using an inverted microscope. Usually 100—500 units (cells or colonies) of the most common species were counted in each sample. Phytoplankton cell volumes were obtained using geometrical formulae and converting the values into biomasses assuming a density of 1 g/cm^3.

On the basis of temporal differences in total arsenic concentrations in the water, the time series were divided into five periods. Within each period, differences were tested using a non-parametric Quade test (CONOVER 1980). In order to avoid "differences" due to counting errors, only species that were counted in more than 5 units per sampling occasion, calculated as an average within each period, were included in the analysis.

Results

Fate of As in the pelagial

The total arsenic concentrations in the lake (A) and in the control (B) were both approximately 1 µg/l (Table 1, Fig. 1). Dosing with As resulted in increased As levels, although the levels were not the same as the theoretical dosages. In the high As corrals (D) and (E), As concentrations fell rapidly for one month after initial dosing and then stayed relatively constant or decreased slowly for the rest of the summer period. These corrals showed very similar concentrations at the beginning of the experiment as long as the concentrations were high and declining but diverged more markedly later when the concentrations stabilized. In the low arsenic corral (C) too, the concentrations decreased rapidly at the beginning of the experiment. After three weeks the concentration in this corral was close to that of the lake and the control corral, although it remained about 0.5 µg/l above those until the end of August. The second As addition in early September only caused a minor increase in the concentrations in the high As corrals (D and E), and these declined rapidly shortly afterwards.

Studying the temporal variation in the concentrations of total As in treatments D and E, five different periods can be distinguished: 1. the pre-treatment period, including two samples; 2. the high As period after addition, June 11 to July 4, including five samples; 3. the relatively stable and quite high As period during July, including six samples; 4. the lower As period during August, 6 samples; and 5. the period after the small As peak in September, 5 samples. These periods will be referred to when discussing other data.

In the lake and in the control, particulate As was 20—30 % of total As throughout the investigated period. The same percentage was also recorded in the other corrals before arsenic dosing, but decreased drastically when As was added and remained low (< 10%) as long as total As was high. The adsorption and uptake of As by particles was thus low.

The majority of particulate As was found in particles smaller than 40 µm (Table 1), usually in the 2—40 µm (phytoplankton) size fraction.

Phytoplankton development in the lake

During the ice-free part of the year, total biomass of phytoplankton in Lake Sörmogen ranged between 380 and 3400 µg WW/l (Fig. 2). A gradual increase took place from early spring and throughout the summer and

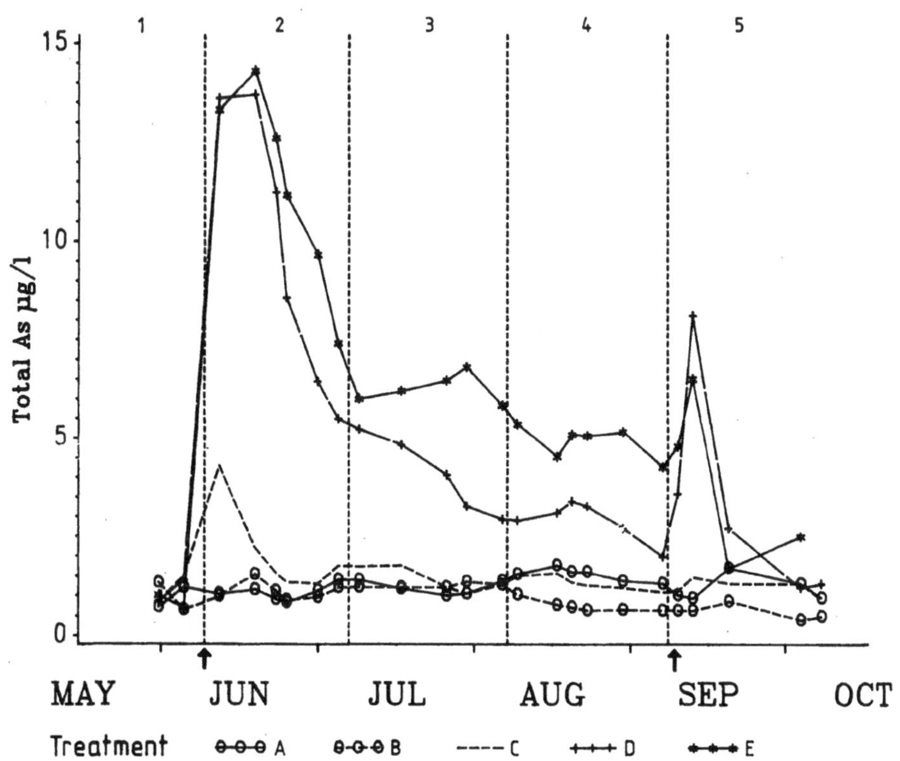

Fig. 1. Total arsenic concentrations in the different corrals (B = control; C = low As; D and E = high As) and in the lake (A). Broken vertical lines delimit the different time periods (see text)

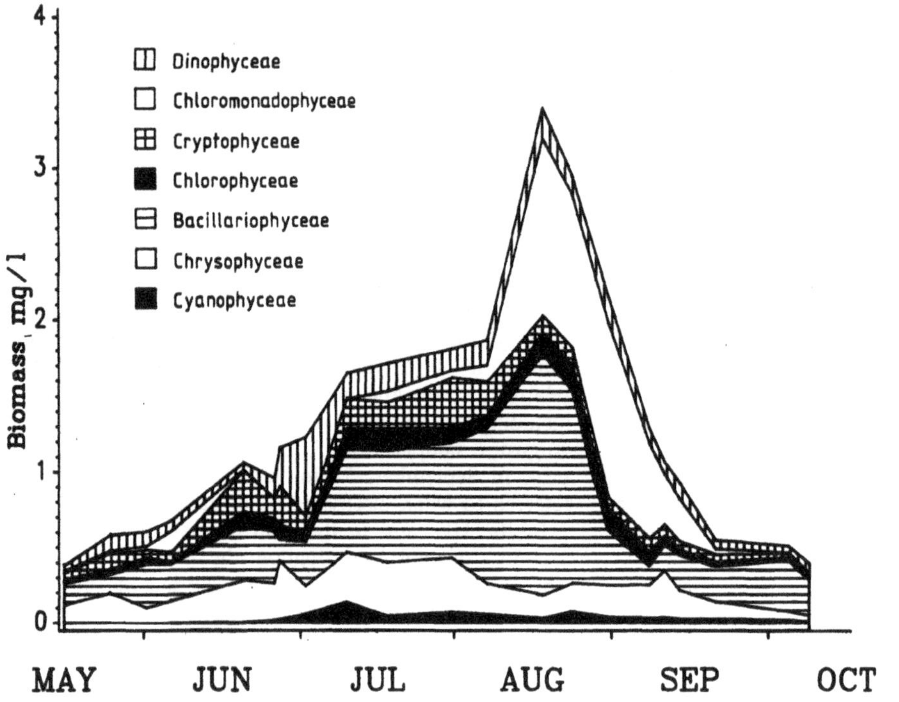

Fig. 2. Total biomass of phytoplankton in Lake Sörmogen, Central Sweden during 1986 and the relative contribution of different algal groups

was followed by a rapid decrease at the beginning of the autumn.

Of the various taxonomic groups, Bacillariophyceae usually dominated (Fig. 2). As an average for the entire season, this group constituted 33% of the total biomass. Chrysophyceae, Cryptophyceae, Chloromonadophyceae and Dinophyceae were also of great importance to total biomass, contributing 20, 15, 11 and 11% respectively.

In terms of the number of taxa, Chrysophyceae were the dominant group followed by Chlorophyceae (Table 2). Chloromonadophyceae, which constituted a considerable proportion of total biomass in late summer, were represented by only one species, *Gonyostomum semen* (Ehr.) Diesing. In all, 87 taxa were included in the biomass estimates. Thirteen of these taxa contributed more than 100 µg WW/l to total biomass on at least one occasion during the growth season and most of these dominant taxa were found throughout the ice-free period.

As regards the number of algal groups and taxa contributing to biomass, Lake Sörmogen accomodates a very diverse phytoplankton community. On the basis of total biomass level and composition, the lake can be classified as a mesotrophic brownwater lake of the *Gonyostomum* type (Rosén 1981).

Table 2. Distribution of phytoplankton taxa among the different algal groups contributing to total biomass in Lake Sörmogen in 1986

Group	Number of taxa
Cyanophyceae	5
Chrysophyceae	32
Bacillariophyceae	11
Chlorophyceae	22
Conjugatophyceae	1
Cryptophyceae	9
Chloromonadophyceae	1
Dinophyceae	6

Effects of As on phytoplankton biomass and composition

A considerable and significant reduction in the total biomass of phytoplankton was found during periods 2, 3 and 4 in the two corrals showing the highest concentrations of As (Fig. 3, Table 3). In the low arsenic corral too, total biomass decreased significantly during periods 2 and 4. The lake and the control corral never showed any significant differences as regards total biomass. The increased concentrations of As during period 5 seem to have had little effect on phytoplankton.

The decrease in total biomass during the high As periods was accompanied by decreases in the number of species counted, while the proportionality between the species, measured by the Shannon diversity index, was usually not affected (Table 3).

If the phytoplankton community is divided by morphotype into motile (i. e. flagellated) and non-motile (i. e. coccoid and trichoid) taxa, it becomes evident that both these groups were negatively affected in the As-dosed corrals (Table 3). Flagellates showed a greater (treatment E) or similar (D) reduction during the initial high As period in the As-dosed corrals but were then usually less affected than the coccoid forms and showed faster recovery during the low As period (4) at the end of the summer. Dividing the coccoid group further, based on their density, into heavy and light forms shows that the heavy forms (i. e. the diatoms) were most affected and showed the lowest degree of recovery.

Dividing the material into taxonomic groups clearly reveals negative effects of As on all the five major groups of algae as well as on Cyanophyceae and usually also on all major contributing taxa within each group (Table 4). Especially among Bacillariophyceae the decrease was obvious for all major contributing species. The only group deviating from the pattern of general decrease was Chlorophyceae, in which two of the species (*Chlorella* sp. and *Oocystis parva*) showed significant increases during the high As period in one of the high As corrals (E)

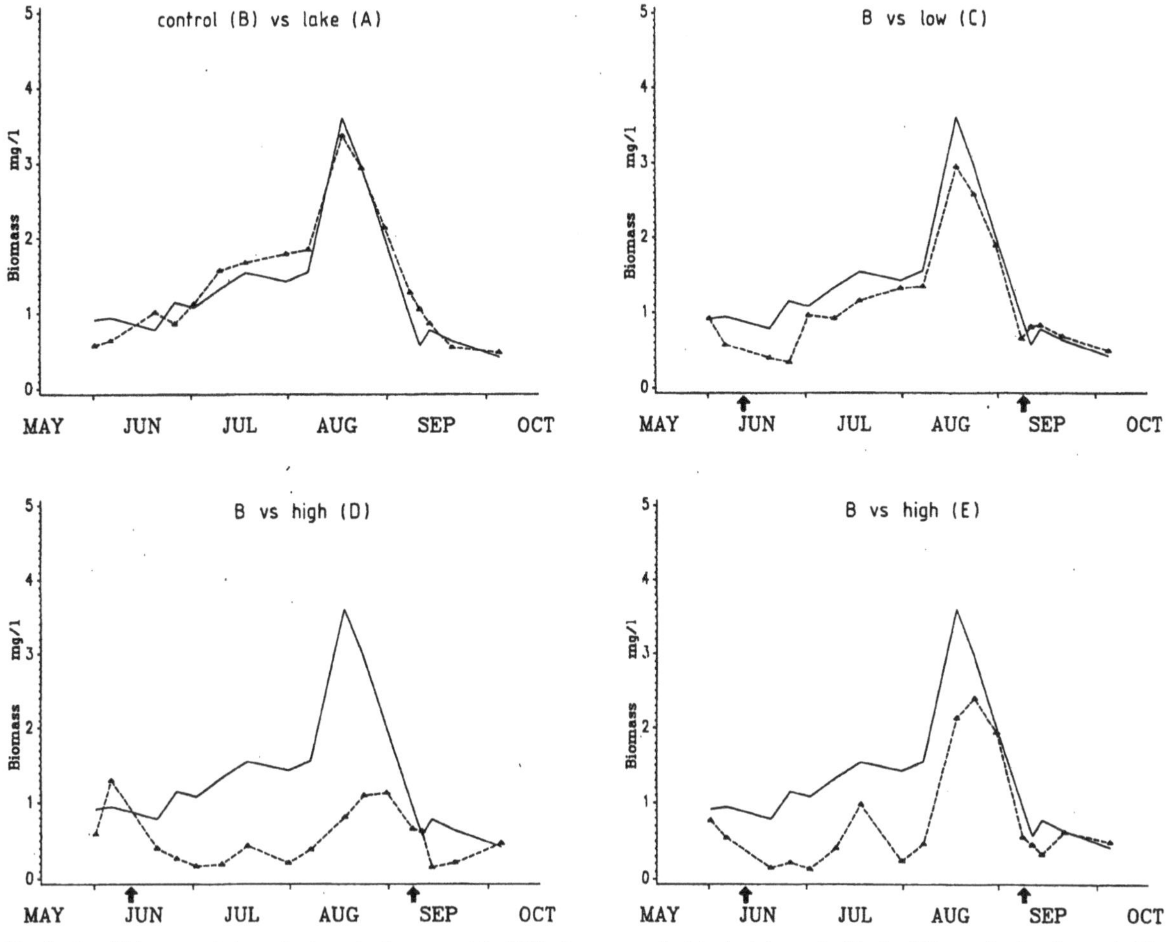

Fig. 3. Total biomass of phytoplankton in the control (solid line), compared with the lake and with the different As-dosed corrals. Black arrows denote initial and additional As dosing

Table 3. Total biomass, number of species, Shannon diversity index and biomass of morphotype-based groups of phytoplankton in the As-dosed corrals and in the lake, calculated as percentages of the values of the same parameters in the control

Treatment	Period	Total biom.	No of species	Shannon index	Flagel-lates	Coccoid (all)	Coccoid (heavy)	Coccoid (light)
E	1	70	104	100	73	58	57	—
(high As)	2	16**	47*	107	12**	29**	7**	95
	3	36**	79	146**	42**	27**	23**	49**
	4	74**	94	123	85	49**	47**	75
	5	79	97	99	88	63**	65	54
D	1	101	100	99	106	83	82	—
(high As)	2	28**	65	61	30	21**	14**	45
	3	21**	68**	128	26**	14**	11**	33**
	4	39**	79**	114	44*	27**	24**	56**
	5	61	88	93	62	58**	54*	77
C	1	80	90	89	85	62	62	—
(low As)	2	56*	66	96	53	65	41	136
	3	81	100	102	86	74	72	84
	4	85**	94	87	80*	95	94	97
	5	117	103	100	129	98	97	103
Lake	1	65	94	93	50	123	123	—
	2	100	106	109	96	111	121	80
	3	118	111	77	98	144	143	153*
	4	102	107	84	91	128	125**	161**
	5	123	95	90	127	116	118	109

* and ** denote which differences were significantly distinguished from zero at the $p < 0.1$ and $p < 0.05$ levels respectively.

— denotes that the species or group was not counted in sufficiently significant amounts to be included in the analysis.

Table 4. Biomass of phylogenetic groups and dominant species of phytoplankton in the As-dosed corrals and in the lake, calculated as percentages of the biomass in the control. For legends to species abbreviations see footnote. Symbols as in Table 3

Treatment	Period	CYANO-PHYCEAE	MINC	AFLA	CHRYSO-PHYCEAE	CH34	CH57	CH80	DIBA	MFAS
E	1	—	—	—	64	77	58	48	43	—
(high As)	2	1**	4**	0	16**	32	19*	7	1*	29
	3	40*	30	—	54	88	51**	50	2**	93
	4	87	53	—	86	94	96	78	71	—
	5	49	94	—	107	92	76	—	111	—
D	1	—	—	—	122	97	108	150	251	—
(high As)	2	44	132	0	47*	48	76	37	6	54
	3	28**	23**	—	33**	75	22**	19*	1**	159
	4	62*	43*	—	54**	92	48*	54	5**	—
	5	69	88	—	58	89	55	—	15**	—
C	1	—	—	—	86	84	52	33	18	—
(low As)	2	31*	40**	18	37*	74	30*	48	5	108
	3	100	82	—	94	78	74**	208	38	109
	4	110	101	—	52**	70	50*	78	23**	—
	5	81	114	—	104	90	64	—	124	—
Lake	1	—	—	—	43	61	48	26	72	—
	2	71	76	65	70	59	70	167	70	83
	3	183**	196	—	111	98	81	160	145	130
	4	143	148	—	87	96	78	77	20**	—
	5	107	203	—	102	106	101	—	63	—

Treat-ment	Period	BACIL-LARIO-PHYCEAE	MAMB	MDVA	CSTE	RLON	TFEN	DIAT	ASTF	CHLORO-PHYCEAE	CHLO	OOCP	KOLI
E	1	57	55	—	—	50	62	—	123	—	—	—	—
(high As)	2	7**	15**	12**	—	9**	4**	3**	1**	192	1000**	178	5**
	3	23**	13**	62	31**	4**	19**	12**	22**	61*	179	87	—
	4	47**	60**	—	—	12**	51**	34**	59	65	—	76	—
	5	65	59	—	—	48	68	322	65	67	—	—	—
D	1	82	85	—	—	65	57	—	138	—	—	—	—
(high As)	2	14**	27**	25**	—	0**	11**	6*	18**	46*	—	33	61
	3	11**	17**	25**	13**	0**	9**	6**	12**	42**	—	58	—
	4	24**	7**	—	—	24**	27**	21**	72	60*	—	100	—
	5	54*	34**	—	—	57	83	154	47	94	—	—	—
C	1	62	67	—	—	40	29	—	42	—	—	—	—
(low As)	2	41	58	41	—	13*	67	20*	38*	246	727**	389	11**
	3	72	20**	83	62	29*	75	94	71	79	—	68	—
	4	94	83	—	—	67*	101	92	105	90	—	76	—
	5	97	69*	—	—	96	116	283	113	109	—	—	—
Lake	1	123	141	—	—	55	62	—	73	—	—	—	—
	2	121	125	122	—	168	116	67	87	81	136	89	48
	3	143	136	141*	100	164	145	187	133	130	—	97	—
	4	125**	66**	—	—	115	169**	101	123	131	—	148	—
	5	118	84	—	—	161**	137**	357	55	121	—	—	—

Treat-ment	Period	CRYPTO-PHYCEAE	CMAR	CSPP	RPUS	KATO	CHLORO-MONADO-PHYCEAE	DINO-PHYCEAE	GFUS	GYLA	GUBB	PINC
E	1	87	—	108	37	79	75	65	—	53	—	112
(high As)	2	10*	0*	3**	—	5*	—	5*	0	9	1*	6*
	3	44	94	26**	38**	—	46**	18**	37	—	16**	2**
	4	50**	—	46	—	—	93	66	107	—	60	20**
	5	78	—	61	—	—	63	76	—	—	—	19**
D	1	155	—	157	53	142	72	91	—	124	—	108
(high As)	2	11*	6*	3**	—	23**	—	11	0	45	2*	4*
	3	33	50	14**	48**	—	8**	10**	5*	—	10**	2**
	4	50**	—	42**	—	—	44	29*	27	—	33**	17**
	5	49	—	63	—	—	91	43*	—	—	—	24**

Table 4. (Continued)

Treatment	Period	CRYPTOPHYCEAE	CMAR	CSPP	RPUS	KATO	CHLOROMONADOPHYCEAE	DINOPHYCEAE	GFUS	GYLA	GUBB	PINC
C	1	89	—	116	47	21	100	87	—	29	—	84
(low As)	2	43	118	22	—	13*	—	111	0	18	43	218
	3	77	75	78*	68	—	131	70**	121	—	63	53*
	4	62*	—	70	—	—	90	71	61	—	98	57
	5	157**	—	155	—	—	167	129	—	—	—	56*
Lake	1	61	—	75	26	64	49	64	—	82	—	52
	2	97	147	106	—	59	—	149	100	50	58	267
	3	84	69	84	73	—	88	86	160*	—	82	55*
	4	121	—	163	—	—	89	84	98	—	145	22**
	5	136*	—	173	—	—	206	92	—	—	—	56**

Legend to species abbreviations in Table 4:
MINC: *Microcystis incerta* (LEMM.) LEMM., AFLA: *Anabaena flos-aquae* RALFS, CH34: *Ochromonas* + *Chromulina* spp. 3—4 µm in diameter, CH57: *Ochromonas* + *Chromulina* spp. 5—7 µm, CH80: *Ochromonas* + *Chromulina* spp. 8—10 µm, DIBA: *Dinobryon bavaricum* IMH., MFAS: *Mallomonas fastigata* ZACH., MAMB: *Melosira ambigua* (GRUN.) MÜLL., MDVA: *Melosira distans* var. *alpigena* GRUN., CSTE: *Cyclotella stelligera* CL. & GRUN., RLON: *Rhizosolenia longiseta* ZACH., TFEN: *Tabellaria fenestrata* (LYNGB.) KÜTZ., DIAT: *Diatoma elongata* (LYNGB.) AG., ASTF: *Asterionella formosa* HASS., CHLO: *Chlorella* sp., OOCP: *Oocystis parva* WEST & WEST, KOLI: *Koliella* sp., CMAR: *Cryptomonas marssonii* (MARSS.) SKUJA, CSPP: *Cryptomonas tetrapyrenoidosa* SKUJA + spp., RPUS: *Rhodomonas lacustris* PASCH. & RUTTN., KATO: *Katablepharis ovalis* SKUJA, GFUS: *Gymnodinium fuscum* (EHR.) STEIN, GYLA: *Gymnodinium* cf. *lacustre* SCHILL., GUBB: *Gymnodinium uberrimum* (ALLM.) KOF. & SWEZY, PINC: *Peridinium inconspicuum* LEMM.
The group Chloromonadophyceae only includes one species: *Gonyostomum semen* (EHR.) DIESING.

and in the low As corral (C). However, as a group, Chlorophyceae generally decreased in the high As corrals.

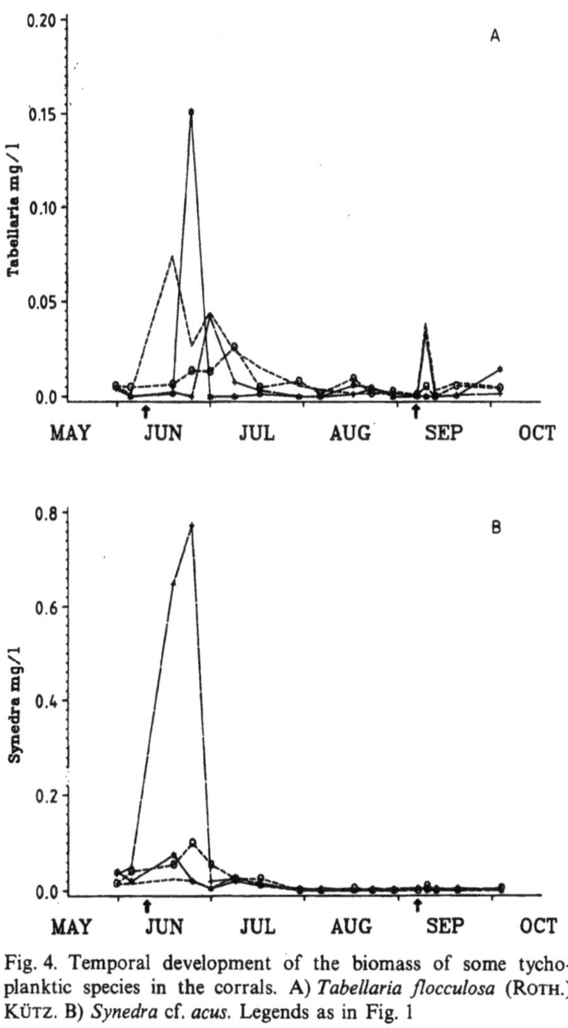

Fig. 4. Temporal development of the biomass of some tychoplanktic species in the corrals. A) *Tabellaria flocculosa* (ROTH.) KÜTZ. B) *Synedra* cf. *acus*. Legends as in Fig. 1

Occurrence of pseudopelagic species

Shortly after initial As dosing, a number of tychoplanktic species began to occur in relatively large amounts in the dosed corrals. A slight increase was also noted in in the control corral, but the general trend were increasing amounts of these species with increasing concentrations of As. In the low As corral (C) and in one of the high As corrals (E), the dominant tychoplanktic species was *Tabellaria flocculosa* (ROTH.) KÜTZ. (Fig. 4A); this species also responded rapidly to the second addition of As in early September. In the other high As corral (D) *Synedra* cf. *acus* became dominant and virtually created a "bloom" (Fig. 4B). Both species disappeared from the water quickly after the peaks, probably due to sedimentation, since few empty cells were recorded in the samples from late June.

Discussion

An experimental set-up such as the one described here has great advantages in its realism and close correspondence to the mother system (cf. Table 1 and LANDNER et al. 1982). It does, however, suffer from its lack of replication, which means that the variation between enclosures cannot be quantified and that statistically rigid conclusions thus cannot be drawn. In the present study it can be argued that the highly significant differences between the control and the treatments were the effect of other factors than arsenate stress or some sort of random phenomenon. However, the reasonable dose-response relationship as well as the increased similarity between the systems when arsenate concentrations decreased strongly indicates that the differences were induced by the treatments (SMITH et al. 1982).

Zooplankton grazing is likely to reduce phytoplankton biomass and the large differences in zooplankton biomass at the start of the experiment (Fig. 5) must be taken into consideration. During period 1 the low As corral (C) and one of the high As corrals (E) were characterized by high zooplankton densities, which in the high As corral persisted throughout period 2. The two high As

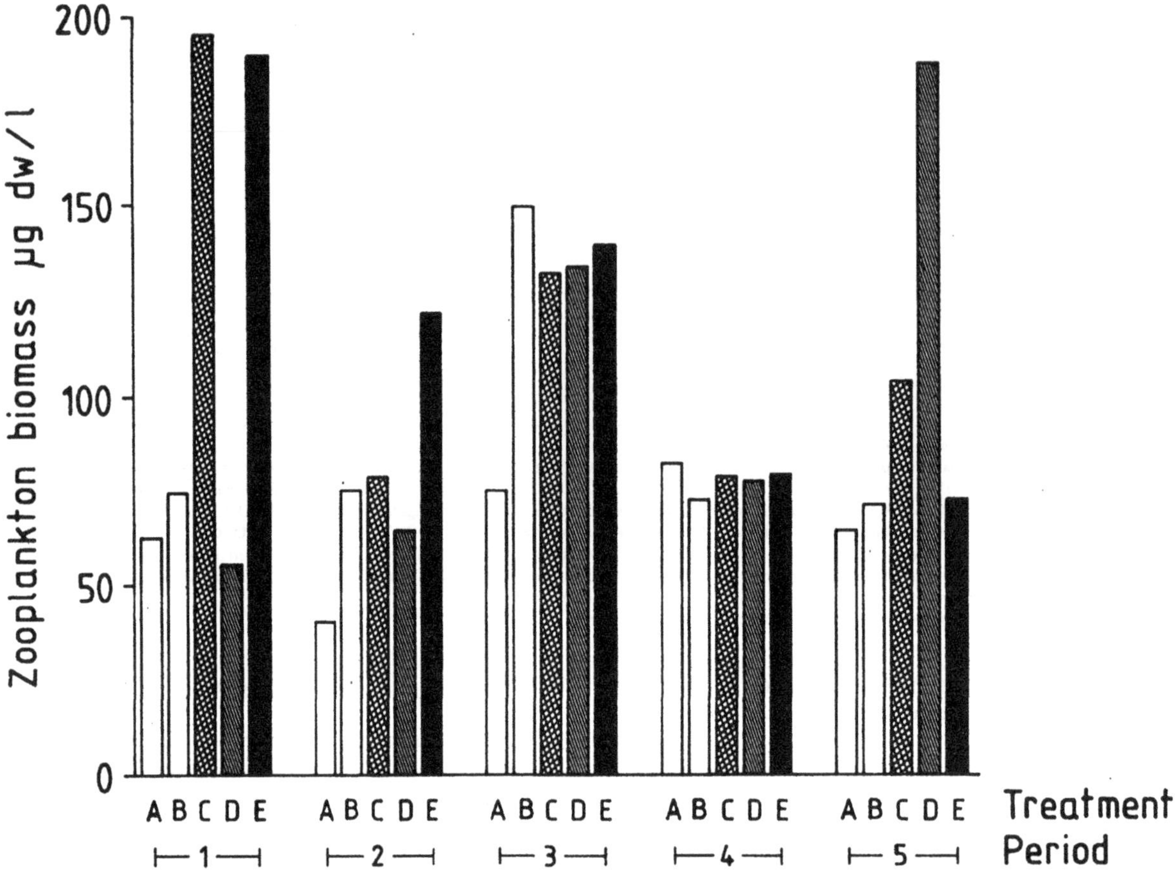

Fig. 5. Average biomass of zooplankton in Lake Sörmogen, Central Sweden (A) and in the corrals (B = control; C = low As; D and E = high As, see Fig. 1) during the different time periods (see Fig. 1)

corrals were thus initially different as regards the grazing aspect, which might explain some of the differences between them but did not affect the general decreasing trend due to arsenate stress. It can thus be concluded that arsenate in these concentrations has a drastic effect on phytoplankton biomass. The results from the low As corral are more complicated since zooplankton biomass before arsenate dosage began was very high and probably caused the reduction in biomass evident in Fig. 3. This is also indicated by the decrease in flagellates as well as the increase of small coccoid green algae (NAUWERCK 1963; PORTER 1976). However, the depression of biomass lasted longer than could be expected from the zooplankton data and the phytoplankton that decreased most as a group were the large and heavy coccoid forms, which are unlikely to be susceptible to grazing. It seems reasonable that the phytoplankton in the low As corral responded to both grazing and to an As concentration of 2 µg/l or twice the background level. This result is similar to those of other studies (PLANAS & HEALEY 1978; SANDERS & VERMERSCH 1982; SANDERS 1986), in which it was concluded that concentrations 2—10 times of background levels in seawater affected algal species. The decrease in total biomass of phytoplankton in our study is consistent with the results of other studies (SANDERS & VERMERSCH 1982; PLANAS & LAMARCHE 1983).

The fate of arsenic in the biotic communities of pelagic ecosystems is still relatively unknown. It is not surprising that much of the particulate As should be found in the phytoplankton size fraction since, as a chemical analogue to phosphate, arsenate is indiscriminately taken up by phytoplankton (SANDERS & WINDOM 1980). The low incorporation of As in the zooplankton size fraction is consistent with the finding that As is not progressively concentrated along a food chain (FERGUSON & GAVIS 1972). It has also been shown that As metabolizes to water-soluble compounds, which can be excreted from animals (NORIN et al. 1987).

The considerable diversity, both as regards the number of algal groups and the number of species contributing to phytoplankton biomass in the test system chosen should, when the system is contaminated by a pollutant, provide a great potential for new species to take over if the pollutant had a selective effect on the species. However, our results do not show As to have any clear selective effect on phytoplankton, since negative effects were found in virtually all common groups and species of algae and since diversity remained largely unaltered. The decrease in the number of species in the As-dosed corrals is, furthermore, a typical artifact caused by the use of a constant sedimentation volume of water, which results in the number of species decreasing with decreasing biomass, due to the extinction of rare species (MARGALEF 1983).

Thus, our results support those of PLANAS & LAMARCHE (1983), who found no correlation between changes in species composition of algae and the presence of As, and contradict those from most of the single-species laboratory tests cited above, as well as those of SANDERS & VERMERSCH (1982) and SANDERS (1986), who found pronounced selective effects of As on mixed populations of phytoplankton in outdoor tank experiments.

The only sign of As having negative selective effects in our study is the weak tendency for heavy non-motile forms (i. e. diatoms) to show a lower degree of recolonization of the water when concentrations of As decreased. This may be partly due to these species being inoculated from the sediments, while many of the other species (e. g. small flagellates, coccoid green algae) depend on a small inoculum in the water (REYNOLDS 1984). Unlike large encysting flagellates (e. g. Dinophyceae and certain Chrysophyceae) and Cyanophyceae which can actively emerge from the sediments, recolonization of water by heavy non-motile diatom species is probably dependent on a great deal of vertical turbulence, which is most likely reduced in the corral systems. Note the clear, but in most cases non-significant, tendency for Bacillariophyceae to be more abundant in the lake than in the control corral (Table 4).

The results also indicate a weak positive effect on certain coccoid green algae. This is in accordance with MAEDA et al. (1985) who showed that the growth of green algae increased with increased As (V) concentrations, but can also be explained with the above mentioned differences in zooplankton biomass in the corrals.

However, if the appearance of tychoplanktic species, and especially that of *Synedra* in corral 2 (treatment B), is taken into consideration, our results become similar to those of SANDERS & VERMERSCH (1982), although our interpretation differs. Neither *Tabellaria flocculosa*, *Synedra* spp. nor *Nitzschia seriata*, the species found by SANDERS & VERMERSCH (op. cit.), are typical members of the truly planktonic flora of lakes or the sea. *Tabellaria flocculosa* is a very common tychoplanktic species in shallow brownwater lakes but is usually found as an attached alga in the littoral region (HUBER-PESTALOZZI 1942; CLEVE-EULER 1951). The majority of the species and varieties of *Synedra* also belong to the littoral flora (HUBER-PESTALOZZI 1942), together with *Nitzschia seriata*, which is a pelagic-neritic species forming blooms of littoral origin (CLEVE-EULER 1955).

The emergence of tychoplanktic species after additions of arsenate thus seems to be a common phenomenon. It is very interesting to note the rapid response of these species to arsenic additions, which makes them early indicators of changes in the environment. Since no data on abundance changes in the corral-wall or sediment communities is at hand, these results merit further investigation. Measurements of the As and phosphorus content of material adhering to the corral walls have, however, been carried out (HEYMAN et al., in prep.). The data obtained showed that arsenate stress, measured as the quotient between As and P, was the same in the material adhering to the walls as in the 2—40 µm particle fraction in the water. WÄNGBERG et al. (pers. comm.) have shown that, compared with phytoplankton, periphytic communities are equally or more sensitive to arsenate. It is reasonable, then, to assume that the tychoplanktic species originate from a periphytic state on the corral walls and that they respond to increased stress by escaping from that environment by changing to a planktonic state. This mechanism may explain the higher sensitivity of periphyton communities found by WÄNGBERG et al. since, even at relatively low concentrations, it may drastically change the community structure.

Acknowledgement

We wish to express our gratitude to ALF LUNDGREN and Dr. RUSSELL BELL for valuable comments on the manuscript. Our thanks also to the members of the project and to the staff at Grimsö wildlife research station for the cooperation during fieldwork. BJÖRN ANDERSSON drew the figures and MARTIN N. NAYLOR made the linguistic revision of the manuscript. This work was supported by the Swedish National Environmental Protection Board (SNV grant 5333195-5 within the project area ESTHER).

References

ANDREAE, M. O., & FROELICH JR., P. N. (1984): Arsenic, antimony and germanium biogeochemistry in the Baltic Sea. Tellus **36B**, 101—117.

BLANCK, H., HOLMGREN, K., LANDNER, L., NOREN, H., NOTINI, M., ROSMARIN, A., & SUNDELIN, B. (1989): Advanced hazard assessment of Arsenic in the Swedish environment. In: LANDNER, L. (ed.), Approaches to advanced hazard assessment of chemicals in the aquatic environment, Chapter 11. Heidelberg (in press).

BOTTINO, N. R., NEWMAN, R. D., COX, E. R., STOCKTON, R., HOBAN, M., ZINGARO, R. A., & IRGOLIC, K. J. (1978): The effects of arsenate and arsenite on the growth and morphology of the marine unicellular algae *Tetraselmis chui* (Chlorophyta) and *Hymenomonas carterae* (Chrysophyta). J. Exp. Mar. Biol. Ecol. **33**, 153—168.

BORG, H. (1983): Trace metals in Swedish natural fresh waters. Developments in Hydrobiology, Hydrobiologia **101**, 27—34.

— HOLM, K., & HOLMGREN, K. (1985): Determination of arsenic with automated hydride generation and flameless atomic absorption spectrometry. Swedish Environmental Protection Board, Report 1985.

BRUNSKILL, G. J., GRAHAM, B. W., & RUDD, J. W. M. (1980): Experimental studies on the effect of arsenic on microbial degradation of organic matter and algal growth. Can. J. Fish. Aquat. Sci. **37**, 415—423.

CHRISTENSEN, E. R., & ZIELSKI, P. A. (1980): Toxicity of arsenic and PCB to a green alga (*Chlamydomonas*). Bull. Envirom. Contam. Toxicol. **24**, 43—48.

CLEVE-EULER, A. (1951): Die Diatomeen von Schweden und Finnland. Kungl. Sv. Vetensk. Akad. Handl. Ser. 4, Bd. 2, No. 1, 158 pp.

— (1955): Die Diatomeen von Schweden und Finnland. Kungl. Sv. Vetensk. Akad. Handl. Ser. 4, Bd. 5, No. 4, 153 pp.

CONOVER, W. J. (1980): Practical nonparametric statistics. 2nd ed. New York.

CONWAY, H. L. (1978): Sorption of arsenic and cadmium and their effects on growth, micronutrient utilization and photosynthetic pigment composition of *Asterionella formosa*. J. Fish. Res. Board Can. **35**, 286—294.

FERGUSON, J. F., & GAVIS, J. (1972): A review of the arsenic cycle in natural waters. Water Res. **6**, 1259—1274.

HEYMAN, U. et al. (in prep.): The use of limnocorrals for testing ecosystem response to arsenic.

HUBER-PESTALOZZI, G. (1942): Das Phytoplankton des Süßwassers. Systematik und Biologie. In: THIENEMANN, A. (ed.), Die Binnengewässer. Bd. XVI, Part 2: 2. Stuttgart.

LANDNER, L. et al. (1982): Systems for Testing and Hazard Evaluation of Chemicals in the Aquatic Environment "ESTHER". A background paper and a programme outline as a basis for the research plan in preparation. Swedish National Environment Protection Board. Bulletin. SNV PM 1631, Stockholm, 81 pp.

MAEDA, S., NAKASHIMA, S., & TAKESHITA, T. (1985): Bioaccumulation of arsenic by freshwater algae and the application to the removal of inorganic arsenic from an aqueous phase. Part II. By *Chlorella vulgaris* isolated from arsenic-polluted environment. Separation Science and Technology **20**, 153—161.

MARGALEF, R. (1983): Limnologia. Barcelona.

NAUWERCK, A. (1963): Die Beziehungen zwischen Zooplankton und Phytoplankton im See Erken. Symb. Bot. Upsal. **17**(5), 1—163.

NORIN, H., CHRISTAKOPOULOS, A., RONDAHL, L., HAGMAN, A., & JACOBSSON, S. (1987): Identification and quantification of arsenocholine and acetylarsenocholine in trace amounts in biological material by use of pyrolysis gas chromatography/mass spectrometry (accepted by Biomedical and environmental Mass Spectrometry).

PLANAS, D., & HEALEY, F. P. (1978): Effects of arsenate on growth and phosphorus metabolism of phytoplankton. J. Phycol. **14** 337—341.

— & LAMARCHE, A. (1983): Lack of effect of arsenic on phytoplankton communities in different nutrient conditions. Can. J. Fish. Aquat. Sci. **40**, 156—161.

PORTER, K. G. (1976): Enhancement of algal growth and productivity by grazing zooplankton. Science **192**, 1332—1333.

ROSÉN, G. (1981): Phytoplankton indicators and their relations to certain chemical and physical factors. *Limnologica* (Berlin) **13**, 263—290.

SANDERS, J. G. (1979): Effects of arsenic speciation and phosphate concentration on arsenic inhibition of *Skeletonema costatum* (Bacillariophyceae). J. Phycol. **15**, 424—428.

— & WINDOM, H. L. (1980): The uptake and reduction of arsenic species by marine algae. Estuarine Coastal Mar. Sci. **10**, 555—567.

— & VERMERSCH, P. S. (1982): Response of marine phytoplankton to low levels of arsenate. J. Plankton Res. **4**, 881—893.

SANDERS, J. (1986): Direct and indirect effects of arsenic on the survival and fecundity of estuarine zooplankton. Can. J. Fish. Aquat. Sci. **43**, 694—699.

SMITH, W., GIBSON, V. R., & GRASSLE, J. F. (1982): Replication in controlled marine systems: Presenting the evidence. In: GRICE, G. D., & REEVE, M. R. (eds.): Marine Mesocosms. Biological and chemical research in experimental ecosystems. New York.

THURSBY, G. B., & STEELE, R. L. (1984): Toxicity of arsenite and arsenate to the marine macroalga *Champia parvula* (Rhodophyta). Env. Toxicol. Chem. **3**, 391—397.

UTERMÖHL, H. (1958): Zur Vervollkommnung der quantitativen Phytoplanktonmethodik. Mitt. Int. Verein. Limnol. **9**, 1—38.

Address of the authors:

PETER BLOMQVIST
ULF HEYMAN
Institute of Limnology
University of Uppsala
Box 557
S-751 22 Uppsala
Sweden

Limnologica (Berlin) **20** (2), 339—346 Juni 1990

Department of Hydrology, University of Uppsala, Uppsala, Sweden

Environmental Hazard Analysis — Contamination of Nutrients, Mercury and Cesium-137 in Natural Waters

By LARS HÅKANSON

With 4 Figures

Key words: Lakes; Coastal areas; Ecological effect; Dose; Sensitivity; Nutrients; Metals; Mercury; Cesium-137.

Abstract

Environmental hazard analysis would mean different things in different contexts. This work summarizes results from some ongoing Swedish research projects on different types of contamination of limnic as well as marine areas. The paper gives a brief theoretical outline on the central concepts of the "meso-scale-type" of environmental hazard analysis, utilizing examples on eutrophication of coastal waters. The concepts are further substantiated in two subsequent parts dealing with radioactive cesium and mercury. The idea is to illustrate that the basic concepts for ("real" world/ "meso scale") environmental hazard analysis can be used for different substances and different aquatic environments. It is important to give clear, quantifiable definitions of the effect, dose and environmental sensitivity parameters, which should be valid for a defined area and for a defined timespan. All other parameters should be compatible and have the same area and time resolution.

Background and Presuppositions

Environmental hazard analysis would mean different things to different persons. So, the presuppositions have to be properly defined to avoid unnecessary confusion. What, for example, is the environment to be considered? E-flasks, microcosms, lakes or the atmosphere, i. e. micro, meso or macro scale? Different concepts and models are often used by scientists working on these different scales.

Physiological concepts and test methods (on, e. g. liver and cell damage, reduced reproduction) predominate among scientists dealing with the mechanisms of the micro world. Correlations of empirical data and mass-balance models dominate among the scientists and many environmental administrators dealing with the "real" world of rivers, bays and lakes (see, e. g. HÅKANSON 1984). As a bridge between the controlled, simplified environments in the laboratory and the natural environment, there is a growing spectrum of very interesting and important test systems in the "micro-meso scale", like combined laboratory/natural systems and microcosms placed in natural aquatic environments (see, e.g. O'NEILL et al. 1982; CAIRNS & PRATT 1987; LANDNER 1988). Large-scale budget calculations and fugacity models dominate in the macro world dealing with the flow of contaminants between air, soils, vegetation, water and biota (see, e. g. MACKEY & PATERSON 1982).

The micro, meso and macro systems of environmental hazard analysis may be more or less sophisticated, the possibilities to validate the results may vary and the concepts, definitions of parameters, etc., may complement each other to varying extents. Many textbooks, scientific journals and academic schools now exist on these topics. It has been evident for more than a decade that, in spite of the fact that the word "environment" is used and abused in so many contexts, "Environmental science" is a flourishing new branch on the tree of science linking biological and abiotic factors and schools. The aquatic branch of this science deals with problems like acidification, contamination (metals and organics) and eutrophication, which in their own obvious right attract the interest of a growing number of concerned citizens, politicians, journalists and a new brand of scientists. It deals with the "diagnosis and medication" of threatened waters and species in analogy with what physicians in the medical faculty are doing for the human body.

This introduction may sound like a prelude to a new textbook advocating a certain school in environmental science. That is not, however, the intention. This work is, instead, meant to summarise results from some ongoing research projects on contamination of lakes and coastal waters. The aim of this paper is to demonstrate that the same general conceptual framework may be used for dose-sensitivity-effect analysis for such diverse substances as nutrients, mercury and radioactive cesium. The concepts can, in fact, be used for any given substance in any defined lake or coastal area. The idea has been first to present the concepts using examples on nutrients in coastal areas, then to demonstrate the ecometric or statistical set-up with Cs-137 as a type substance and, finally, to illustrate how the statistical information may be used for a more detailed causal analysis, using mercury as a type substance.

Brief Theory on Effect-Dose-Sensitivity Concepts

The purpose of this section is to give a short discussion on the fundamental concepts, effect, dose, sensitivity. For a more thorough theoretical account, see HÅKANSON (1984).

The famous VOLLENWEIDER diagram (VOLLENWEIDER 1968 and later versions), and the analysis behind this load diagram, constitute an important base for practically all environmental risk assessments of phosphorus in limnic environments. For marine areas there is no such load diagram available today. In practice, this makes it impossible to conduct relevant evaluations of potential hazards of nutrient emissions to brackish and marine environments (see HÅKANSON et al. 1988a). And it is not possible to make simple adjustments of the VOLLENWEIDER diagram to marine areas, because for such areas nitrogen is often the most limiting nutrient for the bioproduction (see ROSENBERG 1986) and because the complex conditions in the sea often have a profound impact on the conditions in a given coastal area.

To clarify the "road" between nutrient dose (D) and eutrophication effect (E), for a defined coastal area for a defined timespan, one must also account for the environmental sensitivity (Wi):

$$E = f(D, Wi) + R \quad (1)$$

The effect term (E)

In many contexts where environmental hazards are concerned, the term ecological effect is used very loosely. What is actually meant by ecological effect? It is important to give a clear definition of the E-term. It should be a quantitative value for a defined area for a defined timespan. All other dose and sensitivity parameters should be compatible and have the same area and time resolution. In the VOLLENWEIDER model the E-term is defined by concepts like hypertrophy (extremely productive lakes), eutrophy, mesotrophy and oligotrophy, which are related to defined phosphorus concentrations in the water. Many other E-terms are discussed in contexts of eutrophication, e. g. focussed on the bottom fauna community, algae, fish community, primary productivity, water transparency or the oxygen conditions. From the perspective of the environmental authorities, one may advocate that the oxygen status of the water is a critical effect term, which is also quite easy to measure. But then it is not primarily the mean O_2-concentration of the water mass that is requested, but rather the extremely low O_2- concentration. A crucial question is: Where, when, and how often will low O_2-concentrations (<3—4 mg/l) appear in the bottom zone?

The dose term (D)

This term may be given by the load of nitrogen in marine environments, by N and P in brackish waters and by P in fresh waters. To calculate the potential increase of nutrient concentration one must know the background concentration of the nutrient in the water mass, one must quantify the ongoing emissions per time unit as well as the retention time of the water in the given area. The most relevant dose factor may not be the total dose, but rather the biologically available dose. The nutrient load can, e. g. be divided in two posts: Particulate load, which is not directly biologically available, and dissolved load, which is directly available to enhance the primary productivity. The dose concept can be linked to the mass-balance equation. This emphasizes the strict scientific rule system concerning definitions of area and time that should be applied in these contexts.

Environmental sensitivity (Wi)

Different coasts respond differently to one and the same dose. Besides the characteristics that have to be known to determine an increase in concentration from an emission, it is also of great importance to know the bottom dynamic conditions, i. e. if areas of erosion, transportation and accumulation prevail in the given coastal area. If the nutrients are emitted in an area dominated by erosion and/or transportation processes, the possibilities for the O_2-demanding organic material to accumulate will be minimized and the material will be spread to surrounding waters. If they are emitted to a bay dominated by fine sediments and accumulation processes, the areal spread will be smaller. Possible O_2-problems will be accentuated if the bottom water exchange is small. Resuspension of fine material can only, according to definition, occur on areas of erosion and transportation. During storm events, when the wave base is lowered and the areas susceptible to resuspension are comparatively large, there is generally a large input of oxygen (from oxygen saturated surface water) and also an enhanced oxygen consumption from resuspended material. The internal loading of nutrients in coastal areas is also enhanced by resuspension (FLODERUS & HÅKANSON 1989).

Consequently, an area-typical sensitivity factor, which in quantitative terms should describe how a given coast responds to a given nutrient dose, must be regarded as a function of several variables, like coastal volume, water retention time, mean depth, and areas of erosion, transportation and accumulation.

The unaccountable residual term (R)

This term is included to emphasize the importance of quantifying the relationships between dose, sensitivity and effect, and the impossibility in these contexts to know everything to 100%. The R-term may be given by the statistical measure called the degree of explanation (= coefficient of determination, r^2, which is defined as the square of the correlation coefficient, r). It can be used to quantify how well a model can predict empirical data. If the degree of explanation is 100% ($r^2 = 1$), then there is perfect agreement between model data and empirical data. Such models are not obtainable in ecological contexts. It is, however, always important to try to declare how well a given model can predict the reality, since there is a great difference between models that yield a 50% and a 80% degree of explanation.

The concepts used for nutrients in coastal areas would also hold for most contaminants in aquatic environments. This will be illustrated for cesium-137 and mercury.

Empirical Correlations — Cesium-137

Comprehensive investigations are presently being carried out in 41 lakes in northern Sweden (Gävleborg and Västernorrland counties), which received a significant fall-out of cesium from Chernobyl. Subsequently, when discussing cesium-137 in Swedish lakes, we will use the following effect, dose and sensitivity parameters. For a more thorough account of the presuppositions and set-up of the cesium work, see HÅKANSON et al. (1988b).

Effect measured as concentration of cesium-137 (Csa) in small perch (1+, i. e. one year old perch). The perch is caught in September. Collective muscle samples from about 15 fishes from different parts of the lakes have been used. Thus, the Csa-value represents an integrated value for a given lake for the productive season. The dose and sensitivity factors should have the same area and time resolution.

Dose measured as: (1) Primary dose, i.e. fall-out of cesium-137 from the Chernobyl accident of Spring 1986, (2) lake dose measured as mean concentration of cesium-137 in water (from water samples from central lake sites sampled every or every second month) and (3) lake dose of cesium-137 as measured in material collected in sediment traps placed 2 m above the lake bed and 2 m below the lake surface at a central site in the lake during the productive period (June—Sept.).

Table 1. Statistical data for defined limnological parameters (see HÅKANSON et al. 1988b, for further information)

Parameter	n'	n	Mean	S.D.	Min.	Max.
pH	12	41	6.0	0.40	5.1	6.8
alk (meq/l)	12	41	0.07	0.07	0	0.37
cond (mS/m)	12	41	3.0	1.17	1.6	8.2
color (mg Pt/l)	12	41	102	41.4	35	210
totP (μg/l)	12	41	11.7	4.7	5	29
Fe (μg/l)	6	41	629	433	59	
Ca (meq/l)	12	41	0.17	0.07	0.08	0.48
Ca + Mg (meq/l)	12	41	0.23	0.10	0.09	0.64
K (μeq/l)	5	23	11.1	7.1	5.9	36.7
Cssoil (kBq/m²)	1	41	25.2	20.8	2.5	70
Csa (Bq/kg ds)	2	41	29424	60951	514	524045
Csf (Bq/kg ds)	4	15	38521	82565	410	576000
Csw (Bq/l)	3	15	0.314	0.25	0.03	0.79

n' = number of samples per lake; n = number of lakes; Cssoil = cesium-137 in soils; Csa = Cs in 1+perch; Csf = Cs in sediment traps; Csw = cesium in lake water (data from April to Sept. 1987); ds = dry substance or sediments; ws = wet substance.

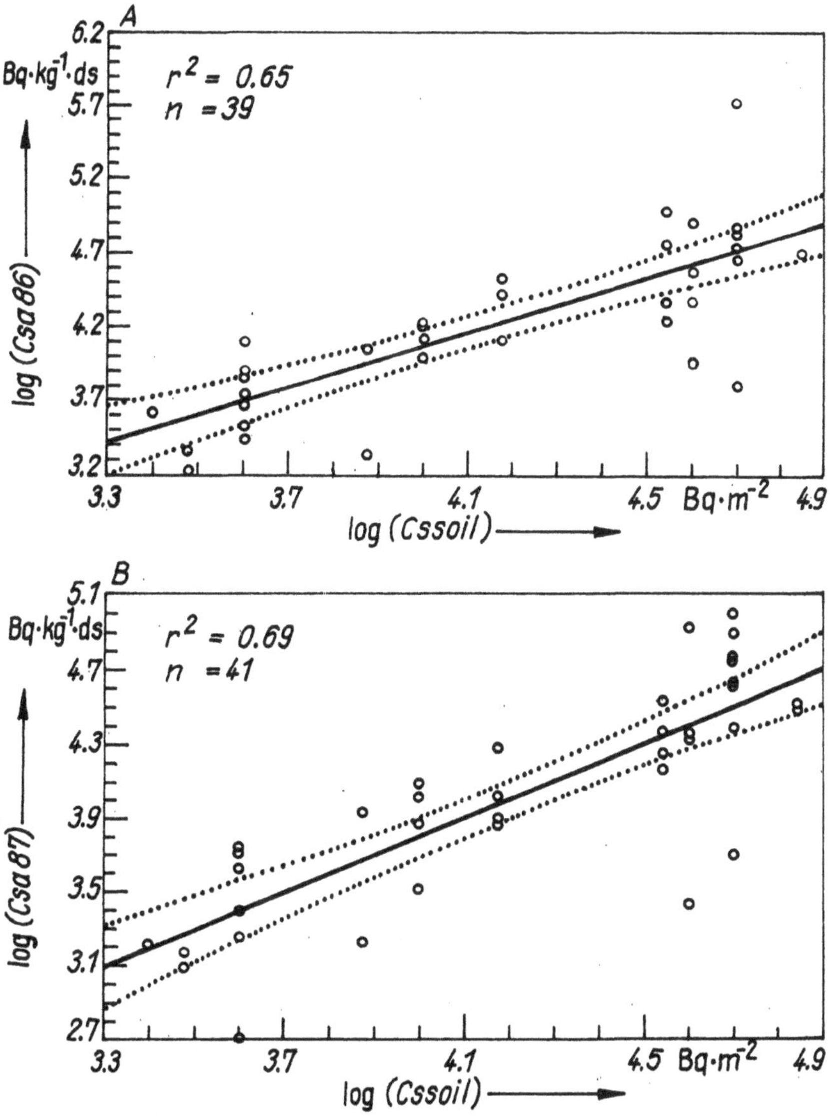

Fig. 1. A. The correlation between cesium-137 in perch 1986 (1+) and the primary dose (Cssoil). B. The correlation between cesium-137 in perch 1987 (1+) and the primary dose (Cssoil)

Sensitivity measured as mean lake values of a broad set of limnological data (pH, transparency, color, Ca, Fe, P, alkalinity, etc.), morphometrical data (volume, depth, bottom types, etc.) and environmental parameters on the drainage area (water discharge, soils, land use, bedrock, etc).

Here we will discuss some multiple regressions to see what factors seem to best explain (from a statistical perspective) the variation in the effect parameter. It should be noted that the trick is not to make multiple regressions per se, but rather to have access to compatible data for the regression analysis. Many factors could potentially affect the concentration of cesium-137 in perch and the idea is to try to quantify what factors seem to be most important. This knowledge may be used directly by environmental authorities to select suitable measures to reduce the environmental hazard. It may also be used by scientists to pinpoint what factors would be worthwhile subjecting to a more thorough causal analysis. The subsequent section on mercury will illustrate how results from statistical exercises can be used for a more thorough causal analysis.

Table 1 summarizes limnological and cesium data from the 41 studied lakes.

Before we discuss the results of the correlation analysis, it may be interesting to give the following general background information on cesium in lakes.

On cesium in lakes

Most studied lakes have perch with Csa-values far higher than the Swedish "blacklisting" limit of 1500 Bq/kg ws [ws = wet substance]. There is a clear linkage between the primary dose (Cssoil) and the Csa-values, see Fig. 1.

The Csa-values have decreased from 1986 to 1987. The average value for all lakes has dropped from 9800 Bq/kg ws to 5040, i. e. with 49%; the max.-value from 131000 to 24500 (= 81%). Generally, the decrease is between 5 and 90%. We have not found that any specific factor can be linked to the decrease in Csa (dCsa) between the two years.

There exist very marked positive correlations (see Table 2) between the given dose parameters and between the dose parameters and the effect parameter.

The concentrations of cesium in sediment-trap material have, on average, decreased by 80%, as compared to 30—50% for the perch. In the sediment traps, the concentrations seem to have decreased relatively more in lakes with a high primary dose. This trend is not apparent for perch.

Table 2. Degree of explanation (r^2) between the dose parameters cesium-137 in water (Csw), cesium in sediment traps (Csy = surface sediment traps and Csb = bottom sediment traps) and the effect parameter, cesium in perch (Csa) for 1986 and 1987

	Csa86		Csa87	
	r^2	n	r^2	n
Csy86	0.82	15	0.90	15
Csb86	0.73	15	0.89	15
Csy87	0.62	15	0.61	15
Csb87	0.44	15	0.48	15
Csw	0.63	15	0.67	15

The following ratios have been determined for the dose parameters and the effect parameter (average conditions):

Csy86/ Csa86	Csb86/ Csa86	Cssed/ Csa86	Csy87/ Csa87	Csb87/ Csa87	Csw/ Csa87
1.40	2.69	0.25	0.51	0.62	0.000085

Thus, in relative terms, the concentrations of Cs-137 in water are very low. These ratios are rather stable throughout the year. The increase in Cs-content in perch relative to the values in sediment-trap material for 1987 are evident just as the change, the levelling effect, between the years concerning cesium in bottom and surface sediment traps.

Correlations

Correlations between cesium parameters and drainage area parameters, Cs-parameters versus each other and Cs-parameters versus limnological parameters show, among other things (HÅKANSON et al. 1988b):

- that there is a cluster with strongly internally correlated lake parameters that include alkalinity, conductivity, K-content and water hardness (= Ca + Mg); correlation coefficient > 0.85;
- that also total-P and pH are linked to this group (r about 0.5); and
- that the Cs-content in perch and the cesium dose to the lake is negatively correlated to this group.

The primary dose (Cssoil) explains 65—68% of the variability in the effect term [log (Csa)]; the degree of explanation increases markedly (to about 80%) if one also accounts for a lake-typical sensitivity factor (any of the cluster factors); the degree of explanation increases further (up to 85% for 1987) when also the phosphorus content is taken into account (see Table 3).

Table 3. Results illustrating how the degree of explanation (r^2) for the variability of the effect parameter (Csa) increases as information on sensitivity parameters (cond., hardness and totP) is supplemented to the primary dose factor (Cssoil)

Parameter	Csa86		Csa87	
	r^2	n	r^2	n
Cssoil	0.65	39	0.68	41
Cssoil, cond	0.79	39	0.83	41
Cssoil, CaMg	0.78	39	0.83	41
Cssoil, CaMg, totP	0.79	39	0.85	41

An equation (2) which (from the given statistical presuppositions) provides the best possible degree of explanation ($r^2 = 0.85$, $n = 41$) has been derived:

$$\log(\mathrm{Csa87}) = 0.92 * \log(\mathrm{Cssoil}) - 1.57 * \mathrm{CaMg} - 0.02 * \mathrm{totP} + 0.75 \quad (2)$$

where: Csa87 = the mean lake Cs-content 1987 in 1+perch in Bq/kg ds (ds = dry substance); Cssoil = the primary dose in Bq/m^2; CaMg = yearly mean water hardness in meq/l; and totP = mean total phosphorus concentration in μg/l.

The relationship between observed and predicted data is illustrated in Fig. 2. At the outset of the project we did not anticipate that it would be possible to obtain a simple formula that would provide such a high degree of explanation (85%), since so many factors could, potentially, affect the Cs-content in perch.

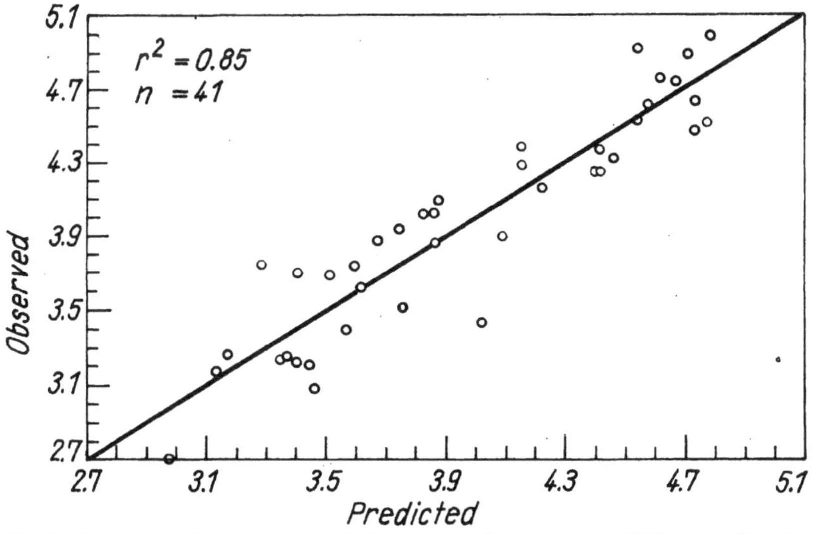

Fig. 2. The relationship between empirically observed data on cesium-137 in perch (1+) in 1987 and data predicted from eq. 2. Logarithmic values in Bq/kg dry weight

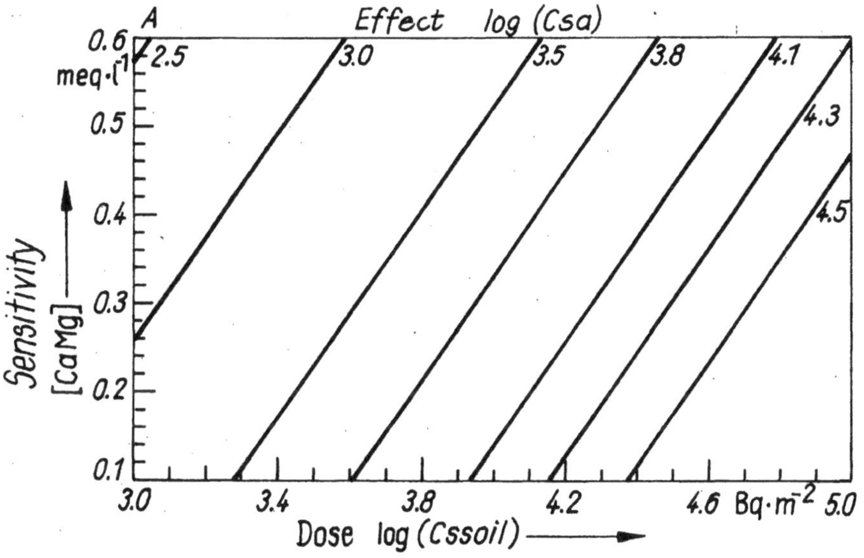

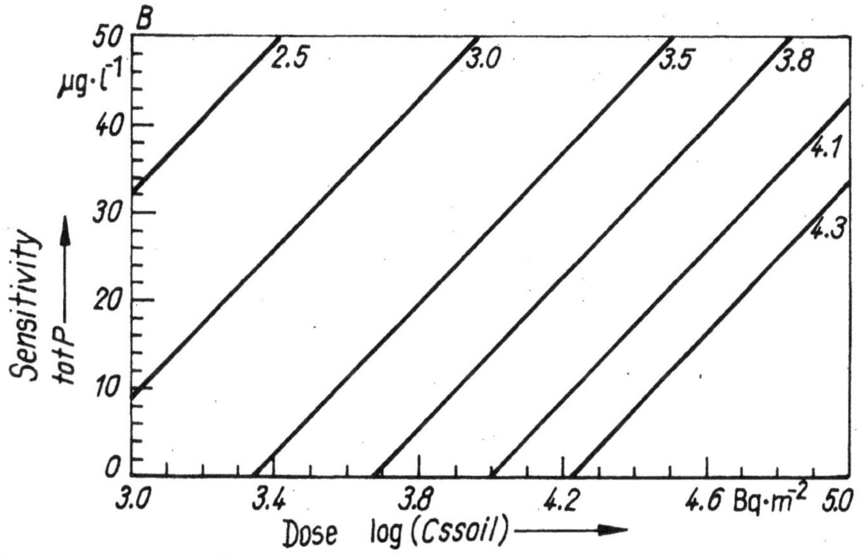

Fig. 3. Dose-sensitivity-effect-diagram.
A. The relationship between the primary dose (Cssoil), lake sensitivity (as given by water hardness, CaMg) and the effect parameter, i. e. cesium-137 in 1 + perch. The concentration of totP is, in this nomogram, held constant at 10 µg/l.
B. The relationship between the primary dose (Cssoil), lake sensitivity (as given by totP-concentration) and the effect parameter, i. e. cesium-137 in 1 + perch. The water hardness is, in this nomogram, held constant at 0.2 meq/l

Eq. (2) can predict values in the given cesium range, but it should not be used uncritically, since it is based on an empirical set of data which only covers certain lake types (see Table 1). In Fig. 3 we have drawn a dose-sensitivity-effect diagram (from eq. 2). From this figure one may, for a hypothetical lake, e. g. with a primary dose of 104.2 (= 15 800 Bq/m²), a hardness of 0.13 meq/l, calculate a probable value on the cesium content in perch (1+) to 104.3 (= 20 000 Bq/kg ds), i. e. about 5000 Bq/kg ws. Eq. (2) may also be used to estimate, e.g. that a measure that would increase the mean yearly hardness by 0.13 meq/l would imply a reduction of Csa by 40%.

Many factors could potentially affect the concentration of Cs-127 in perch. The statistical calculations have revealed that some of these factors may be more important than others, e. g. the "cluster" parameters (cond., alk., Ca + Mg) and the P-concentration of the lake water. The next step would be to ask **why** would these particular factors be important. To illustrate the subsequent causal analysis, mercury will be used as a type substance.

Mechanisms — Mercury

Background

There are today about 50 000 publications on "Mercury as an environmental pollutant" (cf. TAYLOR 1975). Various abiotic and biological factors of natural water systems have been discussed and correlated to high Hg-content in fish (VERTA 1984). Some of these factors, like pH and lake bioproductivity, have a general importance, others are more associated to local contamination. From a recent study, with data from 1456 Swedish lakes (HÅKANSON et al. 1988c), on the factors linked to the content of mercury in 1-kg-pike (FHg), we may conclude that the following were most important to explain the variability in that effect parameter, FHg ($p < 0.005$):

1. Hgsed, i. e. the mercury dose to the lakes as given by the mean concentration of Hg in surficial sediments (linear correlation coefficient, $r = 0.66$).
2. Mean lake water pH ($r = 0.35$); i. e. pH is the most important sensitivity factor as revealed by these data.
3. Distance from point source and water hardness ($r = -0.29$).
4. Lake water alkalinity and conductivity ($r = -0.26$).
5. Water retention time, size of drainage area and lake surface ($r = -0.23$, -0.20 and -0.19, respectively).

Table 4 summarizes the best-fit linear correlations for one to four "independent" variables; the degree of explanation (r^2) increases from 0.43 for the dose-factor, Hgsed, to 0.68 for the four variables Hgsed, cond., pH and lake area. A formula which (from given statistical presuppositions) provides the best possible degree of explanation ($r^2 = 0.78$) for these four variables has been derived:

$$FHg = -0.25 + ((Hgsed + 600)*10^{-3})/(\ln(cond - 0.2)) + (2.8*(6 - \ln(A + 1)))/(pH - 3.2)^3 \quad (3)$$

FHg = mean Hg-content in 1-kg pike in mg/kg ws; Hgsed = Hg-content in surficial sediment in ng/g ds; cond = mean lake water conductivity in mS/m; A = lake area in km²; pH = mean lake water acidity; ln = the natural logarithm.

Table 4. The variables (1—4) which provide the best degree of explanation (r^2) of the variation in FHg and the number of lakes (n)

Variable	r^2	n
Hgsed	0.43	69
Hgsed, cond	0.56	59
Hgsed, cond, pH	0.64	59
Hgsed, cond, pH, A	0.68	57

Eq. (3) can be applied on lakes with the following characteristics: Hgsed should vary between 50—5000 µg/kg ds, pH between 5.5—8, conductivity between 2—20 mS/m and the lake area should be in the range 0.05—50 km². Most lakes in the Northern hemisphere fall within these ranges. It is a statistical formula which may reveal little about real causal relationships. The exponents, e. g. pH³, may have no real significance beside providing the best possible fit to this set of data.

Since many biological, chemical and physical factors are linked to each other in the limnic ecosystem, and because many of these factors seem to be of importance for the Hg-content in fish, it is not difficult to understand why it has been considered hard to find simple mechanisms explaining why certain lakes have high Hg-content in fish and others have low. BJÖRNBERG et al. (1988) presented a theory which explains why it is logical that pH would be so important in these mercury contexts. The approach, with its coupling between effect, dose and regulatory factors, could serve as model not only for Hg but for other substances as well.

Theory

The theory stresses some chemical reactions that take place outside the fish and are linked to pH and Eh (redox potential) of the lake water. The effect term (E, the Hg-content in fish) is related to a Hg-dose (D) and a regulatory factor (Re); the dose term is given by the amount of Hg discharged to the lake with primarily humic matter; the regulatory factor is given by the activity of sulphide, selenide (and telluride), see Fig. 4.

The S^2-activity and its relation to pH and redox potential (Eh) is of special relevance. H_2S dominates at low pH, S^{2-} at high pH. At the pH values normally found in natural waters (4—9), the S^{2-}-activity is **very** low; 12—4 orders of magnitude lower than the total sulphide concentration. H_2S formed through protonization of S^{2-} will ventilate away from acid waters, which will be continuously depleted of S^{2-}. The activity of Hg^{2+} is primarily determined by the S^{2-}-activity. The solubility constant (Ks) of HgS (s) is 10^{-52}. This is an extremely low constant, and implies that practically all Hg will be in the form of HgS (s) when sulphide is available. The sulphide activity may be reduced by a lowering of pH and/or an increase in Eh. Besides sulphide, selenide and telluride display a similar relationship to Hg (Ks = 10^{-58} and Ks = 10^{-70} respectively).

The Hg^{2+}-concentration in natural waters varies within a wide range but is often in the order of about 5 ng/l (HÅKANSON et al. 1988c). This is about 10—11 molar, which is a high concentration in this context. Such high concentrations can only prevail when the S^{2-} (and/or the Se^{2-})-activity is very low (lower than 10^{-41} and 10^{-47}, respectively). Thus, if waters have naturally high S^{2-}-and/or Se^{2-}-activities, e. g., from sulphidic rocks in the drainage area, or if S^{2-} and/or Se^{2-} is added to the waters, then the Hg^{2+}-activity will

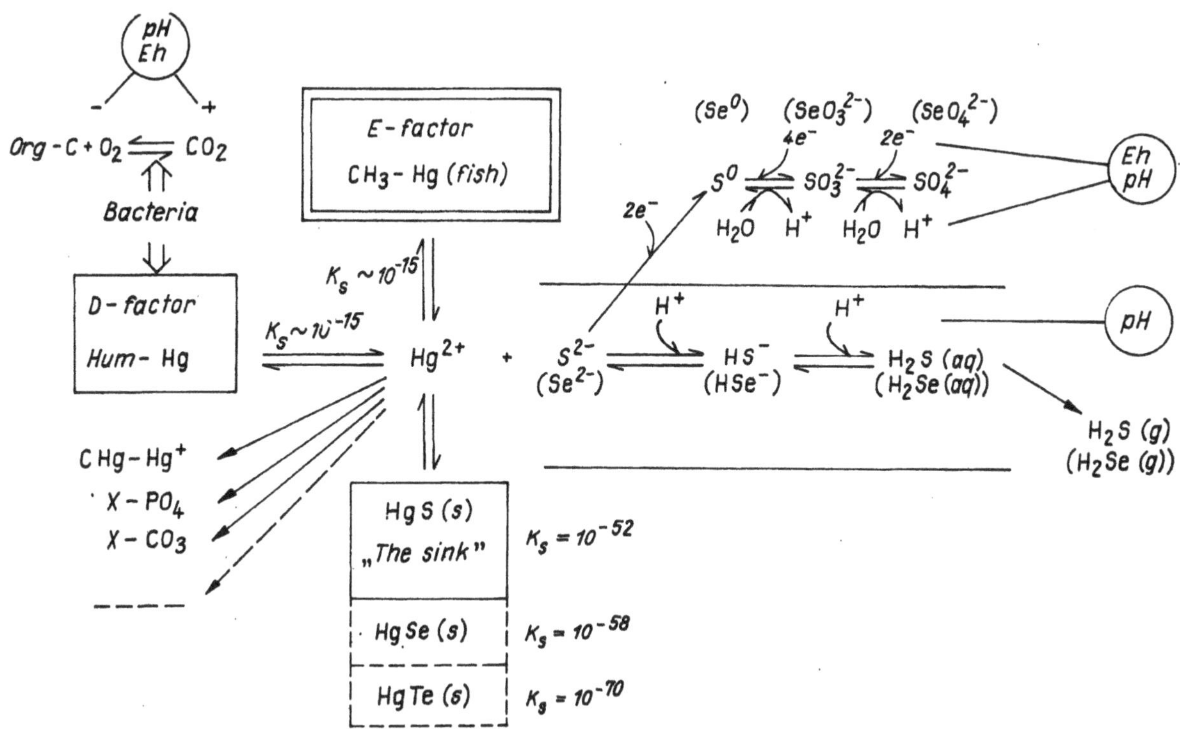

Fig. 4. Diagram illustrating the theory on the mechanisms regulating the relationship between the E-factor (Hg-content in fish), the dose factor (given here as Hum-Hg, i. e. the Hg associated to humic matter) and the regulatory factor (the equilibrium between Hg^{2+} and regulatory ions (S^{2-}, Se^{2-} and Te^{2-}) and the linked relationships with pH and redox potential (Eh). The "sink" illustrates the most effective (Ks < 10^{-52}) trapping of HgS (s), HgSe (s) and HgTe (s). When, e. g. pH decreases, the S^{2-}-activity will decrease very rapidly. This would increase the biologically available fraction, the Hg^{2+}-activity, and the possibilities for Hg^{2+} or CH_3—Hg^+ to enter the fish. An increase in pH will cause the opposite result, i. e. an ultimate decrease in Hg-content in fish, provided that there is sulphide and/or selenide available in the water. The equilibrium constants for all other reactions are much higher than the solubility constants of HgS, HgSe (and HgTe)

be reduced most effectively. This means that it should be possible to treat natural waters with S^{2-} as well as Se^{2-}. In practice, adding S^{2-} would not be ideal since it would be effectively eliminated as H_2S (g) or oxidized (see Fig. 4). This would not, to the same extent, be the case with Se.

A project is presently going on to test this theory more thoroughly. Some results from Lake Öltertjärn can be used to illustrate how rapid and effective the treatment of Se may be. The Se-concentration of the test lake was raised artificially from 0.4 to 2—4 µg Se/l. The increase was achieved by sodium selenite (Na_2SeO_3) slowly leaching from a supporting material (rubber). These data (in mg/kg fish muscle) demonstrate the quick response obtained between 1984 and 1985/86.

	Pike						Perch					
	Hg			Se			Hg			Se		
	Min.	Mean	Max.	Min.	Mean	Max.	Min.	Mean	Max.	Min.	Mean	Max.
	n = 13			n = 3			n = 21			n = 21		
Before	0.74	1.51	2.08	0.50	0.53	0.60	0.10	0.56	1.23	0.12	0.53	0.95
	n = 11			n = 11			n = 46			n = 46		
After	0.34	0.70	1.30	0.13	1.98	5.10	0.03	0.16	0.39	0.31	7.60	22.0

Initially, one may expect that about 70% of the Se added to the lakes would be attached to suspended particles and thus have a limited effect on the Hg-content in fish.

If a lake receives a certain dose of Hg primarily from high molecular weight humic substances, what will happen? There are large numbers of bacteria in the water which use the carbon supplied by the organic material as a source of energy. During the bacterial decomposition, O_2 is consumed and CO_2 (and bacterial biomass) is created. Thus, there exists a link between the oxic conditions, the redox conditions and the activity of S^{2-} and Se^{2-} (i. e., our regulatory ions). When the redox potential decreases, the S^{2-} activity increases and vice versa. But there are also chemical equilibria between the humic Hg-complexes and Hg^{2+}. The equilibrium constants of these reactions are in the order of 10^{10} to 10^{20} (Smith & Martell 1976), which is much higher than the Ks-values for HgS and HgSe. Similarly, there are equilibrium constants between Hg^{2+} and methyl-Hg or Hg-complexes with phosphates, carbonates, etc., which also ought to be in this order of magnitude. Assuming that methyl-Hg in organisms/fish is connected to Hg^{2+} in water via a chain of equilibrium reactions, this means that the presence of S^{2-} or Se^{2-} will cause reduction of Hg-levels or uptake in biota. The exact nature of these equilibrium reactions (inorganic, organic or biologic) will be important primarily for the rate of uptake or depletion of Hg. As experiments have shown, this rate is rapid enough to make addition of selenium a practical remedy for high Hg levels in fish.

The theory does not address many important issues like: Where does the methylation of mercury takes place? Is Hg intake via food more important than Hg intake via gills and skin?

It should be stressed that the theory focusses on chemical reactions taking place outside the cell that are linked to the pH of the water, which has been demonstrated by ample empirical data to be an important sensitivity factor.

References

BJÖRNBERG, A., HÅKANSON, L., & LUNDBERGH, K. (1988): A theory on the mechanisms regulating the bioavailability of mercury in natural waters. Environ. Poll. **49**, 53—61.

CAIRNS, Jr., J. & PRATT, J. R. (1987): Ecotoxicological effect indicies: A rapidly evolving system. Wat. Sci. Tech. **19**, 1—12.

FLODERUS, S., & HÅKANSON, L. (1989): Resuspension, ephemeral mud blankets and nitrogen cycling in Laholmsbukten, SE Kattegatt. Hydrobiologia **176/177**, 61—75.

HÅKANSON, L. (1984): Aquatic contamination and ecological risk. An attempt to a conceptual framework. Wat. Res. **18**, 1107—1118.

— ERVIK, A., MÄKINEN, T., & MÖLLER, B. (1988a): Basic concepts concerning assessments of environmental effects of marine fish farms. Nordic Council of Ministers **90**, 103p.

— ANDERSSON, T., NILSSON, Å., NOTTER, M., & NEUMANN, G. (1988b): Cesium in perch in lakes in northern Sweden after Chernobyl — present situation, relationships, trends. National Swedish Environmental Protection Board, Solna, SNV, Report 3497 (in Swedish, with English summary).

— NILSSON, Å., & ANDERSSON, T. (1988c): Mercury in fish in Swedish lakes. Environ. Poll. **49**, 145—162.

LANDNER, L. (ed.) (1988): Approaches to advanced hazard assessment of chemicals in the aquatic environment (in prep.).

MACKEY, D., & PATERSON, S. (1982): Fugacity revisited. Environ. Sci. Technol. **16**, 654—660.

O'NEIL, R. V., GARDNER, R. H., BARNTHOUSE, L. W., SUTER, G. W., HILDEBRAND, S. G., & GEHRS, C. W. (1982): Ecosystem risk analysis: A new methodology. Environ. Tox. & Chem. **1**, 167—177.

ROSENBERG, R. (ed.) (1986): A review of eutrophication in marine waters surrounding Sweden. National Swedish Environment Protection Board, Solna, SNV Report 3054, 137 p.

SMITH, R. M., & MARTELL, A. E. (1976): Critical stability constants. Vol. 4. New York.

TAYLOR, D. (1975): Mercury as an Environmental Pollutant. 4th ed., Birmingham Imperial Chemical Industries Ltd.

VERTA, M. (1984): The mercury cycle in lakes; some new hypotheses. Aqua Fennica **14**, 215—221.

VOLLENWEIDER, R. (1968): Scientific fundamentals of the eutrophication of lakes and flowing waters, with particular reference to nitrogen and phosphorus as factors in eutrophication. Rep. OECD/DAS/SCI/68.27, Paris.

Address of the author:

LARS HÅKANSON
Department of Hydrology
University of Uppsala
V Ågatan 24
752 20 Uppsala
Sweden

Limnologica (Berlin) **20** (2), 347—359, Juni 1990

Swedish Environmental Research Institute, Stockholm, Sweden

The Role of Sediments as Sink or Source for Environmental Contaminants — a Case Study of Mercury and Chlorinated Organic Compounds

By INGELA ANDERSSON, HELENA PARKMAN & ARNE JERNELÖV

With 3 Figures

Key words: Mercury; Chlorinated organic compounds; Sediment; Transformations; Transports.

Abstract

This is a review-article on the role of lake sediments for the distribution of two groups of compounds which are important environmental contaminants in the Nordic countries — mercury and chlorinated organic compounds. The article deals with distribution of the pollutants and processes occurring within the sediments of significance for transformations of the compounds — transformations that might lead to the formation of more toxic forms. Different mechanisms of transports between sediment and water, such as diffusion, turbulence, gas convection and bioturbation are reviewed. Accumulation in organisms migrating out of the sediments is also discussed as a pathway for toxic compounds from the sediment to the water.

1. Introduction

Sediments play an important role for the distribution of different compounds in aquatic systems. They may serve as a sink which lowers the concentrations in the water column but they may also serve as a source. The most studied flux between sediment and water in limnic systems is the phosphorus flux. In the early fourties MORTIMER (1941, 1942) presented the classical iron-phosphorus model which suggested that when oxic conditions prevail in the water mass phosphorus can associate with iron (III), in one way or another, and precipitate to the sediments. At the same time free iron (II) from deeper reduced sediment layers diffuses upwards and, when reaching oxic surface sediments (Eh > 200 mV), precipitates with (among other substances) phosphorus. In this way sediments could serve as a sink for phosphorus. However, when anoxic conditions appeare in the bottom water, iron(III) will be reduced to iron(II) and subsequently a release of accumulated phosphorus will follow. Under these conditions, the sediments instead serve as source for phosphorus to the water. This model predominated for a long time, but recent research shows a much more complex picture of the exchange of phosphorus between sediments and water. A review article on today's theories has recently been presented by BOSTRÖM et al. (1988). The interest for the exchange between sediments and water was directed towards phosphorus because of its importance for primary production in lakes, but there are also other compounds, e. g. toxic ones, where the knowledge about the flow into or out of the sediment is of great importance. This will be exemplified with a review on two groups of toxic compounds found in amounts of environmental significance in Scandinavian lakes, chlorinated organic compounds and mercury.

Sources, distribution and concentrations

A large number of chlorinated organic compounds are found in the environment — some that have been commercially produced, some that have been formed as byproducts in industrial processes and further some that have been formed in processes occurring in nature. Concerning chlorinated organic compounds, this paper mainly deals with PCBs and the chlorinated organic compounds formed when bleaching pulp with conventional methods (i. e. with chlorine). PCBs have been studied rather intensively during the last decades. The chlorinated material in spent bleach liquors, on the other hand, has been paid attention to more recently. None the less, these compounds may be important environmental contaminants, especially in areas with many pulp industries, like in the Nordic countries.

Before 1967 pulp industries was also one of the main sources for the anthropogenic dispersal of mercury to the environment, but in the seventies and eighties an extensive reduction of industrial and agricultural mercury discharge has been accomplished. Today the biggest source of emission of mercury in Sweden is the incineration of house-hold waste (3300 kg/year) (LINDQVIST et al. 1986). Still, mercury amounts have continued to increase in fish, especially from oligotrophic forest lakes that never have been subjected to direct discharge of mercury. In Sweden almost 10000 lakes ought to be blacklisted according to the mercury concentrations in pike of one kilo (> 1 mg/kg ww) (ANDERSSON et al. 1987). The leaching of earlier deposited airborne mercury from the drainage area is proposed to be the explanation (ANDERSSON et al. 1987; VERTA et al. 1986). In Finland elevated mercury levels have been found in newly impounded reservoirs (e. g. SURMA-AHO et al. 1986) and both increased and decreased mercury concentrations in sediments have been observed as an effect of drainage activity (SURMA-AHO 1986; LIEHU et al. 1986; SIMOLA & LODENIUS 1982). Mercury effectively binds to particles and therefore a very large proportion of total mercury in the lake ecosystem is to be found in the sediments (> 99% calculated by VERTA 1984).

Chloroorganic compounds often show a high lipophilicity and are therefore, like mercury, readily associated with particulate material in water and sediments. This gives the sediment a role as a buffer, which may be benificial during times of high inputs, but in the future, at decreased loading, the sediment is a potential source for the pollutant to the water and organisms. Theoretical models for the partitioning between different compartments of the system can be constructed (e. g. REUBER et al. 1987).

The sediment properties are important for the distribution of hydrophobic pollutants. It has been reported from several investigations (e. g. LYNCH & JOHNSON 1982; KARICKHOFF et al. 1979; STEEN et al. 1978; OLIVER 1987) that organic matter content and particle size of the sediments affect the release of chlorinated organic compounds to the water and to the organisms living in the sediments. When the organic content is high and the sediments mostly contain fine particle size fractions, as is normally the case at accumulation bottoms, the sediments more efficiently sorb the compounds. HÅKANSON et al. (1988) found, in a monitoring program outside pulp mills in the Baltic sea, that elevated levels of EOCl (extractable organic chlorine) were mostly obtained at accumulation bottoms.

Also the vertical distribution of hydrophobic pollutants within the sediments may be influenced by the sediment properties. The surficial sediments normally have a higher organic matter content than lower layers. Mobile, hydrophobic compounds may therefore be concentrated in the surficial sediments. LARSSON (1983) found that the largest amounts of PCBs injected 10 mm below the surface of a sediment core 10 weeks later were found in the uppermost 5 mm of the core. However, a vertical gradient of chlorinated organic compounds found in natural sediments may indicate an increased contamination of recently deposited sediments. Steep vertical gradients between surficial sediments and deeper "pre-industrial" layers have been found in different Nordic areas (SALKINOJA-SALONEN et al. 1981; SALKINOJA-SALONEN 1984; HÅKANSON et al. 1988; PAASIVIRTA et al. 1988).

Accumulation rates for mercury in the sediments of forest lakes have been estimated at between 10 and 50 µg Hg/m^{2}y (REKOLAINEN et al. 1986; LINDQVIST et al. 1984). Two investigations on vertical mercury distribution in sediments from Nordic lakes show elevated concentrations (1.5 to 6.5 times) in the surface layers, compared to deeper "preindustrial" layers (REKOLAINEN et al. 1986; JOHANSSON 1985) (Fig. 1), and concentrations in the surface layers range from 0.02 to 0.60 µg Hg/g dw. In these investigations mercury concentrations were positively correlated with organic content of the sediments.

When analysing chlorinated organic compounds in sediments many different methods are used. This makes it difficult to compare the results from different investigations. However, several investigations show that elevated levels of chloroorganics are found near industries producing bleached pulp with conventional methods. As mentioned, HÅKANSON et al. (1988) found high concentrations of EOCl in surficial sediments in the Bothnian sea outside pulp mills. However, low concentrations

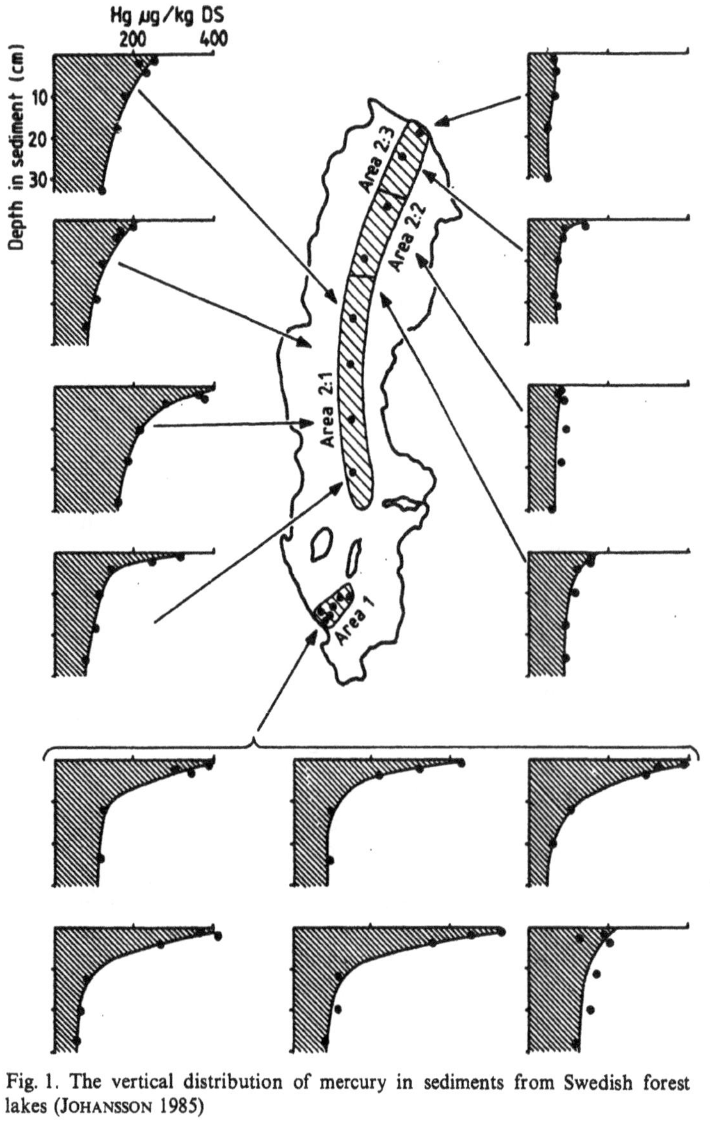

Fig. 1. The vertical distribution of mercury in sediments from Swedish forest lakes (JOHANSSON 1985)

of EOCl were also found in sediments far from any pulp mill. The levels of chloroguaiacols, dioxins and furans in the sediments seem to be correlated with the concentrations of EOCl (KIERKEGAARD & RENBERG 1987). High levels of chloroorganics in sediments have also been found near pulp mills in Lake Mjøsa, Norway (CARLBERG et al. 1987), and in some Finnish lakes (PAASIVIRTA et al. 1988: SALKINOJA-SALONEN et al. 1981). PAASIVIRTA et al. (1988) found that chlorophenolic compounds could be divided into two groups; one, consisting of chlorinated guaiacols and catechols, that only were detected downstream a pulp mill and another, consisting of several chlorinated phenols, that also could be detected upstream the mill. SALKINOJA-SALONEN et al. (1984) likewise found pentachlorphenol and tetrachlorphenol in the sediments of a lake in an unpolluted area. It is possible that the presence of these compounds is due to airborne distribution.

Actual measurements of PCBs in sediments are scarce (SÖDERGREN & LARSSON 1982). In 1980—1982 LARSSON (1984a) studied the sedimentation of PCBs in two southern Swedish lakes and in the southern Baltic (HANÖ-Bight). The concentrations of PCBs in the surface sediments were 0.9 ng/g dw in Hanö Bight, and 7.9 ng/g dw and 22.0 ng/g dw respectively, in the two lakes. The content of PCBs was higher in the sedimenting material than in the surface sediments. Another study in Lake Järnsjön, a small Swedish lake situated downstream a mill based on recycled paper, showed that the surficial organogenic sediments contained 14—100 µg PCBs/g dw. In river Emån 11—42 g of PCBs per day leached from the sediments to the water (LARSSON & RYDING 1988).

Concerning mercury, total mercury and methyl mercury are the fractions that have been considered to be of environmental significance. In the seventies total mercury was measured with neutron activity analysis. This was a reliable but quite expensive method, while methyl mercury analyses were comparatively cheap. A quicker and less expensive method for analysing total mercury was the flameless AAS (Atomic Absorption Spectrophotometry). After the introduction of this method researchers have tended to solely measure total and not methyl mercury in their investigations. Maybe the availability of a convenient technique has influenced our scientific approach to the mercury problem?

In monitoring work it is desirable to get a quick idea of the extent of mercury pollution. For this purpose models to estimate mercury concentrations in aquatic organisms have been constructed from easily measured parameters (e. g. HÅKANSON 1980, 1984; ANDERSSON et al. 1987; LANGSTON 1985). These models are based on total mercury concentrations in the sediments as the pollution factor. Inclusion of other abiotic factors like pH, drainage area and organic content of the sediments have improved the accordance with "true" measured concentrations. VERTA (1984), though, measured higher concentrations than the estimated ones in fish from some brown-water lakes. On the other hand very low concentrations were measured in fish from lakes in a mine district, where the sediments were heavily polluted with mercury and other metals. In this area the lakes with the highest mercury concentrations in the sediments showed the lowest concentrations in pike muscle (LINDESTRÖM & GRAHN 1982).

The biologically important form of mercury is methyl mercury (e. g. JERNELÖV et al. 1975). This form has, in comparison with other mercury compounds, a longer biological half-life and therefore biomagnifies to a higher degree in food chains. Methyl mercury is just a minor part of the total mercury pool in the environment, and is mainly a product of microbial activity. For example, climatic variations affect this activity. Other factors important for the accumulation in organisms are, except for the available mercury and methyl mercury pool, also biological factors like growth rate (biologic dilution), food chains and life cycles of prey organisms (MEILI 1988).

All these circumstances imply the necessity of understanding the biotic as well as the abiotic processes involved in the turnover of pollutants. Not least important is the pool of the pollutant deposited in the sediments. Do the sediments serve as a sink or a source? How is the turnover affected by biological and chemical changes due to, e. g., acidification?

2. Degradation and transformation of chloroorganic compounds

Some chlorinated organic substances, such as PCBs, have very high persistence against degradation and transformation. They have therefore been widely spread in the environment and accumulated in living organisms. Spent bleach liquors from pulp bleaching contain many different compounds. Some of these are degraded or transformed comparatively easily (e. g. MCKAUGE et al. 1988; YAMASHITA et al. 1987) whereas others are more persistent. An investigation outside pulp mills producing bleached chemical pulp showed that the major part of the extractable chlorinated material in the sediments differed from that in spent bleach liquors (DE SOUSA et al. 1988). Only small quantities of chlorinated phenolic compounds (Fig. 2) and high molecular mass material originating from spent bleach liquors could be detected. The discrepancy between the chlorinated compounds in the sediments and in the spent liquors suggests either that the material had been transformed on its way to the sediments or in the sediments, or that the chlorinated material in the sediments only to a minor extent originates from the spent bleach liquors.

2.1. High molecular mass material

When bleaching pulp about 70% of the spent bleach liquor from the prebleaching stage (C) and 95% from the first alkali extraction stage (E) consist of high-molecular mass material, i. e. MW > 1000 (LINDSTRÖM et al. 1981). High molecular mass material (chlorolignin) has been regarded as rather harmless since the molecules are too big to penetrate biological membranes. However, recent investigations have spread light on the degradation and transformation of this material. As a result of the transformation, low-molecular material which may be toxic to living organisms could be formed. ERIKSSON et al. (1985) showed that chlorolignins in sterile solutions were slowly decomposed chemically to yield chlorinated catechols and guaiacols. NEILSON et al. (1983) and ERIKSSON et al. (1985) also showed that chlorinated veratrols are formed when chlorolignins are incubated with bacteria, obtained from pulp mills, or with the white-rot fungus *Sporotrichum pulverulentum*. However both authors pointed out that it was not possible to determine whether the lignin was depolymerized by the bacteria, or if the bacteria methylated low concentrations of

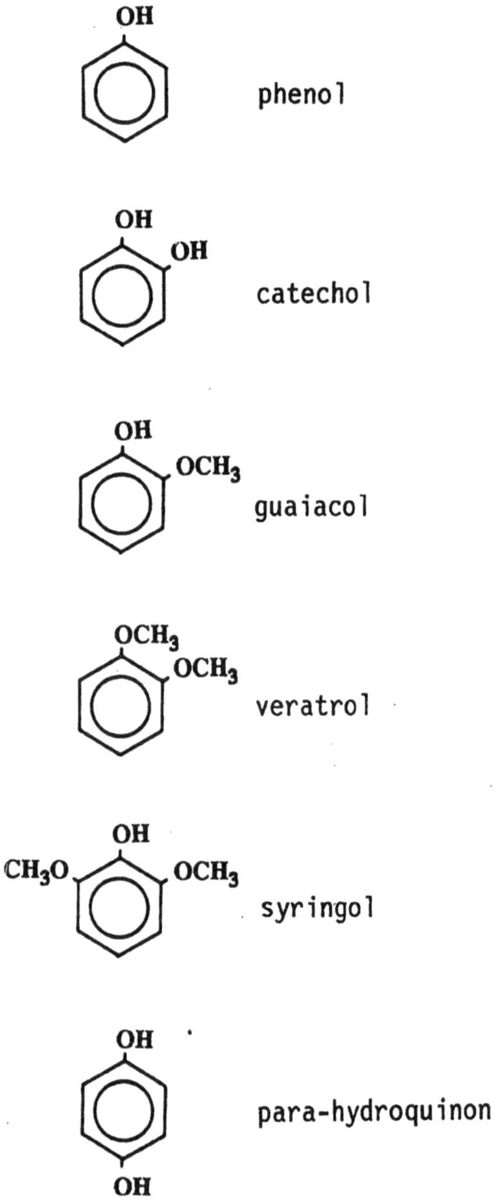

Fig. 2. The structural skeleton of some of the compounds mentioned in the text

chlorinated guaiacols slowly released from the chemically unstable substrate.

ERIKSSON & KOLAR (1985) also investigated the complete degradation of ^{14}C-labeled chlorolignins by measuring evolved $^{14}CO_2$. They found that bacterial mixtures, isolated from aerated lagoons receiving spent bleach liquors, within three months had converted less than 4% of the material, whereas *S. pulverulentum* degraded about 40% of the same material within two months.

2.2. Chlorophenolic compounds

The low-molecular mass material formed when bleaching pulp may be divided into neutral, phenolic and acid compounds. REMBERGER et al. (1986) found that chlorophenolic compounds were readily sorbed to a natural sediment. The degradation and transformation of these compounds have been studied in several investigations.

The bacterium *Rhodococcus chlorophenolicus* has been isolated from a pentachlorophenol enrichment culture inoculated from lake sediment (APAJALAHTI et al. 1986). This bacterium has been shown to degrade several chlorinated phenols, guaiacols and syringols (APAJALAHTI & SALKINOJA-SALONEN 1986; HÄGGBLOM et al. 1985; HÄGGBLOM et al. 1988). It was found that the most toxic guaiacols in general were most readily biodegraded (HÄGGBLOM et al. 1985). The chlorinated guaiacols which carry one methyl substituent and the chlorinated syringols which carry two methyl substituents, were degraded more slowly than the corresponding chlorophenols (HÄGGBLOM et al. 1988). A suggested sequence for the complete dechlorination of pentachlorphenol by *R. chlorophenolicus* has been presented by (APAJALAHTI & SALKINOJA-SALONEN (1987): The degradation is initiated by a hydrolytic and dechlorinative para-hydroxylation which leads to the formation of tetrachlorohydroquinone. This compound is finally converted to 1,2,4-trihydroxybenzene in a reaction involving both hydrolytic and reductive dechlorinations HÄGGBLOM et al. (1988) proposed that the initial attack on tetrachloroguaiacol and trichlorosyringol is a dechlorinative hydroxylation, which leads to the formation of chlorinated para-hydroquinones.

Biological transformations of chlorinated guaiacols, catechols and veratroles added to natural sediments — taken from localities near factories producing fully bleached chemical pulp — have been investigated by REMBERGER et al. (1986). When the samples were incubated aerobically, 4,5,6-chloroguaiacol was O-methylated to the corresponding chloroveratrol with a yield of 25%. However, when incubated anaerobically de-O-methylation of both chloroguaiacols and chloroveratrols took place and resulted in the formation of chlorocatechols. These were not completely stable under anaerobic conditions, but their further fate is not known.

It is plausible that O-methylation serves as a detoxification mechanism for microorganisms (NEILSON et al. 1985). However, the veratrols formed have been shown to be concentrated in fish and they are at least as toxic as their precursors (NEILSON et al. 1984). This implies that not only the compounds found in the bleachery discharge, but also the metabolites formed, must be considered in environmental hazard assessment.

3. Factors affecting the association and mobilization of mercury

3.1. Inorganic mercury

Many authors have concluded that mercury in sediments is associated with **organic matter**. This has been shown, in field studies, with correlations between mercury content and organic matter content (e.g. JOHANSSON 1988; REKOLAINEN et al. 1986) and in laboratory experiments (REIMERS et al. 1974; HÅKANSON 1972; BOURGH & SCHINDLER 1985). JACKSON et al. (1980) found a large fraction of mercury bound to humic substances and mainly to humic acids (MW > 150000). THANABALASINGAM & PICKERING (1985a) observed that as most, at pH 4—4.5, about 98% of the added mercury was adsorbed to humic acids. HÅKANSON (1972) showed an increased rate of mercury bound to deposited organic material with decreased pH and SCHINDLER et al. (1980) observed that acidification caused an increased proportion of ^{203}Hg binding to particulates in a whole-lake experiment. On the other hand JACKSON et al. (1980) suggest that acidification can inhibit the assimilation of humic complexes into the sediments. REIKOLAINEN et al. (1986) found significantly negative correlations between mercury content in the sediments and pH of the water, but at the same time organic matter content in the sediments correlated negatively with pH. In contrast mercury content correlated positively with pH in sediments in another Finnish investigation, based on material from 15 lakes (MYLLYMAA & MURTONIEMI 1986).

Mercury in sediments might also be **associated with inorganic components**, e. g. loosely adsorbed to particles, coprecipitated with iron and manganese oxides or bound to sulphides (or Selenium). When adding mercury into the water above a sediment MILLER (1975) recovered 50—75% in the sediment, in a KOH-insoluble form. This was assumed to be mercury bound to sulphides and adsorbed to oxides and clay particles. JACKSON (1988a) found, when comparing different lakes, that total mercury in sediments gave a highly significant positive linear correlation with S^{2-} and somewhat weaker correlations with organic C, N and iron oxides (extractable with NH_2OHHCl). He also found that lakes with the highest abundance of sand and with the highest Mn/Fe ratios had the lowest total Hg concentrations in the sediments whereas the overlaying water had the highest total mercury concentrations, suggesting release of solubilized Hg from the sediments. Mercury loosely sorbed to the surface of inorganic particulate matter (clays etc.) could be exchanged by other cations, e. g. by H^+ at acidification. It has also been proposed, for other metals, that Ca^{2+} can compete for binding sites when liming lakes (ANDERSSON et al. 1983).

The adsorption of mercury to **manganese oxides** has been studied in laboratory experiments (THANABALASINGAM & PICKERING 1985b). It was concluded that colloidal manganese species could be important for controlling soluble mercury levels. "Uptake was not influenced by pH over the range 5 to 11 but decreased in the 5 to 3 region due to protonation of sites. The amount adsorbed was reduced in the presence of chloride ions... it is proposed that the adsorption mechanism involves migration of the Hg(II) species into the internal structure of the Mn(IV) oxide, where they either undergo ligand exchange with surface hydroxyl groups or are transformed into a solid hydroxy phase."

In 1983 IVERFELDT (1988) measured "reactive" mercury in a vertical profile in the water of fjord Framvaren in Norway and found the highest concentrations at the redox cline. This peak coincided with a peak in dissolved iron concentration which indicates coprecipitation of mercury with **iron oxide** coatings in the oxic surface layer. When reaching the redox cline these coatings undergo reductive dissolution and mercury is released. This relation was also observed in an in situ enclosed sediment-water interface subjected to a sudden change in the redox potential. JACKSON et al. (1980) also found that mercury in the water mass probably was sorbed to peptized colloidal hydrated ferric oxide. Mercury bound to iron oxides might be of importance in oligotrophic lakes, rich in iron, which is a common type of lake in the Nordic countries.

During anaerobic conditions Hg(II) has a high affinity to **sulphide**, with a solubility constant for HgS as low as 10^{-53}. This means that once bound in the HgS form the mercury will not be available as long as reduced conditions are maintained. BJÖRNBERG et al. (1988) present a theory for natural waters which says that the activity of Hg^{2+} is essentially regulated by the activity of S^{2-}. (If selenide or telluride are present they display a similar relationship to Hg). The S^{2-} activity decreases with decreasing pH and increasing redox potential. On the other hand iron sulphides can be solubilized at low pH and the liberated sulphide can precipitate with Hg(II). This was proposed to be the reason for finding lower amounts of total mercury in the **porewater** of acidified systems compared to non-acidified (RAMLAL et al. 1985). In the experiment mentioned earlier MILLER (1975) found higher concentrations at higher pH in the supernate after centrifugation of sediment. In his experiments, the amount of added mercury recovered in the porewater ranged between 0 and 5%, which is in agreement with analyses on "naturally contaminated" sediments (PARKMAN, unpubl. data).

3.2. Methyl mercury

It is well known that the form of mercury most readily accumulated in biota is methyl mercury (e. g. JERNELÖV et al. 1975; RIISGÅRD 1985; WOOD 1984), and it is of importance that we learn to determine the rate of methylation of mercury in sediments. JERNELÖV (1972) once estimated the amount of total mercury in sediments annually converted to methyl mercury to 0.1%. VERTA (1984) constructed, from field data, a model for the mercury distribution in a "typical freshwater lake". He estimated that almost 99% of total mercury would be in the uppermost 5 cm of the sediment, whereas about 1% would be bound to fish. The proportions of methyl mercury, in contrast, would be 49% in sediments and 48% in fish.

In the late sixties and early seventies the interest for speciation of mercury in lakes started, and biotic methylation of mercury was observed (e. g. JERNELÖV 1968). Laboratory as well as field studies were carried out to investigate the processes of mercury methylation. FAGERSTRÖM & JERNELÖV (1972) summarized the knowledge at that time: i). There are two known chemical pathways for the methylation process: 1) a non-enzymatic as well as an enzymatic transfer of methyl groups from methyl cobalamin to inorganic mercury, 2) an enzymatic methylation of mercury bound to homocystein as a "wrong" synthesis of methionin. ii). Mercury is methylated by the organisms as a detoxification process. iii). Mercury can be methylated by both aerobic and anaerobic bacteria and by fungi. iv). The rate of methylation is correlated with the general microbiological activity and is enhanced by the same concentration. v). Formation of methyl mercury is higher the lower the oxygen concentration. vi). The net result of the biological methylation of mercury is either mono- or dimethyl mercury of which the latter is volatile. vii). The overall methylation is not much affected by pH variations between 5 and 9 but so is the relation between the two products formed. At higher pH (> 8) almost all the methylated mercury formed was released as dimethyl mercury. At pH below 5.6 dimethyl mercury was unstable and monomethyl mercury would be formed. viii). When sulphide ions are present methyl mercury can be converted to mercury sulphide and dimethyl mercury. ix). Decompositon of methyl mercury occurs but very little is known about the overall rate of demethylation in nature.

Since then, as for the phosphorus example, we have gained lots of new knowledge showing a more complex picture of interacting factors affecting the methyl mercury production.

Methyl mercury can be formed both biotically and abiotically. **Chemical** (abiotic) **methylation** has been observed, for example, in the presence of humic substances, and especially fulvic acids (NAGASE et al. 1984; WEBER et al. 1985; LEE et al. 1985). This form of methylation is however assumed to be of minor importance compared to biological methylation (e. g. BERMAN & BARTHA 1986b).

Organisms capable of methylating mercury have been found among anaerobes (sulphate reducing bacteria are probably important according to COMPEAU & BARTHA 1985 and KORTHALS & WINFREY 1987), facultative anaerobes and aerobes (JERNELÖV 1987). The toxicity of methyl mercury is considered to be due to its lipid solubility and its high ability to pass through membranes (RIISGÅRD 1985; WOOD 1984; NORDBERG et al. 1985). However, once inside the cell the reactive, inorganic Hg(II) is more devastating, e. g. for the protein synthesis (PAN HOU & IMURA 1982). In view of this the methylation of Hg(II) by decomposing bacteria could be seen as a **detoxification process** (TREVORS 1986; WOOD 1984). In an experiment two strains of *Clostridium cochlearium* were studied. Only one of the strains was able to methylate mercury. Both strains transported mercury into the cells at the same rate but the methylating strain could live in mercury concentrations 40 times higher than concentrations inhibiting the non-methylating strain (PAN HOU & IMURA 1982).

Methyl mercury is lipophilic but presumably associated with small, water-soluble organic molecules. It has, compared to inorganic mercury, less affinity to sulphur groups and can probably associate with single amino acids or short polypeptide chains (JERNELÖV et al. 1975).

Biotic decomposition of methyl mercury, **demethylation**, has been proved for several different microorganisms and is of great importance for the net methyl mercury occurrence in the aquatic environment. In one demethylating process methane and elementary mercury is formed (JERNELÖV et al. 1975).

The effects of pH on the methylation and demethylation processes are discussed in several papers. Abiotic methylation has different pH optima for different humic substances. LEE et al. (1985) found a maximal methyl mercury production at pH 4—4.5 while NAGASE et al. (1984) observed the highest production at pH 2 and 13.5. REIMERS et al. (1974) found that humate organics did not release mercury as pH was increased. However, they became water-soluble, thus releasing a soluble complex with methyl mercury. Adsorbtion of methyl mercury by sulphur containing organic sediments was not affected by pH-fluctuations.

When oxic sediments were acidified, an increase in methyl mercury production was reported, while demethylation was hardly effected at all (XUN et al. 1987). According to a theory of the authors the H^+ ions liberate Hg^{2+} by cationic exchange at low pH. SANGFORS (pers. comm., cited in JERNELÖV 1987) presented an idea that acidification leads to an increased flux of available cobalt from the soil. This might catalyze vitamine B-12 production which could cause a higher rate of biological methylation of mercury. In contrast, in laboratory experiments performed in anoxic milieu RAMLAL et al. (1985) showed acidification to cause a decrease in both demethylation and, to a greater extent, in methylation activity. Also the concentration of total mercury in porewater decreased, which might be due to the dissolution of iron sulphides mentioned above. BAKER et al. (1983) found methyl mercury in sediments incubated at pH 5.5 and 6.5, while no production was detected at pH 3.5, 4.5 and 7.5.

Mercury methylation processes are evidently dependent on **redox conditions**. Several authors have reported higher methyl mercury production (e. g. COMPEAU & BARTHA 1984; KORTHALS & WINFREY 1987; CALLISTER & WINFREY 1986) and net methyl mercury rates (as much as 10 times) (REGNELL 1987) in anaerobic environments. COMPEAU & BARTHA (1984, 1985) conclude that reducing microbiota in sulphide-poor anoxic estuarine sediments are the probable converters of inorganic mercury to monomethyl mercury. The methylation can be severely inhibited, though, if **sulphide** (or selenide), with its high affinity to Hg(II), is present. H_2S effectively affects methyl mercury production by precipitating available mercury as HgS, or by transforming methyl mercury to the volatile dimethyl form (WOOD 1984; BERMAN & BARTHA 1986a). High methylation rates in sulphide-rich sediments have been observed, though, which indicates that the sulphide was in the form of insoluble FeS (FURUTANI & RUDD 1980; RAMLAL et al. 1985). Demethylation appears to be depressed in anaerobic environments (COMPEAU & BARTHA 1984). Thus, aerobic conditions enhance the availability of inorganic mercury but depress the "methylating capability" of the sediments (JACKSON 1988a).

High **salt concentrations** inhibit methylation but not demethylation of mercury (COMPEAU & BARTHA 1984). **Other metal ions** like Fe might have a catalythic effect on chemical methylation (LEE et al. 1985). However, LINDESTRÖM & GRAHN (1982) have proposed antagonistic effects of other metals as the reason for low mercury concentrations in fish from a heavily polluted mine district.

Impoundments that caused flooding of densely forested land had the following effects on lake sediments (JACKSON 1988a): 1) abnormally high rates of methyl mercury production, accompanied by accelerated CO_2 and CH_4 production indicative of intensified heterotrophic microbial activity, 2) elevated concentrations of both methyl and total mercury, 3) elevated organic C concentrations accompanied by O_2-depletion and reducing conditions (a favourable environment for microbial methylation). It was also found that C-rich terrestrial organics played a much greater role than N-rich autochthonous organic matter in stimulating mercury methylation. JACKSON (1988a) concludes that methylation rates are limited not by the total abundance of inorganic Hg but by environmental variables affecting the activities of methylating microbes and the availability of the mercury. [The available inorganic Hg pool can be seen as comprising all free dissolved Hg(II) and all weakly complexed and adsorbed Hg(II).]

It has been shown, in **laboratory experiments**, that the rate of methylation of mercury, in systems with natural sediments, is very much dependent on which inorganic mercury compound is added (CRAIG & MORETON 1985; WEBER et al. 1985a) or if the sediment is "naturally" contaminated (LINDESTRÖM 1986). For instance the rate of methyl mercury produced was 50 times higher when Hg was added in the form of $Hg(OAc)_2$ compared to an adding of $HgCl_2$. The difference is much less for longer incubation times. pH variations have different effects when different compounds are added. This shows the difficulty in drawing general conclusions about mercury transformations in natural systems from experimental systems.

3.3. Seasonal variation

Seasonal variations in mercury methylation activity were measured in water and sediment samples from a vertical profile in a Canadian oligotrophic forest lake (KORTHALS & WINFREY 1987). The methylation/demethylation rate (M/D) was highest in the sediment surface layer where M/D was <1 in May. The rate increased to a peak (>2) in July/August after which it decreased again. The peak in the middle of summer was due to a sudden drop in demethylation activity which coincided with oxygen deficit in the hypolimnion water. The rapid decrease in methylation activity and consequently the M/D rate was simultaneous with the fall-turnover. The variation in temperature could explain 30% of the variance in methylation activity, while demethylation activity not at all was correlated with temperature. In the vertical sediment profile methylation covariated with microbial activity. It was also observed that the greatest potential for mercury methylation in the lake was found in the sediments of the littoral/profundal transition zone. This area is of importance as it serves as spawning and feeding regions of many fishes.

In a similar investigation on methyl mercury production (FURUTANI & RUDD 1980), a big peak was found in the middle of May and two smaller peaks at the end of June and the end of August, respectively. These peaks coincided with increased bacterial activity in the water mass. The authors concluded that "the cyclic nature of methylation could be caused by the cyclic supply of phytoplankton-derived bacterial substrates or periodic resuspension of sediments". This pattern with seasonal variation in methyl mercury formation rates correlates with a peak in early summer for mercury concentrations observed in mesoplankton from Swedish forest lakes (MEILI & PARKMAN 1988).

In sedimenting material from the Norwegian Lake Tyrifjord the highest amount of total mercury was found in spring, but this maximum was not found for any other measured parameter (e.g. C, P, N, Fe, Chl.a). Amounts of coal and, e.g., iron were highest in August when mercury concentration was very low (ABRAHAMSEN 1983). On the other hand JACKSON (1988a) showed that 'available' mercury (extractable with $CaCl_2$) in the sediments of a lake was more abundant in August than in June.

4. Transports

4.1. Diffusion

Pollutants that desorb from the sediments can diffuse from the pore water to the overlaying water. Because of the high affinity of mercury to particulate matter the diffusion of mercury from sediments is probably of minor importance. Most chloroorganic compounds, likewise, show a high affinity for particulate matter. FISHER et al. (1983) found, when investigating four different PCBs, that the apparent diffusivities for the PCBs were very small. They concluded that PCB-contaminated sediments quickly can be removed from diffusional communication with overlaying water by covering with uncontaminated material, provided that the sediments remain undisturbed.

LARSSON (1983) studied the transport of three PCBs in aquatic model systems composed of "undisturbed" sediment cores with overlying water. He found that the lowest chlorinated and most water-soluble of the compounds used was easier desorbed than the others, and that it was transported to the water both in systems without macroinvertebrates and in systems fixed with $HgCl_2$. OLIVER (1985), on the other hand, found little variation in the amount of long-term desorption from sediments of 20 chlorinated organic compounds, despite their wide range of physical and chemical properties.

4.2. Current action

In lakes laminar and turbulent currents are induced by outlet and inlet flows, wind and wave action, temperature and density differences, etc.. These currents are important for the stratification and aeration of the sediments (e. g. JACKSON 1988a, b) and for the resuspension and transport of particle bound toxic compounds in the sediments. The vertical and horizontal distribution of the compounds are highly dependent on the sedimentation velocity of their "carrier particles". Despite the fact that chloroorganic compounds and mercury are very different from the chemical point of view, they both show high affinity (as discussed above) to organic materials, which act as "carrier particles". In areas of Lake Vättern, affected by diffuse contamination, a high correlation was found between the distribution of PCBs and mercury in the sediments (HÅKANSON & JANSSON 1983). JOHANSSON (1988) and PHILLIPS et al. (1987) found positive correlations between mercury in sediments and water depth "presumably because deeper waters contain finer, more highly organic sediments". LINDESTRÖM (1986) observed the highest concentrations of mercury in surface sediments in the deepest parts of the mercury-contaminated Kattfjorden, even though shallower areas shoved higher values for loss of ignition. This indicates that mercury is bound to and transported with fine sized particles.

OLSSON et al. (1978) found higher levels of PCB in roach from Lake Roxen during spring than in other seasons. This was explained as an effect of erosion of PCB-containing sediments in the feeder streams, due to the increased water flow during spring.

4.3. Gas convection

In anaerobic systems gas formed in the sediments by microbial activity may be of great importance for the transfer of materials across the sediment/water interface. SÖDERGREN & LARSSON (1982) showed that PCBs injected below the sediment surface in anaerobic model systems were transported to the overlying water. Rising gas bubbles transfer particles and hydrophobic compounds from the sediments to the water. The chemical and physical conditions in the sediments are also changed due to gas production.

Microbial activity causes production of volatile elementary- and dimethyl-mercury. The disappearance of those from the system can be hastened by transportation with other gases like methane. Higher mercury rates in the air over lakes than over nearby land were reported by SCHROEDER & FANAKI (1987). In Swedish forest lakes XIAO et al. (1988) found higher fluxes of mercury from water to air during daytime than during nights.

4.4. Bioturbation

Zoobenthos, such as oligochaetes, chironomids, some molluscs and some fishes, release compounds to the water by their feeding and burrowing in the sediment. They also change the chemical and physical conditions in the sediment which in turn affects the release of different compounds from the sediment. For instance, they might influence the effect of acidification by mixing acidic bottom water into the pore water. DAVIS (1974) showed that bioturbation by tubificids caused an increase in the redox potential in lake sediments, whereby the oxidized layer was deepened 0.3—1.6 cm. pH increased in the deeper layers (2—14 cm) and slightly decreased in the surface layer. GRANÉLI (1979) reported a sharp increase in iron leakage from sediments when *Chironomus plumosus* was present.

Bioturbation may also stimulate bacterial decomposition of organic material in the sediment, which could lead to an increased release of hydrophobic compounds previously adsorbed to the organic material (SÖDERGREN & LARSSON 1982).

Tubificids have been reported to mix sediments down to a depth of about 10 cm (KARICKHOFF & MORRIS 1985; FISHER et al. 1980). They ingest fine sediment particles and egest them at the sediment surface as fecal pellets. Substantial amounts of sorbed pollutants can be transported within the sediment by this process (KARICKHOFF & MORRIS 1985). Grain size might be of importance for the intensity and effect of bioturbation, which is also seasonally dependent because of generation cycles and temperature-dependent activity (NICHOLS 1974; GOLDHABER et al. 1977).

BODDINGTON et al. (1979) found in experiments that tubificid worms caused a significant density-dependent clearance of added mercury from sediments. JERNELÖV (1970) showed in a laboratory experiment that *Anodonta* could mobilize mercury from 9 cm sediment depth. Sediments were spiked at different depths with inorganic mercury and the mobilization caused by *Anodonta* was measured as methyl mercury uptake in fish living in the overlying water. Tubificids caused an equal uptake in fish with spikes from 0 to 2.5 cm, whereas sediments with no macrobenthos only resulted in uptake from sediments spiked in the surface layer. In other experiments fishes were allowed to swim over a sediment and some fishes were prevented from direct contact with the sediment by a net (LINDESTRÖM 1986). The bioturbation caused by fish, in the system without a net, resulted in a decrease of mercury in sediments but, despite this, the uptake in fish was lower in this system. This was probably due to competition by suspended particles for the released mercury. LINDESTRÖM also calculated the importance of bioturbation on Lake Vänern, a lake recovering from mercury discharge. The most important organisms in this respect were bream, roach and oligochaetes. In shallow areas they might mix the whole surface layer to a depth of 1—4 cm. Thus, bioturbation could effectively counteract a covering of contaminated sediments.

Chaoborus is a genus that might be of great importance for bioturbation in the mercury-sensitive forest lakes. These midge larvae often occur in great numbers and they perform diurnal migrations into and from the sediments, which can cause a considerable disturbance of the sediment surface.

SÖDERGREN & LARSSON (1982) found that both chironomids and tubicifids increased the amount of suspended particles in the water in model systems comprising sediment and water. Thus PCBs attached to these particles were released from the sediments. The release was greatest when both groups of animals were held together. This might be a synergistic effect — tubificides transporting PCBs from deeper sediment layers to the surface where chironomid activity leads to a further transport. When three different PCBs were compared it was found that bioturbation lead to a higher transport from sediment to water via particles for compounds with a higher degree of chlorination than for those with a lower (LARSSON 1983).

4.5. Accumulation in organisms

Benthic organisms accumulate chemicals from sediments. Some of these organisms then migrate out of the sediments (e. g. emerging chironomids) and further some are subjected to predation by organisms living in the overlying water. The accumulated chemicals thus can be transported out of the sediments.

FOWLER et al. (1978) compared the accumulation of PCBs from sediments and water by the benthic worm *Nereis diversicolor*. They found that the accumulation was much slower from sediments than from water. However, the sediment was suggested to contribute the bulk of these compounds to the worms, due to higher relative concentrations of PCBs in the sediment.

The bioaccumulation of a chemical is governed by several factors. Firstly, the molecule must be small enough to penetrate biological membranes. OPPERHUIZEN et al. (1985) suggested a loss of membrane permeability for hydrophobic molecules with a width over 9.5 Å. This means that many of the chlorinated organic compounds originating from pulp bleaching cannot be accumulated if not degraded.

Further, concerning organic compounds, the lipophilicity of the compounds studied is of importance. MACKAY (1982) suggests that a directly proportional relationship exists between the bioconcentration factor in water organisms and the lipophilicity of a compound — measured as the octanol/water partition coefficient (K_{ow}). OLIVER & NIIMI (1985) also give a formula for this relationship. However, deviations between the measured bioconcentration factors and those predicted from the K_{ow}'s are quite frequent.

OLIVER (1987) studied the uptake of chlorinated hydrocarbons from sediments by oligochaete worms. The highest concentration factors were obtained for chemicals with log K_{ow}'s about 6 (e. g. many PCBs). For compounds with low K_{ow}'s the bioconcentrations were generally low (Fig. 3). Compounds with log K_{ow}'s less than 5 were also found to have biological half-lifes of less than 5 days. If this is the general figure the uptake of compounds such as chlorophenolic ones — with log K_{ow}'s about 3—5

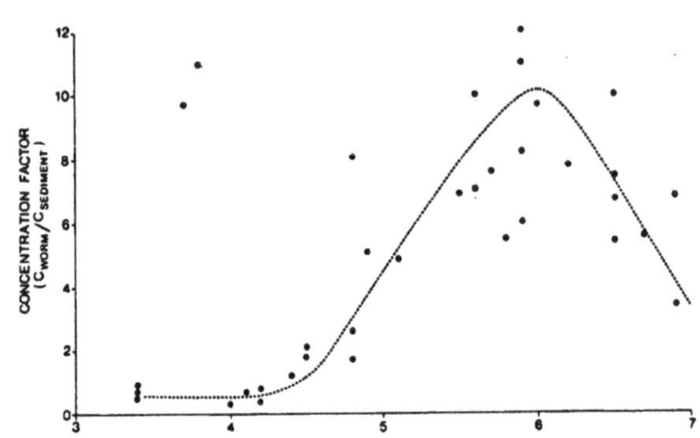

Fig. 3. Concentration factor [chemical concentration in worm (ng/g dry weight)/chemical concentration in sediment (ng/g dry weight)] vs. log octanol-water partition coefficient, K_{ow} (OLIVER 1987)

(XIE et al. 1984) — from sediments by benthic organisms would be rather small.

If the accumulation of hydrophobic compounds is governed mainly by partition processes the fat content of the organisms would be important. However, PAASIVIRTA et al. (1980) investigated chlorophenols in organisms from Finnish lakes. He did not find any clear connection between chlorophenol contents and fat content of the animals.

McLEESE et al. (1980) reported that animal size affects the accumulation of PCBs. Small individuals of *Nereis virens* and *Crangon septemspinosa* were found to accumulate more PCBs than large ones.

Concerning mercury, methyl mercury is the form most readily accumulated and SURMA-AHO et al. (1986) found the mean ratio of methyl mercury (to total) to be 51% for settled solids, 86% for zoobenthos and 95% for fish. As methylation is favoured by anaerobic, and demethylation by aerobic (see 3.2) conditions it might be possible that animals burrowing in anaerobic layers in sediments are exposed to higher methyl mercury concentrations than animals in the oxidized sediment surface layer.

Organisms living in sediments with high organic content and in sediments consisting of fine particle size fractions have been reported to show the lowest concentration factors of chlorinated organics (e. g. OLIVER 1987; LYNCH & JOHNSON 1982). LANGSTON (1985) proposed that high organic content also reduced availability of sediment-bound mercury to organisms. However, JACKSON (1988b) found that mercury concentrations in chironomids were positively related to the abundance of carbon in the organic fraction of the sediments and inversely correlated with the concentration of hydrated Fe and Mn oxides. Further, the pH in the sediments may affect the lipophilicity of weak acids, such as chlorophenols, which might in turn affect the bioaccumulation.

The transport of PCBs from sediment to air by chironomids was investigated by LARSSON (1984b). He found that chironomid larvae readily accumulated PCBs from sediments. When the chironomid larvae metamorphosed to adults the PCBs were concentrated 3.8 times (fresh weight basis). In an experiment with mercury it was found that chironomid larvae accumulated mercury very effectively, but emerging adults only contained about 30% of the mercury concentrations in the larvae (ROSSANO et al. 1986). However, this was a short-time (30 days) experiment and only total mercury was measured. Total annual transfer of mercury from sediments by emerging and predated chironomids was calculated to 0.04 resp. 0.08% of the total mercury in the sediments (BISSONETTE 1976).

4.6. Exchange by plants

Aquatic plants may be of importance for the exchange of chemicals between sediments and water. MORTIMER (1977) found in experiments that sediments with *Elodea* accumulated more mercury from water than sediments without plants. This indicates a transport of mercury to sediments via plants. Also results from ERIKSSON & MORTIMER (1975) support this theory. However, in a field investigation from three polluted and one control area in Finland, LODENIUS (1980) found no significant correlation between mercury content in aquatic plants and sediments.

In laboratory experiments with a two compartment system (leaves and water/roots and sediment) BRIX & LYNGBY (1984) studied the translocation of labelled mercury in *Zostera marina*. After 6 days of incubation less than 0.04% of the total amount of mercury absorbed was translocated in the plant. (The corresponding value for phosphorus was 4%). The transport from leaves to roots was somewhat larger than the transport in the reverse direction. No direct release fom the plant to the surrounding water was detected in this study.

Uptake via the roots of a macrophyte (*Spartina alterniflora*) has been reported to represent a potential pathway for the mobilization of PCBs from sediments (MROZEK & LEIDY 1981). The uptake was suggested to be selective against the lower chlorinated compounds of a PCB mixture.

5. Research in Nordic countries

5.1. Chlorinated organic compounds

In the beginning of the present decade a joint biological/chemical project on the effects of effluents from pulp industries was initiated by the Swedish Environmental Protection Board. The project, called "Environment/Cellulose", consisted of several subprojects and engaged researchers mainly in Sweden but also in Norway. The results of the project have already been reported (SÖDERGREN et al. 1988). In 1987 "Environment/Cellulose II" started. Several of the sub-projects initiated concern chloroorganics in sediments. For instance, there are projects dealing with transport and distribution models, transformations of chloroorganic compounds, uptake and effects in benthic animals, chemical characterization of the chloroorganic material in the sediments and synthesis of compounds found in the sediments. Besides this big project there are several other minor projects in Sweden and Norway dealing with similar problems. In Finland research has been carried out on the distribution of chloroorganics in lake sediments and on the degradation and transformations of chlorinated organic material (e. g. SALKINOJA-SALONEN 1984; PAASIVIRTA 1988; HÄGGBLOM et al. 1988). Studies on leaching of PCBs from lake sediments have been carried out in Sweden (e.g. LARSSON & RYDING 1988).

5.2. Mercury

In 1984 a coordinated research project on the occurrence and turnover of mercury in Swedish environments started. The project was initiated by the research council of the Swedish Environmental Protection Board in cooperation with some branches of industry. "The purpose is to describe and quantify those mercury fluxes in the air, ground and water that are most significant to the enrichment of mercury in the food chains" (LINDQVIST et al. 1986). One of the four subareas is dealing with turnover in the aquatic system, emphasizing biological factors. Studies on factors regulating methylation and demethylation of mercury in an aquatic model ecosystem is also in progress (REGNELL 1987).

A parallel project is the "Liming/Mercury"-project (HÅKANSON 1986) with the purpose to document mercury levels in fish, before and after liming or liming in combination with other treatments like fertilization, selenium treatment and intensive fishing. Both these projects continued until autumn 1989, followed by an international symposium on mercury in the environment.

In Norway a big regional investigation on sediments from some hundred lakes has just been initiated by NIVA

(KJELLBERG, pers. comm.). In the lakes vertical and horizontal distribution in sediments of mercury, among other metals, is going to be outlined. The biggest lake, Lake Mjøsa, has already been investigated (ROGNERUD 1985). Another project which included mercury was reported 1983 (ABRAHAMSEN 1983). This dealt with turnover and sedimentation of organic material and heavy metals in Lake Tyrifjord.

A joint project was undertaken in Finland (1982—1984) to study the relative impact of different anthropogenic sources of mercury on mercury concentrations in fish in small forest lakes not subjected to any direct mercury contamination (VESIHALLITUS 1986). Special interest was directed towards newly impounded reservoirs and, connection with that, humic substances as mercury carrier. Also effects of ditching on, e.g., mercury in sediments have been studied (SURMA-AHO 1986; LIEHU et al. 1986; SIMOLA & LODENIUS 1982). Furthermore, a regional investigation on metals and other abiotic parameters, in sediments from 42 lakes, was reported in 1986 (MYLLYMAA & MURTONIEMI 1986).

In Denmark mercury studies have mainly been accomplished in streams and marine environments (e. g. KIÖRBOE et al. 1983; KRISTENSEN 1982).

References

ABRAHAMSEN, H., HONGVE, D., HINDAR, A., & SKOGHEIM, O. K. (1983): Omsetning og sedimentasjon av organisk materiale og tungmetaller i Tyrifjorden 1979. Tyrifjordundersökelsen, Drammen, Fagrapport Nr. **21** (in Norwegian).

ANDERSSON, P., BORG, H., & HOLM, K. (1983): Lösligheten av metaller i sediment vid olika pH och kalcium tillsats — laboratorieförsök. Nat. Swed. Environm. Prot. Board, Report **1772** (in Swedish).

ANDERSSON, T., NILSSON, Å., HÅKANSON, L., & BRYDSTEN, L. (1987): Kvicksilver i svenska sjöar. Nat. Swed. Environm. Prot. Board, Report 3291 (in Swedish).

APAJALAHTI, J. H. A., KÄRPÄNOJA, P., & SALKINOJA-SALONEN, M. S. (1986): *Rhodococcus chlorophenolicus* sp. nov., a chlorophenol-mineralizing actinomycete Int. J. Syst. Bacteriol. **36**, 246—251.

— & SALKINOJA-SALONEN, M. S. (1986): Degradation of polychlorinated phenols by *Rhodococcus chlorophenolicus*. Appl. Microbiol. Biotechnol. **25**, 62—67.

— — (1987): Complete dechlorination of tetrachlorohydroquinone by cell extracts of pentachlorophenol-induced *Rhodococcus chlorophenolicus*. J. Bacteriol. **169**, 5125—5130.

BAKER, M. D., INNISS, W. E., MAYFIELD, C. I., WONG, P. T. S., & CHAU, Y. K. (1983): Effect of pH on the methylation of mercury and arsenic by sediment microorganisms. Environm. Techn. Letters **4**, 89—100.

BERMAN, M., & BARTHA, R. (1986a): Control of the methylation process in a mercury polluted aquatic sediment. Environm. Poll. (Series B) **11**, 41—53.

— (1986b): Levels of chemical versus biological methylation of mercury in sediments. Bull. Environm. Contam. Toxicol. **36**, 401—404.

BISSONETTE, P. (1976): Extent of mercury and lead uptake from lake sediments by chironomids. In: Biological Implications of Metals in the annual Henford Life Sciences symposium at Richland, Wash., **42**, 609—622.

BJÖRNBERG, A., HÅKANSON, L., & LUNDBERGH, K. (1988): A theory on the mechanism regulating the bioavailability of mercury in natural waters. Environm. Poll. **49** (1), 53—62.

BODDINGTON, M. J., DE FREITAS, A. S., & MILLER, D. R. (1979): The effect of benthic invertebrates on the clearance of mercury from sediments. Ecotoxicology and environmental safety **3**, 236—244.

BOSTRÖM, B., ANDERSEN, J. M., FLEISCHER, S., & JANSSON, M. (1988): Exchange of phosphorus across the sediment-water interface. Hydrobiologia **170**, 229—244.

BOURGH, A. C. M., & SCHINDLER, P. W. (1985): Control of the trace metals in natural aquatic systems by the adsorptive properties of organic matter. In: LEKKAS, T. D. (ed.), Int. Conf. Heavy metals in the environment, Athen, Vol. 1, 97—99.

BRIX, H., & LYNGBY, J. E. (1984): The influence of eelgrass (*Zostera marina* L.) on the exchange of nutrients and heavy metals between sediment and water. In: HENRIKSEN, K. (ed.), 12th Nordic Symposium on Sediments, Skallingen, Denmark, 41—47.

CALLISTER, S. M., & WINFREY, M. R. (1986): Microbial methylation of mercury in upper Wisconsin river sediments. Environm. Techn. Letters **29**, 453—465.

CARLBERG, G. E., KRINGSTAD, A., MARTINSEN, K., & NASHAUG, O. (1987): Environmental impact of organochlorine compounds discharged from the pulp and paper industry. Paperi ja Puu, 337—341.

COMPEAU, G. C., & BARTHA, R. (1984): Methylation and demethylation of mercury under controlled redox, pH and salinity conditions. Appl. Environm. Microbiol. **48**(6), 1203—1207.

— — (1985): Sulfate-reducing bacteria: Principal methylators of mercury in anoxic estuarine sediment. Appl. Environm. Microbiol. **50** (2), 498—502.

CRAIG, P. J., & MORETON, P. A. (1985): The role of speciation in mercury methylaton in sediments and water. Environm. Poll. (Series B) **10**, 141—158.

DAVIS, R. B. (1974): Tubificids alter profiles of redox potential and pH in profundal lake sediment. Limnol. & Oceanogr. **19**, 342—346.

ERIKSSON, C., & MORTIMER, D. C. (1975): Mercury uptake in rooted higher aquatic plants; laboratory studies. Verh. Int. Verein. Limnol. **19**, 2087—2093.

ERIKSSON, K. E., & KOLAR, M. C. (1985): Microbial degradation of chlorolignins. Environm. Sci. Technol. **19**, 1086—1089.

— — LJUNGQUIST, P. O., & KRINGSTAD, K. P. (1985): Studies on microbial and chemical convertions of chlorolignins. Environm. Sci. Technol. **19**, 1219—1224.

FAGERSTRÖM, T., & JERNELÖV, A. (1972): Some aspects of the quantitative ecology of mercury. Water Res. **6**, 1193—1202.

FISHER, J. B., LICK, W. J., MCCALL, P. L., & ROBBINS, J. A. (1980): Vertical mixing of lake sediments by tubificid oligochaetes. J. Geophys. Res. **85**, 3997—4006.

— PETTY, R. L., & LICK, W. (1983): Release of polychlorinated biphenyls from contaminated lake sediments: flux and apparent diffusivities of four induvidual PCBs. Environm. Poll. (Series B) **5**, 121—132.

FOWLER, S. W., POLIKARPOV, G. G., ELDER, D. L., PARSI, P., & VILLENEUVE, J. P. (1978): Polychlorinated biphenyls: accumulation from contaminated sediments and water by the polychaete *Nereis diversicolor*. Marine Biol. **48**, 303—309.

FURUTANI, A., & RUDD, J. W. M. (1980): Measurement of mercury methylation in lake water and sediment samples. Appl. Environm. Microbiol. **40**(4), 770—776.

GOLDHABER, M. B., ALLER, R. C., LOCHRAN, J. K., ROSENFELD, J. K., MARTENS, C. S., & BERNER, R. A. (1977): Sulfate reduction, diffusion and bioturbation in Long Island Sound sediments. Report of the foam group. Am. J. Science **277**, 193—237.

GRANÉLI, W. (1979): The influence of *Chironomus plumosus* larvae on the exchange of dissolved substances between sediment and water. Hydrobiologia **66** (2), 149—159.

HÅKANSON, L. (1972): Sambandet mellan kvicksilverförekomst och sedimentologisk miljö i Ekoln, del 3. Transport och deposition av kvicksilver. Univ. Uppsala, UNGI, Rapport **15**, 1—56 (in Swedish).

— (1980): The quantitative impact of pH, bioproduction and Hg-contamination on the Hg-content of fish (pike). Environm. Poll. (Series B) **1**, 285—304.

— (1984): Aquatic contamination and ecological risk. Water Res. **18**(9), 1107—1118.

— (1986): Projektplan, Kalkning-Kvicksilver, 1985—1989. Nat. Swed. Environm. Prot. Board, Report **3097** (in Swedish).

— & JANSSON, M. (1983): Principles of lake sedimentology. Berlin, Heidelberg.

— JONSSON, P., JONSSON, B., & MARTINSEN, K. (1988): Distribution of chlorinated organic substances from pulp mills. Wat. Sci. Tech. **20**, 25—36.

HÄGGBLOM, M. H., APAJALAHTI, J. H. A., & SALKINOJA-SALONEN, M. S. (1985): Metabolism of chloroguaiacols by *Rhodococcus chlorophenolicus*. Appl. Microbiol. Biotechnol. **24**, 397—404.

— — — (1988): Hydroxylation and dechlorination of chlorinated guaiacols and syringols by *Rhodococcus chlorophenolicus*. Appl. Environm. Microbiol. **54**, 683—687.

IVERFELDT, Å. (1988): Mercury in the Norwegian fjord Framvaren. Mar. Chem. **23**, 441—456.

JACKSON, T. A. (1988a): The mercury problem in recently formed reservoirs of northern Manitoba (Canada): effects of impoundment and other factors on the production of methyl mercury by microorganisms in sediments. Can. J. Fish. Aquat. Sci. **45**, 97—121.

— (1988b): Accumulation of mercury by plankton and benthic invertebrates in riverine lakes of northern Manitoba (Canada). Can. J. Fish. Aquat. Sci. **45**, 1744—1757.

— KIPPHUT, G., HESSLEIN, R., & SCHINDLER, D. W. (1980): Experimental study of trace metal chemistry in soft-water lakes at different pH levels. Can. J. Fish. Aquat. Sci. **37**, 387—402.

JERNELÖV, A. (1968): Laboratorieförsök rörande kvicksilvrets omvandling mellan olika förekomstformer. Vatten **1**, 53—56 (in Swedish).

— (1970): Release of methyl mercury from sediments with layers containing inorganic mercury at different depths. Limnol. & Oceanogr. **15** (6), 958–960.

— (1972): Environmental mercury contamination. Ann. Arbor Mich., 174 p.

— (1987): Ecological aspects of microbial alkylation of heavy metals. Swed. Environm. Research Inst., Report B-875.

— LANDNER, L., & LARSSON, T. (1975): Swedish perspectives on mercury pollution. J. Wat. Poll. Cont. Fed. **47**(4), 810—822.

JOHANSSON, K. (1985): Mercury in sediments in Swedish forest lakes. Verh. Int. Verein. Limnol. **22**, 2359—2363.

— (1988): Heavy metals in Swedish forest lakes — factors influencing the distribution in sediments. Acta Univ. Ups., Comprehensive Summaries of Uppsala Dissertations from the Faculty of Science, Uppsala, **144**.

KARICKHOFF, S. W., BROWN, D. S., & SCOTT, T. A. (1979): Sorption of hydrophobic pollutants on natural sediments. Water Res. **13**, 241—248.

— & MORRIS, K. R. (1985): Impact of tubificid oligochaetes on pollutant transport in bottom sediments. Environm. Sci. Technol. **19**, 51—56.

KIERKEGAARD, A., & RENBERG, L. (1987): Halter av klorguajakoler i sediment utanför Iggesund. Nat. Swed. Environm. Prot. Board, Special Analytical Laboratory, Report 87-02 (in Swedish).

KIÖRBOE, T., MÖHLENBERG, F., & RIISGÅRD, H. U. (1983): Mercury levels in fish invertebrates and sediment in a recently recorded polluted area (Nissum Broad, western Limfjord, Denmark). Mar. Poll. Bull. **14**(1), 21—24.

KRISTENSEN, P. (1982): Time-dependent variation of mercury in a stream sediment and the effect upon mercury content in *Gammarus pulex* (L.). Water Res. **16**, 759—764.

KORTHALS, E. T., & WINFREY, M. R. (1987): Seasonal and spatial variations in mercury methylation and demethylation in an oligotrophic lake. Appl. Environm. Microbiol. **53** (10), 2397—2404.

LANGSTON, W. J. (1985): The use of organisms and sediments as indicators of environmental contamination with special reference to mercury, in the Mersey estuary. In: LEKKAS, T. D. (ed.), International Conference, Heavy metals in the environment, Athen, Vol. 2, 374—377.

LARSSON, P. (1983): Transport of 14C-labelled PCB compounds from sediment to water and from water to air in laboratory model systems. Water Res. **17**, 1317—1326.

— (1984a): Sedimentation of polychlorinated biphenyls (PCBs) in limnic and marine environments. Water Res. **18**, 1389—1394.

— (1984b): Transport of PCBs from aquatic to terrestrial environments by emerging Chironomids. Environm. Poll. (Series A) **34**, 283—289.

— & RYDING, S. O. (1988): Kontaminerade sediment som spridningskälla för PCB i Emån. Swedish Environmental Research Institute, Publication B-896 (in Swedish).

LEE, Y. H., HULTBERG, H., & ANDERSSON, I. (1985): Catalytic effect of various metal ions on the methylation of mercury in the presence of humic substances. Water, Air and Soil Pollution **25**, 391—400.

LIEHU, A., SANDMAN, O., & SIMOLA, H. (1986): Effects of peatbog ditching in lakes: Problems in paleolimnological interpretation. Hydrobiologia **143**, 417—424.

LINDESTRÖM, L. (1986): Gäddans kvicksilverhalt i norra Vänernsedimentens, fabrikens och andra faktorers betydelse för utvecklingen under 1974—1984. Nat. Swed. Environm. Prot. Board, Report 3181.

— & GRAHN, O. (1982): Antagonistic effects to mercury in some mine drainage areas. Ambio **11** (6), 359—361.

LINDSTRÖM, K., NORDIN, J., ÖSTERBERG, F. (1981): Chlorinated organics of low and high relative molecular mass in pulp bleachery effluents. In: KEITH, L. H. (ed.), Advances in the identification and analysis of organic pollutants in water, Vol. 2, pp. 1039—1058 Ann Arbor, M1.

LINDQVIST, O., JERNELÖV, A., JOHANSSON, K., & RODHE, H. (1984): Mercury in the Swedish environment, global and local sources. Nat. Swed. Environm. Prot. Board, Report 1816.

— JOHANSSON, K., & TIMM, B. (1986): Occurrence and turnover of mercury in the environment. Nat. Swed. Environm. Prot. Board. Report 3265.

LODENIUS, M. (1980): Aquatic plants and littoral sediments as indicators of mercury pollution in some areas in Finland. Ann. Bot. Fennici **17**, 336—340.

— & SEPPÄNEN, A. (1984): Kvicksilvrets binding till humus i vatten. Vatten **40**, 302—307.

LYNCH, T. R., & JOHNSON, H. E. (1982): Availability of a hexachlorobiphenyl isomer to benthic amphipods from experimentally contaminated natural sediments. In: PEARSON, J. G., FOSTER, R. B., & BISHOP, W. E. (eds.), Aquatic toxicology and hazard assessment: Fifth conference, ASTM STP 766, pp. 273—287. American Society for Testing and Materials.

MACKAY, D. (1982): Correlation of bioconcentration factors. Environm. Sci. Technol. **16**, 274—278.

McKAUGE, A. B., KOLAR, M. C. & KRINGSTAD, K. P. (1988): Nature and properties of some chlorinated, lipophilic, organic compounds in spent liquors from pulp bleaching. 1. Liquors from conventional bleaching of softwood kraft pulp. Environm. Sci. Technol. **22**, 523—526.

McLEESE, D. W., METCALFE, C. D., & PEZZACK, D. S. (1980): Uptake of PCBs from sediment by *Nereis virens* and *Crangon septemspinosa*. Arch. Environm. Contam. Toxicol. **9**, 507—518.

MEILI, M. (1988): Bioaccumulation of mercury in aquatic ecosystems — A biological approach. In: ASTRUC, M., & LESTER, J. N. (ed.), Heavy Metals in the Hydrological Cycle, pp. 249—256. Lisbon.

— & PARKMAN, H. (1988): Seasonal mercury accumulation patterns in mesoplankton. Verh. Int. Verein. Limnol. **23**, 1639—1640.

MILLER, R. W. (1975): The role of humic acids in the uptake and release of mercury by freshwater sediments. Verh. Int. Verein. Limnol. **19**, 2082—2086.

MORTIMER, C. H. (1941): The exchange of dissolved substances between mud and water, I. J. Ecol. **29**, 280—329.

— (1942): The exchange of dissolved substances between mud and water, II. J. Ecol. **30**, 147—201.

MORTIMER, D. C. (1977): Interactions between aquatic plants and sediments, invertebrates and fish with respect to transport and accumulation of mercury. Ottawa River Project, final report **1**.

MROZEK, E., & LEIDY, R. B. (1981): Investigation of selective uptake of polychlorinated biphenyls by *Spartina alterniflora* LOISEL. Bull. Environm. Contam. Toxicol. **27**, 481—488.

MYLLYMAA, U., & MURTONIEMI, S. (1986): Metals and nutrients in the sediment of small lakes in Kuusamo, north-eastern Finland. Publications of the Water Research Institute, National Board of Waters and Environment, Finland, Nr. **69**, 33—48.

NAGASE, H., OSE, Y., SATO, T. & ISHIKAWA, T. (1984): Mercury methylation by compounds in humic material. The Science of the Total Environment **32**, 147—156.

NEILSON, A. H., ALLARD, A. S., HYNNING, P. Å., REMBERGER, M. & LANDNER, L. (1983): Bacterial methylation of chlorinated phenols and guaiacols: formation of veratroles from guaiacols and high-molecular-weight chlorinated lignin. Appl. Environm. Microbiol. **45**, 774—783.

— REILAND, S., REMBERGER, M., TÄRNHOLM, A., VIKTOR, T., & LANDNER, L. (1984): Tri- and tetra-chloroveratrole, metabolites produced by bacterial O-methylation of tri- and tetra- chloroguaiacol: An assessment of their bioconcentration potential and their effects on fish reproduction. Can. J. Fish. Aquat. Sci. **41**, 1502—1512.

— & REMBERGER, M. (1985): Biodegradation and transformation of recalcitrant compounds. In: The Handbook of Environmental Chemistry, Vol. 2/Part C, pp. 29—86. Berlin, Heidelberg, New York.

NICHOLS, F. H. (1974): Sediment turnover by a deposit-feeding polychaete. Limnol. & Oceanogr. **19**, 945—950.

NORDBERG, G. F., GOYER, R. A., & CLARKSON, T. W. (1985): Impact of effects of acid precipitation on toxity of metals. Environm. Health Perspectives **63**, 169—180.

OLIVER, B. G. (1985): Desorption of chlorinated hydrocarbons from spiked and anthropogenically contaminated sediments. Chemosphere **14**, 1087—1106.

— (1987): Biouptake of chlorinated hydrocarbons from laboratory-spiked and field sediments by oligochaete worms. Environm. Sci. Technol. **21**, 785—790.

— & NIIMI, A. J. (1985): Bioconcentration factors of some halogenated organics for rainbow trout: limitations in their use for predictions of environmental residues. Environm. Sci. Technol. **19**, 842—849.

OLSSON, M., JENSEN, S., & REUTERGÅRD, L. (1978): Seasonal variation of PCB levels in fish — an important factor in planning aquatic monitoring programs. Ambio **7**, 66—69.

OPPERHUIZEN, A., VELDE, E. W. v. d., GOBAS, F. A. P. C., LIEM, D. A. K., STEEN, J. M. D. v. d., & HUTZINGER, O. (1985): Relationship between bioconcentration in fish and steric factors of hydrophobic chemicals. Chemosphere **14**, 1871—1896.

PAASIVIRTA, J., KNUUTINEN, J., MAATELA, P., PAUKKU, R., SOIKKELI, J., SÄRKKÄ, J. (1988): Organic chlorin compounds in lake sediments and the role of the chlorobleaching effluents. Chemosphere **17**, 137—146.

— SÄRKKÄ, J., LESKIJÄRVI, T., & ROSS, A. (1980): Transportation and enrichment of chlorinated phenolic compounds in different aquatic food chains. Chemosphere **9**, 441—456.

PAN-HOU, H. S., & IMURA, N. (1982): Involvement of mercury methylation in microbial mercury detoxication. Arch. Microbiol. **131**, 176—177.

PHILLIPS, G. R., MEDVICK, P. A., SKAAR, D. R., & KNIGHT, D. B. (1987): Factors affecting the mobilization, transport and bioavailability of mercury in reservoirs of the upper Missouri River Basin. Nat. Tech. Inf. Serv., Springfield, Technical report 10.

RAMLAL, P. S., RUDD, J. W. M., FURUTANI, A., & XUN, L. (1985): The effect of pH on methyl mercury production and decomposition in lake sediments. Can. J. Fish. Aquat. Sci. **42**, 685—692.

REGNELL, O. (1987): Conversion and partitioning of radiolabelled mercury chloride in aquatic model systems. Submitted to Can. J. of Fisheries and Aquatic Sciences.

REIMERS, R. S., KRENKEL, P. A., EAGLE, M., & TRAGITT, G. (1974): Sorption phenomenon in the organics of bottom sediments. In: KRENKEL, P. A. (ed.), Heavy metals in the aquatic environment, pp. 117—129, Ottawa.

REKOLAINEN, S., VERTA, M., & LIEHU, A. (1986): The effect of airborne mercury and peatland drainage on sediment mercury contents in some Finnish forest lakes. Publications of the Water Research Institute, National Board of Waters, Finland, Nr. 65, 11—20.

REMBERGER, M., ALLARD, A. S., & NEILSON, A. H. (1986): Biotransformations of chloroguaiacols, chlorocatechols, and chloroveratroles in sediments. Appl. Environm. Microbiol. **51**, 552—558.

REUBER, B., MACKAY, D., PATERSON, S., & STOKES, P. (1987): A discussion of chemical equilibria and transport at the sediment — water interface. Environm. Toxicol. Chem. **6**, 731—739.

RIISGÅRD, H. U. (1985): Grundrids af kvicksölvs ökotoxikologi i akvatisk miljö. Vand & Miljö **4**, 163—167 (in Danish).

ROGNERUD, S. (1985): Kvikksölv i Mjøsas sedimenter — Arealprofiler og vertikalprofiler av antropogent kvikksölv. Norsk Institutt for Vannforskning (NIVA), Rapport 0-82105.

ROSSANO, B., GAGGINO, G. F., & MARCHETTI, R. (1986): Accumulation of mercury in larvae and adults, *Chironomus riparius* (MEIGEN). Bull. Environm. Contam. Toxicol. **37**, 402—406.

SALKINOJA-SALONEN, M., SAXELIN, M. L., PERE, J., JAAKKOLA, T., SAARIKOSKI, J., HAKULINEN, R. & KOISTINEN, O. (1981): Analysis of toxicity and biodegradability of organochlorine compounds released into the environment in bleaching effluents of kraft pulping. In: KEITH, L. H. (ed.), Advances in the identification and analysis of organic pollutants in water, Vol. 2, pp. 1131 1164, Ann Arbor, MI.

— VALO, R., APAJALAHTI, J., HAKULINEN, R., SILAKOSKI, L., & JAAKKOLA, T. (1984): Biodegradation of chlorophenolic compounds in wastes from wood-processing industry. In: KLUNG, M. J., & REDDY, C. A. (eds.), Current Perspectives in Microbial Ecology. American Society for Microbiology, pp. 668—676. Washington, DC.

SCHINDLER, D. W., HESSLEIN, R. H., WAGEMANN, R. W., & BROEKKER, W. S. (1980): Effects of acidification on mobilization of heavy metals and radionuclides from the sediments of a freshwater lake. Can. J. Fish. Aquat. Sci. **37**, 373—377.

SCHROEDER, W. H., & FANAKI, F. H. (1987): Air-water exchange of mercury. In: LINDBERG, S. E., & HUTCHINSON, T. C. (eds.), Heavy metals in the environment, Vol. 1, pp. 36—40, New Orleans.

SIMOLA, H., & LODENIUS, M. (1982): Recent increase in mercury sedimentation in a forest lake attributable to peatland drainage. Bull. Environm. Contam. Toxicol. **29**, 298—305.

SOUSA, F. DE., STRÖMBERG, L. M., & KRINGSTAD, K. P. (1988): The fate of spent bleach liquor material in reciving waters: Characterization of chloroorganics in sediments. Wat. Sci. Tech. **20**, 153—160.

STEEN, W. C., PARIS, D. F., & BAUGHMAN, G. L. (1978): Partitioning of selected polychlorinated biphenyls to natural sediments. Water Res. **12**, 655—657.

SURMA-AHO, L., PASSAVIRTA, J., REKOLAINEN, S., & VERTA, M. (1986): Organic and inorganic mercury in the food chain of some lakes and reservoirs in Finland. Chemosphere **15** (3), 353—372.

SÖDERGREN, A., BENGTSSON, B. E., JONSSON, P., LAGERGREN, S., LARSSON, Å., OLSSON, M., & RENBERG, L. (1988): Summary of results from the Swedish project 'environment/cellulose'. Wat. Sci. Tech. **20**, 49—60.

— & LARSSON, P. (1982): Transport of PCBs in aquatic laboratory model ecosystems from sediment to the atmosphere via the surface microlayer. Ambio **11**, 41—45.

THANABALASINGAM, P., & PICKERING, W. F. (1985a): The sorption of mercury (II) by humic acids. Environm. Poll (Series B) **9**, 267—279.

— (1985b): The sorption of mercury (II) by manganese(IV)oxide. Environm. Poll. (Series B) **10**, 115—128.

TREVORS, J. T. (1986): Mercury methylation by bacteria. J. Basic Microbiol. **26** (8), 499—504.

VERTA, M. (1984): The mercury cycle in lakes; some new hypotheses. Aqua Fennica **14** (2), 215—221.

— REKOLAINEN, S., MANNIO, J., & SURMA-AHO, K. (1086): The origin and level of mercury in Finnish forest lakes. Publications of the Water Research Institute, National Board of Waters, Finland, Nr. 65, 21—31.

Vesihallitus — National Board of Waters, Finland (1986): Publications of the Water Research Institute, Nr. 65.

WEBER, J. H., REISINGER, K., & STOEPPLER, M. (1985a): Methylation of mercury(II) by fulvic acid. Environm. Techn. Letters **6**, 203—208.

— (1985b): Methyl transfer to mercury(II) from fulvic acid. In: LEKKAS, T. D. (ed.), International Conference, Heavy metals in the environment, Athen, Vol. 2, 510—512.

WOOD, J. M. (1984): Alkylation of metals and the activity of metal-alkyls. Toxicol. Environm. Chem. **7** (3), 245—255.

XIAO, Z., MUNTHE, J., SCHROEDER, W. H., & LINDQVIST, O. (1988): Mercury fluxes over soil and lake surfaces. Report OOK 88:10, Dept. of Inorganic Chemistry, Univ. of Göteborg, Sweden.

XIE, T. M., HULTHE, B., & FOLESTAD, S. (1984): Determination of partition coefficients of chlorinated phenols, guaiacols and catechols by shake-flask GC and HPLC. Chemosphere **13** (3), 445—459.

Xun, L., Campbell, N. E. R., & Rudd, J. W. M. (1987): Measurements of specific rates of net methyl mercury production in the water column and surface sediments of acidified and circumneutral lakes. Can. J. Fish. Aquat. Sci. **44**, 750—757.

Yamashita, M., Kinae, N., Tomita, I., & Kimura, I. (1987): Effects of pH and temperature on degradation of chloroacetones that are mutagenic. Bull. Environ. Contam. Toxicol. **39**, 549—554.

Address of the authors:

Ingela Andersson
Helena Parkman
Arne Jernelöv
Swedish Environmental Research Institute
Box 21060
S-10031 Stockholm
Sweden

Limnologica (Berlin) **20** (2), 361—374, Juni 1990

Rezensionen

DOROTHY F. SOULE & G. S. KLEPPEL (eds.): **Marine Organisms as Indicators.** New York, Berlin, Heidelberg, London, Paris, Tokyo: Springer-Verlag 1988. XII, 342 S., 79 Abb. Geb. DM 98,—

Indikatororganismen spielen in der Ökologie der Gegenwart eine große Rolle, obwohl keinesfalls klar ist, inwieweit tatsächlich in allen Fällen direkte Beziehungen zwischen dem Auftreten eines Organismus und der gegebenen Umweltsituation bestehen. Der vorliegende Symposiumsband versucht, auf einem ganz bestimmten Teilgebiet, dem der marinen Ökologie, Antworten zu geben auf die Frage nach der ökologischen Bedeutung von Indikatororganismen. Es wird dabei deutlich, daß für die Interpretation biologischer Daten eine genaue Kenntnis der Taxonomie der Organismen eine unabdingbare Voraussetzung ist. Die Herausgeber empfinden es als schmerzlich, daß gegenwärtig taxonomische Untersuchungen sich nicht gerade eines guten Rufes erfreuen. Sie fordern sie aber als Voraussetzung für eine moderne Ökologie. Sie stellen außerdem fest, daß ohne eine detaillierte Kenntnis der Lebensgeschichte, der Physiologie und der Wachstumsbedingungen keine Bioindikation möglich ist.

Ausgehend von diesen allgemeinen Überlegungen vermitteln die 18 Spezialisten in insgesamt 11 Beiträgen ihre Vorstellung von der Bedeutung mariner Organismen als Indikatoren. Von einem neuen Zeitalter einer dynamischen biologischen Ozeanographie sprechen die Beiträge von HAURY & PIEPER sowie von SVEJKOVSKY et al., wenn sie biologische und physikalische Ozeanographie integrieren. HOURIGAN et al., KLEPPEL, STEPHENS et al., WENNER, SMITH et al. und REICH versuchen zu belegen, daß in vielen Fällen Organismen die einzigen effektiven Methoden liefern, um ökologische oder toxikologische Bewertungen vorzunehmen, und das trotz der enormen technologischen Fortschritte der Gegenwart. Im Grunde genommen ist der vorliegende Sammelband ein Appell für eine Biologie, die als Lehre von den Lebewesen verstanden wird. Solche Appelle sind in jüngster Zeit mehrfach geäußert worden. In diesem Falle ist es besonders interessant, daß es Ozeanographen sind, die sich für eine organismische Biologie stark machen. Unabhängig von diesen grundsätzlichen Erörterungen und Absichtserklärungen sind die vorgestellten Ergebnisse etwa über Pigmente als ernährungsdynamische Indikatoren, Fischvergesellschaftungen als Indikatoren für Umweltveränderungen in küstennahen Gebieten, in Korallenriffen oder im Ozean sowie über Toxizitätstests von grundsätzlicher Bedeutung. Der Sammelband wirft mehr Fragen auf als er beantwortet und dürfte gerade deswegen großes Interesse beanspruchen.

S. J. CASPER, Jena

UWE ARNDT, WILLFRIED NOBEL & BIRGIT SCHWEIZER: **Bioindikatoren.** Möglichkeiten, Grenzen und neue Erkenntnisse. Stuttgart: Verlag Eugen Ulmer 1987. 388 S., 36 Farbfotos, 139 Abb., 102 Tab. Kst. DM 68,—

Die Beurteilung von Schadstoffbelastungen der Umwelt mit Hilfe von Bioindikatoren gewinnt immer mehr an Bedeutung. Das vorliegende Buch setzt sich zum Ziel, die bis zum Jahre 1986 erschienene umfangreiche Literatur zu diesem Themenkreis kritisch zu sichten und ihre Ergebnisse und Probleme in zusammenfassender Schau allen Interessenten nahe zu bringen. Das ist, wie man den Bearbeitern bescheinigen kann, durchaus gelungen, wenngleich einige Bereiche auf Grund fehlender Sachkunde der Autoren nur stark verkürzt behandelt werden konnten bzw. ganz ausgeklammert werden mußten. Da zu diesen Lücken auch die Verschmutzung von Meeren und Ozeanen und die Eutrophierung von Gewässern gehören, liegt der Schwerpunkt des Buches auf den terrestrischen Ökosystemen, während auf die aquatischen Ökosysteme nur etwa ein Sechstel des speziellen Teils entfällt. Nachdem im Abschnitt 7 das Saprobiensystem zur Bestimmung der Wassergüte vorgestellt wird, wobei auch neuere Ergänzungen und Entwicklungen, aber auch die Probleme bei der Eichung der Arten zur Sprache kommen, werden im Abschnitt 8 (S. 272—293) die pflanzlichen Organismen und im Abschnitt 9 (S. 294—314) die tierischen Organismen zur Erfassung von Schadstoffbelastungen in aquatischen Ökosystemen abgehandelt. Damit ist das Buch auch für den Limnologen durchaus von Interesse, insbesondere auch durch die Vorstellung von Ergebnissen, die in nicht allgemein verbreiteten Arbeiten veröffentlicht wurden. Allerdings hat man gerade bei diesen Kapiteln den Eindruck, daß Publikationen außerhalb der BRD unterrepräsentiert sind.

Angesichts der wachsenden Fülle der Literatur über Bioindikation kann und will das vorliegende Buch nur eine Zwischenbilanz sein. Als solche aber ist es, trotz der genannten Einschränkungen, für jeden, den dieser Themenkreis beschäftigt oder berührt, von großem Wert.

H.-D. KRAUSCH, Potsdam

R. THOMAS, R. EVANS, A. HAMILTON, M. MUNAWAR, T. REYNOLDSON & H. SADAR: **Effects of In-Situ Contaminants** (Developm. in Hydrobiol. **39**). Dordrecht, Boston, London: Dr. W. Junk Publishers 1987. 272 S., mit zahlr. Abb. u. Tab. Geb. Dfl. 250.00

Die Internationale Joint Commission (IJC) wurde 1909 zum Schutz und zur Bewirtschaftung der Großen Amerikanischen Seen gegründet, sie zählt damit zu den ältesten Vereinigungen zum Schutz von Gewässern. Die Arbeitstagung in Aberrystwyth 1984 beschäftigte sich mit den Auswirkungen anthropogener Verunreinigungen von Sedimenten sowie den ökologischen und politischen Folgen. Band 39 der "Developments in Hydrobiology" veröffentlicht in sehr ansprechender Aufmachung und ausgezeichnetem Druck 22 Beiträge dieser Tagung.

Ausgehend von der Sedimentation-Resuspension und dem damit verbundenen Kreislauf der Elemente werden Xenobiontics, d. h. Spurenmetalle und organische Schadstoffe (PCB, CSB, AHS, DDT usw.) in ihrer Wirkung auf benthische Organismen und in der Nahrungspyramide über die Fische bis zum Menschen untersucht. Von besonderem Interesse sind die Beiträge, in denen die Auswirkungen erläutert werden, die ein gestörtes Ökosystem auf die Gesellschaft, auf das ethische Empfinden, auf den "social context" hat.

Auch Beispiele für die anthropogene Belastung von Gewässern in Europa werden vorgestellt: die Hg-Belastung in einem englischen Flußsystem (J. PARKS & A. HAMILTON), die Verunreinigung im Elbe-Ästuar (L. TENT) und im Golf von Neapel (V. DOMIANI) sowie die Bleibelastung durch den Bergbau in Wales (B. E. DAVIS).

Die Tagung ist beispielgebend dafür, daß Umweltprobleme über die lokale Situation bzw. den Rahmen einer Fachwissenschaft hinaus komplex beurteilt werden können, das trägt zur Entscheidungsfindung im gesellschaftlichen und politischen Bereich bei.

G. PROFT, Jena

R. I. JONES & V. ILMAVIRTA: **Flagellates in Freshwater Ecosystems** (Developm. in Hydrobiol. **45**). Dordrecht, Boston, London: Kluwer Academic Publishers 1988. VI, 279 S., 216 Abb., 37 Tab. Geb. Dfl. 260.00/US $ 145.00/UK £ 80.00

Vom taxonomischen Standpunkt aus gesehen ist noch immer nicht klar, ob die Flagellaten einen eigenen Status beanspruchen können. Während die Zoologen sie meistens als Ordnung in ihren taxonomischen Schemata führen, haben ihnen die Algologen oft einen höheren Rang und eine besondere Stellung zugebilligt. Auf diese Probleme will der vorliegende Band nicht eingehen. Er sieht die Flagellaten vielmehr vor dem Hintergrund ihrer Rolle in den aquatischen Ökosystemen und hier besonders in ihren Abhängigkeiten von den trophischen Gegebenheiten und von ihrer Umwelt überhaupt. Die Erkenntnis, daß diese beweglichen Kleinlebewesen unter Umständen die ihnen zusagenden Räume selbst aufsuchen können und daß sie außerdem ihre eigenen saisonalen Verteilungsmuster besitzen, macht sie zu besonderen Objekten der Hydrobiologie. Es war daher erfreulich, daß sich die IVL entschloß, dem Vorschlag der Abhaltung eines eigenen Symposiums über Flagellaten in Süßwasserökosystemen zuzustimmen. Es fand im Juni 1986 in der Biologischen Station Lammi der Universität Helsinki in Finnland statt und wurde von Teilnehmern aus 9 Ländern besucht.

Der vorliegende Band stellt eine Auswahl der seinerzeit vorgestellten Arbeiten dar. Er gliedert sich in 5 Schwerpunktgebiete: In die Taxonomie und Lebensgeschichte der Flagellaten, in ihre räumliche Verteilung, in ihr jahreszeitliches Auftreten, in ihre Abhängigkeit vom Trophiestatus der Gewässer und schließlich in die Beziehungen zwischen Umwelt und Flagellaten. Alle Beiträge können großes Interesse beanspruchen. Eine große Rolle spielen Goldalgen, *Ceratium*, *Cryptomonas*, *Gonyostomum* und *Chlamydomonas*. Sogar Neubeschreibungen von Arten finden sich (z. B. *Ceratium rhomvoides* HICKEL). Der allgemeine Tenor ist die Erkenntnis, daß ohne das exakte Wissen vom Wesen und von der Entwicklung der einzelnen Sippen keine Aussagen über die Ökologie gemacht werden können. Daß dies in vielen Fällen auf Schwierigkeiten stößt, zeigen vor allem auch Arbeiten, die sich mit Organismenresten aus See-Sedimenten beschäftigen. Hier ist der Rückschluß auf die rezenten Vertreter besonders schwierig. Ausführlich diskutiert wird auch die Bedeutung von Vertikalwanderungen, wobei die Abhängigkeit von den Größenverhältnissen zu einer kritischen Einschätzung führt. Insgesamt aber herrscht die Auffassung vor, daß in kleinen Gewässern die Bewegungsfähigkeit der Flagellaten zu einer besseren Versorgung mit Nährstoffen führen kann. Aufschlußreich sind auch Untersuchungen über das jahreszeitliche Auftreten der Flagellaten in Gebirgsseen, wobei interessante Unterschiede in der Höhenverteilung beobachtet werden konnten. Daß Flagellaten einerseits Bakterien fressen können, andererseits aber nährstoffreiche Beute für räuberische Organismen sein können, spielt in den Nahrungsnetzen eine wichtige Rolle.

Insgesamt gesehen ist das Spektrum der vorgestellten Arbeiten breit gefächert. Die Herausgeber kommen zu dem Ergebnis, daß es in Zukunft sich lohnen wird, sich mit den Flagellaten in den Süßgewässern intensiv zu befassen und schlagen vier zu bevorzugende Arbeitsrichtungen vor: Verbesserung der taxonomischen Untersuchungsmethoden unter Einschluß quantitativer Beobachtungen; integrierte Feld- und Laborstudien über die Bedeutung von Vertikalwanderungen; Kulturen zur Ermittlung der physiologischen Toleranzen einer größeren Zahl von Flagellaten und Erforschung des Verhältnisses von Autotrophie und Heterotrophie bei denjenigen Flagellaten, die beide Ernährungsweisen besitzen.

Ein informativer, gut gedruckter Band, geeignet, das Interesse vieler Hydrobiologen auf die Flagellaten zu richten.

S. J. CASPER, Jena

M. MUNAWAR (ed.): **Limnology and Fisheries of Georgian Bay and the North Channel Ecosystems** (Developm. in Hydrobiol. **46**). Dordrecht, Boston, London: Dr. W. Junk Publishers 1988. 222 S., 6 Schwarz-weiß-Fotos, 118 Abb., 57 Tab. Dfl. 225.00 /US $ 125.00/ £ 67.00

Dieser Band der Developments in Hydrobiology faßt die Resultate eines Symposiums über die Georgian Bay und das North Channel System zusammen, das während der 27. Konferenz über die Großen Seen abgehalten wurde.

Das behandelte Gebiet, im Norden des Lake Huron gelegen, wird in einem einleitenden Kapitel von SLY & MUNAWAR vorgestellt. Dabei gehen die Autoren von der Geologie des Gebietes aus, behandeln die Vegetation, Geschichte, Erschließung und Industrie und geben einen Überblick über die Limnologie und Fischwirtschaft.

Die folgenden Artikel befassen sich mit speziellen Problemen der Limnologie dieses Teils der Großen Seen in Nordamerika.

So legt BENETT in seinem Beitrag zur physischen Limnologie die Auswertung der Meßdaten von Zeitserien des Temperaturregimes, der Bodentopographie, des Einzugsgebietes und der Hydrographie vor. THOMAS beschreibt die Verteilung und Zusammensetzung der Oberflächensedimente von etwa 3 cm Dicke. Dabei wurden 3 Hauptkomponenten dieser rezenten Sedimente festgestellt: Geschiebelehm, glaciallacustriner Clay und Schlamm. Dazu kommt die nordöstliche ufernahe Zone mit einer Mischung aus Muttergestein und den genannten Hauptkomponenten in kleinräumig wechselnden Mengenanteilen. Die Hauptelemente und -mengen sowie die Schwermetallkonzentrationen wurden auf ihre Verteilung in den Sedimenten untersucht. SLY & SANDILANDS ergänzen dies hinsichtlich der geologischen und Umweltrelevanz der Sedimentverteilung in einem Gebiet steiler submerser Hänge mit z. T. 70 m dicken glaciolacustrinen Sedimenten aus Clay. Die Verteilung von organischem Kohlenstoff sowie Kupfer, Blei, Quecksilber und anderen Elementen wird dabei berücksichtigt. Gleichzeitig wird die Bodenfauna — vor allem Chironomiden, Oligochaeten und Amphipoden — in Abhängigkeit von Partikelzusammensetzung, Tiefe und Schluffanteilen der Sedimente untersucht.

WEILER beschreibt die Hauptfaktoren der chemischen Limnologie — Phosphor, Stickstoff und Chlorid — der oligotrophen Georgian Bay und des oligotrophen, aber stark phosphorbelasteten North Channels. Die Untersuchung der Phosphoreinträge in die südliche Georgian Bay und ihre Auswirkungen auf das Phytoplankton ist Gegenstand der Arbeit von NICHOLLS, ROBINSON, TAYLOR & CARNEY. Dabei wurden höhere Phytoplanktonwerte in der Nähe von Siedlungen und Flußmündungen gefunden, die gleichzeitig das sensible Reagieren des Phytoplanktons auf Nährstoffeinträge dokumentieren. Diese Untersuchungen von MUNAWAR & MUNAWAR zum Phytoplankton sowie zur Struktur und der Reaktion auf anthropogene Einflüsse der Georgian Bay bestätigen dieses sensitive Reagieren. Auch der folgende Artikel von MUNAWAR, MUNAWAR, MCCARTHY & DUTHIE zum Phytoplankton des North Channels (Zusammensetzung des Phytoplanktons, Bedeutung des Ultraplanktons sowie die 14-C-Primärproduktion) bestätigt diese Ergebnisse. Die Zooplanktonstruktur in ihrer Abhängigkeit vom Phytoplankton stellen SPRULES, MUNAWAR & JIN für die beiden Untersuchungsgebiete vor, finden aber keine grundsätzlichen Unterschiede zu den beiden angrenzenden Lake Superior und Lake Huron. Die Untersuchung der benthischen Makroalgen wurde von SHEATH, HAMBROOK & NERONE durchgeführt, wobei *Cladophora glomerata* und *Chara globulis/vulgaris* die dominanten Arten darstellen. Fädige Algen stellt JACKSON vor, der dabei besonderes Gewicht auf die Metallkonzentrationen innerhalb der Algen legte.

ROSS & MUNAWAR stellten in ihrer Arbeit nach Messung der Zooplanktonfiltrationsraten fest, daß nach den gefundenen Werten und dem Vergleich mit der Zooplanktonbiomasse das Algenfutter für das Zooplanktonwachstum nicht ausreicht. Den am Ende der Nahrungskette stehenden Fischen, den Auswirkungen der kommerziellen Fischerei sowie der Restaurierung der Salmonidenbestände widmet sich der Beitrag von HENDERSON & PAYNE. JOHNSON's Arbeit zur Effektivität der Seeneunaugenkontrolle (*Petromyzon marinus*), sowie den Auswirkungen der Predation auf den Fischbestand schließt den Band ab.

Für einen Teil der „Großen Seen" Nordamerikas werden somit die Untersuchungsergebnisse zusammengefaßt und einem breiten

Leserkreis zugänglich gemacht. Das Buch stellt einen gewissen Abschluß dar, macht aber gleichzeitig die vielen offenen Fragen deutlich.

D. RONNEBERGER, Neuglobsow

TH. CAPPENBERG & C. L. M. STEENBERGEN (eds.): **Proceedings of the Third International Workshop on the Measurement of Microbial Activities in the Carbon Cycle in Aquatic Ecosystems** (Arch. Hydrobiol., Beih. Ergebn. Limnol. 31). Stuttgart: E. Schweizerbart'sche Verlagsbuchhandlung (Nägele und Obermiller) 1988. 390 S., 111 Abb., 74 Tab. Brosch. DM 174,—

Nach vorangegangenen Veranstaltungen 1977 und 1983 am Max-Planck-Institut für Limnologie in Plön wurde der 3. Workshop in Utrecht vom Limnological Institute of the Royal Netherlands Academy of Arts and Sciences ausgerichtet. SIL und ICOME unterstützten das Treffen, an dem 64 Wissenschaftler aus 16 Ländern teilnahmen. Die an 4 Tagen präsentierten 45 Beiträge sind in diesem Band zusammengestellt.

Entsprechend dem Hauptanliegen der Veranstalter, einen Meinungsaustausch über die gegenwärtig verfügbaren Methoden zur Bestimmung bakterieller Aktivitäten im C-Kreislauf zu organisieren, findet man hierzu vielfältige Anregungen. Unverkennbar ist aber auch der bereits erzielte Erfahrungszuwachs bei der Quantifizierung bakterieller Leistungen. Auch werden diese, wie die Herausgeber einleitend betonen, zunehmend in strukturelle und funktionelle Analysen des gesamten Ökosystems eingeordnet. Dies zeigt sich gegenwärtig besonders deutlich bei der Betrachtung der bakteriellen Biomasseproduktion und des heterotrophen Mikrozooplanktons oder der mikrobiellen Prozesse an der Sediment-Wasser-Grenzfläche. Fortschritte bei der Vervollkommnung oder Entwicklung von Methoden zu erzielen, heißt in diesem Zusammenhang nichts weniger als Kochrezepte zu entwickeln. Vielmehr geht es um außerordentlich sensible Techniken und eine sinnvolle Kopplung von Labor- und Standortmessungen. Die damit verbundene Spezialisierung setzt freilich ausreichend große Arbeitsgruppen und Austauschmöglichkeiten voraus, wenn komplexe Fragestellungen auf dem jetzt zugänglichen Niveau bearbeitet werden sollen.

Die Beiträge sind zu 5 Komplexen zusammengefaßt: Coupling of laboratory experiments and field measurements (6 Vorträge); Measurement of biomass, activity and taxonomy in relation to both structure and function of microbial communities (13 Beiträge); Comparison of different methods in assessment microbial activity and biomass (9 Beiträge); Role of heterotrophic microzooplanktonic grazing activities of the microbial community on the turnover of organic matter (9 Beiträge); Aerobic and anaerobic carbon mineralization at the sediment-water interface (9 Beiträge).

Mit diesen repräsentativen Schwerpunkten ermöglicht der Tagungsbericht wiederum eine zuverlässige Orientierung über den Stand der Gewässermikrobiologie und ihre Entwicklungsrichtung.

H.-D. BABENZIEN, Neuglobsow

Verhandlungen der Gesellschaft für Ökologie. Band XVI (Jahrestagung Gießen 1986). Hrsg. L. STEUBING. Göttingen: Gesellschaft für Ökologie 1987. 512 S., 329 Abb., 92 Tab. Kart. DM 80,— (für Mitglieder der Gesellschaft für Ökologie kostenfrei)

Der vorliegende Band enthält die Vorträge der 16. Jahrestagung der Gesellschaft für Ökologie 1986 in Gießen, die ebenso wie die 1. Tagung der Belastung und Belastbarkeit von Ökosystemen gewidmet war. So befassen sich 4 Beiträge mit allgemeinen Überblicken und Darstellungen zu diesem Thema, 7 mit Waldökosystemen, 9 mit Agrarökosystemen, 7 mit Stadtökosystemen, 7 mit natürlichen Streßfaktoren, 29 mit anthropogenen Streßfaktoren und 6 mit Fragen des Naturschutzes; nur 4 Vorträge beinhalten freie Themen, und der Anhang enthält zur Einführung in die Exkursion in den Frankfurter Stadtwald eine Darstellung des Verfahrensganges und der ersten Ergebnisse der forstlich-ökologischen Beweissicherung im Raum der Startbahn 18 West des Frankfurter Flughafens. Obwohl das Generalthema auf terrestrische Ökosysteme eingeengt war, beschäftigen sich einige Arbeiten auch mit Gewässern. So untersuchten BAUER-DUBAU, SPECK & GUNKEL die Sukzession von Phyto- und Zooplankton von Kleingewässern in landwirtschaftlich genutzten Gebieten, und MOMPER & REDMANN prüften in ihrer Arbeit über Bioindikation von Schwermetallen mit Makrophyten die Eignung von *Ceratophyllum demersum*, *Elodea canadensis*, *Fontinalis antipyretica* und *Myriophyllum spicatum* als aktive Monitoren für Zn-, Pb- und Cd-Belastungen von Fließgewässern. Den Ursachen der variablen Photosyntheseeffizienz in einem Braunkohlerestgewässer in Hessen gingen DIETRICH, ARNDT & SCHWANTES nach. Die Folgen von Maßnahmen der Gewässerunterhaltung wurden von A. STÖCKMANN dargelegt, wobei die Autorin nachweisen kann, daß die starken Veränderungen der Ufer und des Profils eines kleinen Fließgewässers in Hessen drastische Folgen für die Besiedlung hatten. Das Vorkommen gefährdeter Köcherfliegenlarven (Trichoptera) am Vogelsberg untersuchte R. BURKHARDT. Als freies Thema legten GRUNERT & MROWKA die Ergebnisse ihrer Forschungen über Schwimmaktivität und Territorialverhalten einer tropischen Cichlidae-Art vor.

Der Band zeigt, daß den Problemen der gegenwärtigen Umweltbelastungen von den Ökologen in der BRD in vielfältiger Weise nachgegangen wird. Die hier vorgelegten Arbeiten vermitteln sowohl durch ihre Ergebnisse als auch durch ihre Methoden und Fragestellungen dort wie anderswo zahlreiche neue Erkenntnisse und weiterführende Anregungen.

H.-D. KRAUSCH, Potsdam

W. C. KERFOOT & A. SIH (eds.): **Predation, Direct and Indirect Impacts on Aquatic Communities.** Hanover, London: University Press of New England 1987. 386 S., 115 Abb., 66 Tab. Hard cover. US $ 72.00

In einer Arbeit des amerikanischen Limnologen J. SHAPIRO aus dem Jahre 1978 findet sich die Textstelle: "A few years ago we had a saying: Lakes are complex". Trotz dieser Bemerkung schienen die Vorstellungen der meisten Limnologen über das Ökosystem See damals eher einfach und wohlgeordnet. Das lag einerseits am konkreten Entwicklungsstand der Forschung, andererseits an der speziellen Ausrichtung des diesbezüglichen Teils des Internationalen Biologischen Programms (Produktivität der Binnengewässer).

Seit Mitte der siebziger Jahre nun, aber besonders im vergangenen Jahrzehnt, wurde eine Vielzahl von Erkenntnissen gesammelt, die erkennen lassen, auf welchen z. T. verschlungenen Wegen Organismen oder auch Organismengruppen miteinander in Wechselwirkung treten.

Der vorliegende Band hat sich hauptsächlich einem speziellen Typ dieser Wechselwirkungen gewidmet, den indirekten Effekten. KERFOOT & SIH sagen dazu im Vorwort: "The central theme of this volume (is), that the mere presence of a predator can alter the relationship between two or more competitors in a variety of important ways. . .".

Insgesamt werden zu diesem Thema 24 Beiträge vorgestellt, von denen 6 auf einem Symposium gebracht wurden, das schon 1984 in Fort Collins (Colorado, USA) stattgefunden hat. Die übrigen 18 Arbeiten wurden von anderen kompetenten Autoren erbeten. Die Hauptthemen sind: (1) Direct interactions, (2) Types of indirect interactions, (3) Food web dynamics, (4) Behavioral and morphological responses, (5) Local and regional extinctions: Historical ramifications of predation. Zur Illustration des Inhaltes sollen einige Arbeiten vorgestellt werden:

In einem Übersichtsartikel spricht W. J. O'BRIEN über die Probleme planktivorer Fische bei der Nahrungsbeschaffung. Räuber und Beute befinden sich hier in einem Prozeß ständiger Koevolution effektiveren Beuteerwerbs und wirksamerer Schutzmechanismen. W. C. KERFOOT widmet sich in einem Beitrag den sogenannten Kaskadeneffekten. Organismen, die im Wettbewerb um eine Nahrungsquelle unterlegen sind, ermöglichen einem Räuber oft auf indirekte Art das Überleben, indem der Räuber sich vom Konkurrenzgegner ernährt.

M. J. VANNI untersucht den indirekten Einfluß der Fische auf das Zooplankton. Er bietet Phytoplankton an, das aus Enclosuren mit und ohne Fischbesatz gewonnen wurde, und stellt unterschiedliche Ernährungswerte fest.

Ebenfalls in einem Übersichtsartikel referiert A. SIH über den Einfluß von Räubern auf die Lebensweise ihrer Beutetiere. *Notonecta*-Arten, z. B., die mit Fischen vergesellschaftet sind, werden von diesen schwerer aufgefunden, und bewegen sich deutlich weniger als ihre Verwandten, die nur fischlose Lebensräume besiedeln.

Besonders interessant ist ein Beitrag von SCRIMSHAW & KERFOOT, in dem chemische Räuberabwehr bei Wasserinsekten behandelt wird. Weit verbreitet sind solche Mechanismen bei Käfern, Wanzen und Wassermilben. Schließlich versuchen KERFOOT & LYNCH in einer paläolimnologischen Arbeit, das Auftreten kleiner Cladoceren im Mesozoikum mit der Evolution zooplanktonfressender Fische in Verbindung zu bringen.

Insgesamt enthält der Band eine Menge von neuen Erkenntnissen, die für das bessere Verständnis der Vorgänge im Seenökosystem von Bedeutung sind. Darüber hinaus wird an zahlreichen Stellen explizit zu weiteren Untersuchungen angeregt.

P. KASPRZAK, Neuglobsow

ULRICH MANIAK: **Hydrologie und Wasserwirtschaft**. Eine Einführung für Ingenieure. Berlin, Heidelberg, New York, London, Paris, Tokyo: Springer-Verlag 1988. XV, 576 S., 211 Abb. Brosch. DM 74,—

Oberirdische Wasservorräte, ihre quantitative Erfassung und Nutzung, die hydrologischen Berechnungsverfahren und wasserwirtschaftliche Bemessungsmethoden sind die Schwerpunkte dieses Hochschultextes, der für „Studierende des Bauingenieurwesens" bestimmt ist. Die gediegene ausführliche Darstellung und die vielschichtige Problemsicht der Wassermengenwirtschaft sind für den studierenden Ingenieur und jeden Leser sehr anregend. Interessante mathematische und graphische Auswertungen werden vorgestellt und werden sicher in die Praxis hinein wirken, wobei manche Sachgebiete noch ausbaufähig wären. Alarmierend ist die dabei immer wieder bewußtwerdende Lückenhaftigkeit unserer Meßmethoden und Meßnetze, selbst in Mitteleuropa. Die vom Autor eigentlich selbst geforderte ganzheitliche Betrachtung bei wasserwirtschaftlichen Problemen scheint aber aus dem Blick geraten zu sein, denn „... Grundwasser und Wassergüte werden weitgehend ausgeklammert". Bei der detaillierten Darstellung mancher Theorien und Modellierverfahren hätten auch für die Gewässergütefragen wenigstens Hinweise Platz gefunden. Der Nachholebedarf bei der Einbeziehung gewässergütewirtschaftlicher Kenntnisse und Zusammenhänge in das Ingenieurwissen stellt doch ein Hauptproblem unserer Zeit dar.

L. KREY, Jena

LEONHARD A. HÜTTER: **Wasser und Wasseruntersuchung.** Methodik, Theorie und Praxis chemischer, chemisch-physikalischer und bakteriologischer Untersuchungsverfahren. 3. neu bearb. u. erw. Aufl. Frankfurt a. M., München, Aarau, Salzburg: Verlag Diesterweg/Salle—Sauerländer 1988. XII, 448 S., 51 Abb., 42 Tab. Brosch. DM 68,—

Das gut eingeführte Laborbuch, das für Studentenpraktika und auch als kleines Nachschlagewerk zu gebrauchen ist, erscheint innerhalb einer relativ kurzen Zeitspanne von 4 Jahren nun in seiner 3. Auflage. Unter Beibehaltung der bewährten Konzeption (vgl. *Limnologica* **19** (2), S. 189), hat der Umfang um fast 100 Seiten zugenommen, und es wurden neben Umgestaltungen des Textes die Abschnitte Gewässerschutz, Badegewässer, Wasseraufbereitung und Abwasserbehandlung bedeutend erweitert. Die Umweltprobleme unserer Zeit führten zur Aufnahme der Abschnitte über die Gewässerversauerung und die Elemente Aluminium und Bor.

Wasserrechtliche Bestimmungen und Gesetze deutschsprachiger Länder wurden auf den neuesten Stand gebracht. Damit bleibt das gut handhabbare Buch mit seiner großen Materialfülle ein guter Begleiter für den Wasseranalytiker.

G. PROFT, Jena

W. FRESENIUS, K. E. QUENTIN & W. SCHNEIDER (eds.): **Water Analysis**. A Practical Guide to Physico-chemical, Chemical and Microbiological Water Examination and Quality Assurance. With Contributions by F.-J. BIBO, H. BIRKE, H. BÖHM, W. CZYSZ, H. GORBAUCH, H.-J. HOFFMANN, H.-H. RUMP & W. SCHNEIDER. Translated from the German by A. GLEDHILL, R. HOLLAND & T. J. OLIVER. Berlin, Heidelberg, New York, London, Paris, Tokyo: Springer-Verlag 1988. XXV, 804 S., 178 Abb., 55 Tab. Hard cover DM 148,—

Die vorliegende Publikation „Water Analysis" der „Deutschen Gesellschaft für Technische Zusammenarbeit (GTZ) GmbH" ist für Projekte der Zusammenarbeit mit Entwicklungsländern konzipiert. Diesem Anliegen entsprechend werden Methoden dargestellt, die in einfach ausgerüsteten Laboratorien Anwendung finden können.

Mit beeindruckender Ausführlichkeit und Genauigkeit werden drei Themen — Theorie der Arbeitstechniken, Methoden und Rezepte, Sammlung nationaler und internationaler Standards — in sechs Kapiteln behandelt.

Obwohl ein Minimum an technischer Ausrüstung angestrebt wurde, können Analysentechniken, die komplizierte und oft teure Geräte erfordern, wie Atomabsorptionsspektrometrie, Gaschromatographie, HPLC und Massenspektrometrie nicht unerwähnt bleiben. Die für verschiedene Einschätzungen notwendigen Konzentrationsnachweise von Schwermetallen, einigen organischen Stoffen, Radionukliden u. a. lassen sich nicht anders erbringen. In die theoretischen Grundlagen dieser Techniken wird im Kapitel 2 (125 Seiten) eingeführt.

Die Beschreibung der Methoden erfolgt auf ca. 500 Seiten in drei Kapiteln (Inorganic Parameters; Organic Parameters; Biological Analysis). Die Anleitungen sind nach folgendem Schema gegliedert: Theorie; Geräte und Chemikalien; Durchführung; Berechnung; Besonderheiten. Skizzen zum Geräteaufbau, Beispiel-Diagramme bzw. Tabellen illustrieren die Beschreibungen hervorragend. Hervorzuheben ist die Vielfalt der behandelten mikrobiologischen Untersuchungen. Den Abschluß des Buches bildet das Kapitel zur statistischen Auswertung der Daten und ihre Einschätzung anhand bestehender Standards der Wassergüte/Wasserhygiene.

Die Literatur ist auf Basisliteratur, Monographien und einige Zeitschriften beschränkt, im Text werden vollständige Zitate angeführt. Unter Anwendung der dargestellten Techniken ist ein Laboratorium in der Lage, aktuelle Zustandsanalysen von Seen, Abwasser oder Fließgewässern zu erarbeiten. Ohne produktionsbiologische Bezüge bleiben diese Daten Momentaufnahmen, die kaum eine Aussage über die Entwicklung der Wassergüte zulassen.

Mit kaum erweitertem Instrumentarium sind Untersuchungen der Sedimente, der Primärproduktion des Phytoplanktons, der bakteriellen heterotrophen Aktivität u. a. durchführbar. Bei einer späteren Auflage sollten diese Methoden Berücksichtigung finden.

„Water Analysis" stellt neben den etablierten deutschsprachigen Methodenbüchern eine Bereicherung dar und sollte seinen Platz in allen Laboratorien der Wasseruntersuchung finden.

P. CASPER, Neuglobsow

G.-J. KRAUSS & G. KRAUSS: **Hochleistungs-Flüssigchromatographie.** SERVA-Praxishefte, Band 2. Heidelberg: SERVA FEINBIOCHEMICA GmbH & Co. 1986. 115 S., 25 Abb. u. 12 Trennprofile, 13 Tab. Brosch.

Im Jahre 1984 behandelten die Autoren im Band 1 der Reihe SERVA-Praxishefte Grundlagen der Chromatographie. Im vorliegenden Band 2 wird die Hochleistungs-Flüssigchromatographie — HPLC-Technik vorgestellt.

Neben den Hauptkapiteln „Apparative Ausstattung" (30 Seiten) und „Chromatographisches System" (35 Seiten) enthält das Heft eine Einführung, ein Literaturverzeichnis, ein Sachverzeichnis sowie 9 Applikationsbeispiele einschließlich ausgewählter Trennprofile und spezieller Literatur.

Der Band vermittelt die Grundlagen der Hochleistungs-Flüssigchromatographie, er zeigt die Stellung dieser Technik im System der chromatographischen Verfahren und verweist auf Vor- und Nachteile der behandelten Technik. Die Autoren verzichteten weitgehend darauf, mögliche kommerzielle Produkte und Bezugsquellen aufzuführen. Hervorzuheben sind die sehr guten Abbildungen und Tabellen (z. B. Abb. 19: Mischbarkeit von Lösungsmitteln; Abb. 24: Entscheidungsschema für die Wahl des Trennungsprinzips; Tab. 11: UV-Absorption verschiedener funktioneller Gruppen; Tab. 12: Optimierungsmöglichkeiten). Die Arbeiten zu diesem Heft wurden 1984 abgeschlossen, die Literatur bis zu diesem Zeitpunkt ist berücksichtigt.

Obwohl die HPLC-Technik in den vergangenen fünf Jahren eine rasche Entwicklung erfuhr, bleibt dieses Heftchen als Einführung in die Technik bzw. als Grundlagendarstellung aktuell. Es kann allen Wissenschaftlern empfohlen werden, die vor der Aufgabe stehen, Nachweise oder Separationen von Verbindungen durchzuführen, die mit anderen Methoden nicht möglich sind.

P. CASPER, Neuglobsow

Archiv für Hydrobiologie, Suppl.-Band 76, Heft 3. Stuttgart: E. Schweizerbart'sche Verlagsbuchhandlung (Nägele u. Obermiller) 1987. 117 S., 37 Abb., 36 Tab. Brosch. DM 90,60

Das vorliegende Heft enthält 4 monographische Beiträge. EVA PIP & CAROLE SUTHERLAND-GUY untersuchten aquatische Makrophyten in dem in 323 m Meereshöhe im Präkambrischen Schild Kanadas an der Grenze der Provinzen Manitoba und Ontario gelegen Shoal-See. Der erste, von beiden Autorinnen gemeinsam verfaßte Teil der Studie behandelt Diversität, Biomasse und Stoffwechselstatus in Abhängigkeit von Wassertiefe und Lichtintensität (S. 197—222). Insgesamt werden für das untersuchte Areal 26 Arten von Wasserpflanzen angegeben; es muß jedoch bedauert werden, daß einige Sippen (*Sparganium, Sagittaria, Nitella, Chara, Drepanocladus*) nicht bis zur Art bestimmt worden sind. Näher untersucht wurden allerdings nur 6 Arten (*Ceratophyllum demersum, Myriophyllum exalbescens, Potamogeton zosteriformis, P. foliosus, Elodea canadensis, Najas flexilis*), von denen 3 auch in Mitteleuropa vorkommen. Im 2., allein von E. PIP verfaßten Teil der Studie (S. 223—235) werden jahreszeitliche und lokale Chlorophyll-Konzentrationen in Abhängigkeit von Temperatur und Wasserchemismus der oben genannten 6 Wasserpflanzen untersucht.

Den cyclopoiden Copepoden *Mesocyclops leuckartii* und *Cyclops scutifer* in Randbiotopen von 2 sauren Seen im südlichen Norwegen galten vierjährige Untersuchungen von ST. SANDOY & J. P. NILSSEN. Einer dieser Seen wurde aufgekalkt, so daß sein pH-Wert von 4.7 auf 7.5 anstieg, was auf beide Arten einen positiven Effekt hatte (S. 236—255). Die umfangreiche Studie von P. C. DALL über die Ökologie der Egelfauna im flachen und steinigen Litoral des Esrom-Sees in Dänemark behandelt Lebenszyklen, Reproduktion, Mortalität, Wachstum und Nettoproduktion von 6 nichtparasitären Egel-Arten (S. 256—313).

Alle 4 Beiträge bieten eine Fülle von Daten und neuen Erkenntnissen zur Ökologie von Binnengewässern im nördlichen Randbereich der temperaten Zone der Holarktis.

H.-D. KRAUSCH, Potsdam

Decheniana 141. Verhandlungen des Naturhistorischen Vereins der Rheinlande und Westfalens. Bonn: Selbstverlag des Naturhistorischen Vereins 1988. 345 S., 94 Abb., 85 Tab. Brosch.

Der Band 141 der Decheniana zeichnet sich nicht nur durch seinen Umfang aus, sondern auch durch die große Anzahl wertvoller Arbeiten aus allen Bereichen, die von dieser Schriftenreihe gestreift werden. Neben 25 floristischen, pflanzensoziologischen, faunistischen, biogeographischen und geologischen Beiträgen, vorwiegend aus Nordrhein-Westfalen und Rheinland-Pfalz, enthält dieses Heft auch eine Reihe interessanter limnologischer Arbeiten.

So berichtet J. H. JUNGBLUTH über die Situation der Flußperlmuschel *Margaritifera margaritifera* in der ehemaligen Preußischen Rheinprovinz (Nordrhein-Westfalen, Rheinland-Pfalz und Saarland). Besonderes Interesse dürfte die Arbeit von HARBERS et al. finden, in der eine umfangreiche Bestandsaufnahme zu „Fauna und Siedlungsdichten — insbesondere der Mollusken — auf der Sohle des Rhein-Herne-Kanals" vorgelegt wird. Dieser Kanal, bereits heute eine der meistbefahrensten westdeutschen Wasserstraßen, wird zu einer Europawasserstraße ausgebaut. Die Abhängigkeit der Siedlungsdichte vom Ausbauzustand, vom Nutzungstyp und von der Substratkombination wird diskutiert.

Hingewiesen sei auch auf die Arbeit von THIESMEIER et al., einer ökologischen Grundlagenerhebung. Es wird der Versuch gemacht, alle noch bestehenden Bachläufe und Bachrelikte der Großstadt Bochum zu erfassen. Der besondere Wert dieses ohnehin interessanten Vorhabens liegt darin, daß die gewonnenen Daten und das Kartenmaterial den entsprechenden Ämtern die Möglichkeit geben sollen, durch Renaturierungsmaßnahmen zur Verbesserung und Wiederherstellung des Bachnetzes beizutragen.

Auch wenn nur drei Arbeiten kurz umrissen werden konnten — das gesamte Heft weist eine solche Vielzahl wichtiger Arbeiten auf, daß es jedem Biologen sehr empfohlen werden kann.

A. LANGE, Jena

Beiträge zur Meereskunde. Hrsgeg. von KLAUS VOIGT unter Mitwirkung von HANS-JÜRGEN BROSIN, LUTZ BRÜGMANN, DIETER LANGE, WOLFGANG MATTHÄUS, DIETWART NEHRING, KLAUS STRIGGOW und MANFRED STURM (Schriftl.). Berlin: Akademie-Verlag.
Heft 58 (1988): 75 S., 50 Abb., 10 Tab. Brosch. M 20,—
Heft 59 (1989): 82 S., 34 Abb., 18 Tab. Brosch. M 20,—

Heft 58 hat fast ausschließlich Ostsee-Untersuchungen zum Inhalt, wobei auch limnologisch interessante Themen behandelt werden: H.-D. BIRR legt Untersuchungen zu den Strömungsverhältnissen im Strelasund dar, H. U. LASS behandelt theoretische Untersuchungen zum Wasseraustausch zwischen Nord- und Ostsee, H.-J. BROSIN et al. bringen interessante Informationen und Graphiken über kleinräumige Wasseraustauschstrukturen in der südöstlichen Ostsee. Hydrochemisch-hydrobiologische Spezialfragen des Chlorophyllgehaltes der Ozeane stellt D. LOMMATZSCH vor, wobei ein Biospektrometer MKS-BS aus der Satellitenerkundung auf seine Anwendbarkeit vom Schiff aus erfolgreich getestet wurde. Wissenschaftliche Kurzinformationen zu verschiedenen Detailfragen beschließen das informative Heft.

Heft 59 bringt u. a. von A. BACHOR Untersuchungen zum Salz- und Nährstoffhaushalt des Unterwarnow-Ästuars, von E. HAGEN & R. SCHEMAINDA Profilschnitte zum Unterstrom vor Nordwest-Afrika, sowie von D. NEHRING Phosphat- und Nitrat-Probleme und -Trends im Sediment-Bodenwasserbereich in der zentralen Ostsee. Auswertungen von G. SAGER & R. SAMMLER über den Makrelenhai sowie von M. SCHARLIPP über meßtechnische Vergleiche an Tiefseekippthermometern runden das thematisch sehr vielseitige Heft ab.

L. KREY, Jena

S. SKRESLET (ed.): **The Role of Freshwater Outflow in Coastal Marine Ecosystems.** Berlin, Heidelberg, New York: Springer-Verlag 1986. XI, 453 S. Hard cover DM 198,—

Das Buch ist in der Reihe der NATO-ASI-Serie erschienen. Ihm liegen die Vorträge und Ergebnisse eines Workshops in Bödo, Norwegen, aus dem Jahre 1985 zugrunde, an dem 34 Wissenschaftler aus 9 Ländern teilgenommen haben.

Insgesamt 27 Vorträge zu physikalischen, chemischen und biologischen Problemen des Süßwasserstroms in Küstengewässern sind in den Kapiteln „Invited Contributions, Inshore Waters, Marginal Seas und Shelf Waters" geordnet. Aus den unterschiedlichsten Brackwassergebieten, Ästuarien, Fjorden und Küstengewässern der Erde werden Fragen der Durchmischungs- und Strömungsprozesse, der Frontenbildung, des Nährstoffeintrags der Küstenfischerei, der Planktonverteilung und -produktion sowie der Ökosystemmodellierung behandelt. Völlig unberücksichtigt blieben Probleme der Schwermetallanreicherung in Ästuarien und mikrobiologische Fragen, einschließlich des Eintrags pathogener Mikroorganismen in die Küstengewässer.

Alle Beiträge sind übersichtlich gegliedert, straff formuliert und haben einen hohen Informationsgehalt. Zahlreiche Literaturzitate ergänzen den Text und ermöglichen eine weitergehende Information. Am Schluß des Bandes befinden sich die Berichte der Arbeitsgruppen, ein ausführlicher zusammenfassender Bericht über den Workshop sowie Empfehlungen für eine internationale Forschungskooperation. Ein Stichwortverzeichnis ermöglicht eine rasche Orientierung.

Das Buch kann einem breiten Leserkreis der Ozeanographie, Limnologie, Fischerei und der wasserwirtschaftlichen Praxis empfohlen werden.

H. HÜBEL, Kloster a. Hiddensee

Biologia Acuatica. Schriftenreihe des Instituts für Limnologie „Dr. Raul A. Ringulet" der Universität La Plata. Hrsgeg. von HUGO L. LOPEZ. **Heft 13** (1988). 140 S. Brosch. US $ 5.00

Nachdem das im Jahr 1984 erschienene Heft 6 der Reihe „Biologia Acuatica" bereits eine Zusammenstellung der argentinischen limnologischen Literatur von 1979—1983 gibt, wird diese Bibliographie mit dem Heft 13 vervollständigt, das Arbeiten über argentinische Gewässer umfaßt, die zwischen 1961 und 1978 erschienen sind.

Die Bibliographie enthält folgende Gebiete: Chemie (2 S.), Geologie (4 S.), Gefäßpflanzen (12 S.), Wirbellose (außer Insekten; 27 S.), Plankton (12 S.), Insekten (16 S.), Fische (23 S.), Kriechtiere (9 S.), Vögel (8 S.) und Säugetiere (3 S.).

Die hier zitierten Arbeiten sind vorwiegend in argentinischen bzw. lateinamerikanischen Zeitschriften veröffentlicht.

A. LANGE, Jena

DON MCALLISTER & EDWARD KOTT (eds.): **On Lampreys and Fishes:** A Memorial Anthology in Honor of VADIM D. VLADYKOV (Developm. in Environm. Biol. of Fish. **8**). Dordrecht, Boston, Lancaster: Dr. W. Junk Publishers 1988. 160 S., 56 Schwarz-weiß-Fotos, 16 Abb., 22 Tab. Geb. Dfl. 165.00 /US $ 83.00/ £ 48.00

Das Buch ist dem Andenken VADIM VLADYKOVS (1898—1986) gewidmet, einem bedeutenden Ichthyologen, der seit 1930 in Kanada arbeitete.

Nach einer ausführlichen Biographie VLADYKOVS von MCALLISTER und der Auflistung der 290 Publikationen sowie der Neubeschreibungen folgt eine Reihe von ichthyologischen Originalarbeiten. Dabei bildet die letzte Arbeit VLADYKOVS über die Urogenitalpapillen der holarktischen Lampreten, zusammen mit KOTT und REMAND verfaßt, auf Grund ihrer Bedeutung für taxonomische Zwecke den Anfang. Es wird dabei herausgestellt, daß nur die männlichen Urogenitalpapillen nichtparasitischer Lampreten groß genug und sicher unterscheidbar sind, um als taxonomisches Merkmal Verwendung zu finden. Umfangreiches Material von 30 holarktischen Lampretenarten wurden dabei mitberücksichtigt. Im nächsten Beitrag geben WILLIAM, BEAMISH & MUDLAND eine Übersicht über die Metamorphose von *Ichthyomyzon greeleyi*, einer in den Bergflüssen von Ohio und Tennessee lebenden Lampretenart. Eindrucksvolle Zeichnungen geben einen Überblick über die Entwicklung aller Larvenstadien bis zum adulten Tier, wobei besonders auf die Veränderungen in der Kopfregion eingegangen wird. Als Vergleich werden in einer Tabelle die Entwicklung parasitischer und nichtparasitischer Larven verglichen. LANTEIGNE gibt einen ausgezeichneten, mit vielen Photos belegten und mit einem Bestimmungsschlüssel versehenen Überblick zur Identifikation der Larven der Gattung *Ichthyomyzon*. KOTTELAT & CHU XIN-LUO beschreiben in ihrer Revision der Gattung *Yunnanilus* sechs neue Arten. Diese Fischgruppe, bisher aus China und Burma bekannt, ist an die unterschiedlichsten Habitatstrukturen und Futterarten angepaßt, wobei z. T. extreme Biotope besiedelt werden. Von den bisher bekannten 14 Arten werden in einer Bestimmungstabelle alle zugänglich gemacht und taxonomisch neu- oder nachbeschrieben. Photos der einzelnen Arten und eines Teils ihrer Habitate runden die Arbeit ab. RENAUD & MCALLISTER beschreiben im folgenden Artikel den taxonomischen Status von *Rhinichthys cataractae smithi* aus den heißen Quellen Albertas als endemische Unterart. Ausführliche taxonomische Vergleichsdaten und eine Bestimmungstabelle der Gattung sowie eine Darstellung der verschiedenen Hypothesen der ökologischen und räumlichen Abgrenzung der Unterart vervollständigen die Arbeit. Aus dem Iran beschreibt COAD einen neuen Zahnkarpfen als *Aphanius vladykovi*. Dabei werden auch hier ausführliche Vergleichsmaße von verwandten Arten dargeboten sowie die zoogeographischen Besonderheiten und Habitate beschrieben.

Einige kleinere Beiträge von RENAUD & HOLČIK zur Problematik der Synonymie von *Lampetra (Eudontomyzon) gracilis* und *Eudontomyzon danfordi*, von COAN & PAPAHN über Haiattacken in iranischen Flüssen sowie von MCCAULEY & THOMSON zur Thermoregulation von *Cyprinodon nevadensis* in heißen Quellen schließen sich an. Die abschließende Arbeit von RUBEC zur Erhaltung und Bewirtschaftung philippinischer Korallenriffe wendet sich in erster Linie den zerstörenden Folgen der destruktiven Fischereimethoden und des massenhaften Korallen- und Muschelabbaues für aquaristische Zwecke zu. Diese Methoden haben eine ähnlich katastrophale Auswirkung wie die Abholzung der tropischen Regenwälder.

D. RONNEBERGER, Neuglobsow

KURT O. VIETS: **Die Milben des Süßwassers** (Hydrachnellae und Halacaridae [part.], Acari). **Teil 2: Katalog** (Sonderbände des Naturwissenschaftlichen Vereins in Hamburg, Band **8**). Hamburg, Berlin: Verlag Paul Parey 1987. 1012 S. Geb. DM 280,—

Nur wenige Jahre nach dem Erscheinen des 1. Teils (Bibliographie; Sonderbände Naturwiss. Ver. Hamburg, Band 6, 1982) liegt nunmehr sowohl vom wissenschaftlichen Inhalt als auch von der praktischen Handhabe her ein modernes Katalogwerk vor, dessen Nutzen für die zukünftige Entwicklung der Wassermilbenforschung nicht annähernd überschaubar ist. Süßwassermilben bilden mit ihren über 4000 Arten (47 Familien, 295 Gattungen) eine zwar verwandtschaftlich heterogene, aber ökologisch sehr einheitliche Gruppe, deren Formenfülle gerade auch in den letzten Jahrzehnten zu einer unübersehbaren Anzahl von Einzelpublikationen geführt hat. Der Verfasser hat im Anschluß an das berühmte Werk seines Vaters KARL VIETS „Die Milben des Süßwassers und des Meeres" (VEB Gustav Fischer Jena, 1955 und 1956) eine Aufgabe bewältigt, die sicherlich derjenige am besten ermessen wird, der sich jemals dem mühsamen Unterfangen der Katalogisierung einer Tiergruppe unterzogen hat.

Mit 818 Seiten nimmt der Artenkatalog den größten Teil des Buches ein. In ihm sind die Namen aller Taxa bis hinunter zur Unterart erfaßt und alphabetisch angeordnet. Aufgenommen sind auch alle Homonymie-Fälle, die in neuerer Zeit entstanden sind, ohne Rücksicht darauf, ob die dazu notwendigen Neubenennungen bereits erfolgten oder nicht. Arten älterer Gattungen, die nicht mehr in das alte ursprüngliche Genus gehören, werden in der alten Gattung mit einem Hinweis auf ihre Neuzuordnung genannt. Enthalten sind auch die Namen von noch nicht beschriebenen Arten (Nomina nuda). Hinter jedem einzelnen Taxon ist die Literatur nach 1956 zitiert. Zitate aus dem alten Katalog von K. VIETS werden daher nicht wiederholt. Den entsprechenden Zugang findet man durch den Verweis „K. VIETS 1956b".

Bibliographie und Katalog widerspiegeln den aktuellen Stand der Hydrachnellen-Forschung. Die gründliche Aufbereitung aller erfaßbaren nomenklatorischen, taxonomischen und bibliographischen Informationen sowie die kritischen Ergänzungen durch den Autor erlauben ein schnelles und sicheres Arbeiten. Beide Bände werden daher eine weit in die Zukunft reichende unentbehrliche Grundlage für den Fachkarologen sein. Dies dürfte mehr als eine bloße Anerkennung für die bewundernswerte Einzelleistung eines Autors sein, der seit Jahrzehnten das Profil der Wassermilbenforschung bestimmend beeinflußt.

M. MORITZ, Berlin

DIETER MATTHES: **Tierische Parasiten** — Biologie und Ökologie. Braunschweig, Wiesbaden: Friedrich Vieweg & Sohn Verlagsgesellschaft mbH 1988. 253 S., 139 Abb. Geb.

„Parasitologien" gibt es schon viele, und der Zugang zu diesem theoretisch interessanten wie praktisch wichtigem Gebiet der biologischen Forschung ist sehr verschieden. Der Vorzug dieses Parasitologie-Lehrbuches ist, daß ein erfahrener Hochschullehrer „einen wirklich lesbaren Text, unterstützt durch gute Bilder" vorlegt. Es ist ein kurz gefaßter Überblick über die Biologie und Ökologie tierischer Parasiten — für den Anfänger eine sehr gute Hilfe zur Einarbeitung in dieses umfangreiche Wissensgebiet, für den Biologen, Ökologen und Mediziner gleichfalls eine gewinnbringende Lektüre. So dargelegt kann Parasitologie für den Studenten sehr interessant sein und zum Studium anregen.

Zunächst wird der Parasitismus den nichtparasitären Vergesellschaftungen (Symbiosen, Karposen) gegenübergestellt, anschließend werden die mannigfaltigen parasitischen Beziehungen in folgender Weise gegliedert: Brutparasitismus, Sozialparasitismus (bei Ameisen), Raumparasitismus, Kleptoparasitismus, Ektoparasitismus, vom Ekto- zum Endoparasitismus, vom kommensalen Synphorismus zum Endoparasitismus und schließlich der Endoparasitismus (umfangreichstes Kapitel) selbst. Diese zunehmend engeren Beziehungen zwischen Wirt und Parasit werden an Hand von vielen Beispielen überzeugend dargestellt. Abgerundet wird das Buch durch ein Kapitel zum Larval-Parasitismus.

Besonders auffallend sind die eigenwilligen Überschriften für die kleineren Abschnitte, die zum Lesen anregen werden. Auf 14 Seiten wird wichtige, weiterführende Literatur genannt, und ein 16seitiges Register unterstützt die Orientierung.

G. Schäller, Jena

Heinz Stolp: **Microbial Ecology: Organisms, Habitats, Activities** (Cambridge Studies in Ecology). Cambridge, New York, New Rochelle, Melbourne, Sydney: Cambridge University Press 1988. 308 S., 50 Abb. Paperback. £ 12.95/US $ 22.50

Diese Einführung in die Mikrobenökologie ist in erster Linie für Studenten geschrieben, die über Grundkenntnisse in der allgemeinen Mikrobiologie verfügen. Sie entspricht in ihrem Anliegen der Notwendigkeit, Mikroorganismen in die Ökosystemanalyse einzubeziehen, also die Vielfalt ihrer Leistungen zu bestimmen und die Abhängigkeiten von den konkreten Standortbedingungen verstehen zu lernen. Nach einer Einführung, die kurz auf Aufgaben und Stellung der Mikrobenökologie eingeht, bieten 2 Kapitel eine Kurzfassung allgemeiner mikrobiologischer Grundlagen. Es sind Stichworte wie Ernährungstypen, Wachstum, Stoffwechsel, Kulturtechniken sowie eine knappe Kennzeichnung der taxonomischen Hauptgruppen enthalten. Dann werden die natürlichen Biotope Boden und Wasser sowie Extrembiotope mit ihren spezifischen Besonderheiten vorgestellt. Ein weiteres Kapitel informiert über die Wirkung einzelner Umweltfaktoren auf das Verhalten der Mikroorganismen und ihre Adaptationsstrategien. Es folgt eine Darstellung der verschiedenen Möglichkeiten von Interaktionen zwischen Mikroorganismen bzw. Mikro- und Makroorganismen. Schließlich wird die Rolle der Mikroorganismen im Kreislauf der wichtigsten biogenen Elemente behandelt, wobei auch auf den mikrobiellen Abbau von Fremdstoffen, z. B. Pestiziden, eingegangen wird. Das abschließende Kapitel ist der Orientierung auf die Fachliteratur zu grundlegenden methodischen Aspekten gewidmet. In einem kurzen Epilog weist der Autor auf die dringlichen Aufgaben hin, die für die Mikrobenökologie in Hinblick auf die Umweltqualität entstanden sind. Das Buch ist mit einem angemessenen Sachwortverzeichnis ausgestattet. Die Literatur ist kapitelweise geordnet und verweist auf wichtige Lehrbücher und wichtige, meist zusammenfassende, Einzelarbeiten. Man kann auf der Grundlage dieses Buches durchaus einen hinlänglichen Überblick über Grundfragen der Mikrobenökologie erhalten. Daher wird es wohl vor allem da willkommen sein, wo es um eine erste Orientierung geht.

H.-D. Babenzien, Neuglobsow

Alexander J. B. Zehnder (ed.): **Biology of Anaerobic Microorganisms** (Wiley Series in Ecological and Applied Microbiology). New York, Chichester, Brisbane, Toronto, Singapore: John Wiley & Sons, Inc. 1988. 872 S., 186 Abb., 79 Tab. US $ 77.50

Die Kenntnisse über anaerobe Bakterien haben sich im letzten Jahrzehnt in ungewöhnlicher Weise erweitert. Die Besiedlung extremer Habitate, biochemisch/physiologische Besonderheiten und eine noch nicht voll überschaubare Fülle potentieller Leistungen gehören zu den Gründen, die diesen Bakteriengruppen ein fast spektakuläres Interesse gesichert haben. Dem Herausgeber dieses bemerkenswerten Buches ist es gelungen, mit 18 weiteren prominenten Autoren eine ganz aktuelle Übersicht vorzulegen. Die einzelnen Beiträge sind in sich abgerundete Darstellungen der physiologisch zu definierenden Gruppen. Sie enthalten also taxonomische, ökologische und biochemische Aspekte bis hin zu biotechnologisch interessanten Leistungen, wobei wohl durchgängig in der physiologischen Interpretation ökologischer Phänomene ein Vorzug des Buches gesehen werden kann. Es ist in folgende Kapitel gegliedert: Geochemistry and Biogeochemistry of Anaerobic Habitats (A. J. B. Zehnder & W. Stumm), Microbiology, Physiology, and Ecology of Phototrophic Bacteria (M. T. Madigan), Molecular Mechanisms of Bacterial Photosynthesis (J. Amesz & D. B. Knaff), Ecology of Denitrification and Dissimilatory Nitrate Reduction to Ammonium (J. A. Tiedje), Dissimilatory Reduction of Oxidized Nitrogen Compounds (A. H. Stouthammer), Microbial Reduction of Manganese and Iron (W. C. Ghiorse), Anaerobic Microbial Degradation of Cellulose, Lignin, Oligolignols, and Monoaromatic Lignin Derivates (P. J. Colberg), Anaerobic Hydrolysis and Fermentation of Fats and Proteins (M. J. McInerney), Acetogenesis (J. Dolfing), Microbiology and Ecology of Sulfate- and Sulfur-Reducing Bacteria (F. Widdel), Dissimilatory Reduction of Sulfur Compounds (J. LeGall & G. Fauque), Biogeochemistry of Methanogenic Bacteria (R. S. Oremland), Biochemistry of Methane Production (G. D. Vogels, J. T. Keltjens & Ch. van der Drift), Principles and Limits of Anaerobic Degradation: Environmental and Technological Aspects (B. Schink). Ein ausführlicher Index erleichtert den schnellen Zugang zur Klärung einzelner Sachfragen. Mit insgesamt über 3000 Literaturzitaten (bis 1987!) ermöglicht dieses Buch eine wohl gegenwärtig einmalige geschlossene Information über Zusammenhänge und Fakten, die heute zum wesentlichen Grundlagenwissen der allgemeinen Mikrobiologie gerechnet werden müssen. Nicht weniges davon ist aus ökologischen Forschungsthemen hervorgegangen. Auch insoweit handelt es sich um ein richtungsweisendes Buch, das den Fortschritt der Mikrobenökologie markiert.

H.-D. Babenzien, Neuglobsow

Exkursionsflora für die Gebiete der DDR und BRD. Begründet von W. Rothmaler. Bd. 3: Atlas der Gefäßpflanzen. 6., völlig neu bearb. Aufl. Hrsgeg. von R. Schubert, E. Jäger & K. Werner. Berlin: Volk und Wissen, Volkseig. Verlag 1987. 752 S., 2814 Abb., Geb. M 30.—

Mit großem Interesse nehmen wir Kenntnis von dem Erscheinen der 6. Auflage des Bandes 3 der Rothmaler'schen Exkursionsflora, da es sich bei ihm um eine vollständige Neubearbeitung handelt. Es ist von vielen Seiten angemerkt worden, daß die bisherigen Auflagen des Atlasses der Gefäßpflanzen nicht mehr dem wachsenden Niveau des Bestimmungswerkes entsprachen. Es soll gleichzeitig betont werden, daß die 14 Wissenschaftler und 17 Graphiker, die für die neue Konzeption verantwortlich zeichnen, mit Umsicht und Können ans Werk gegangen sind und in mehr als 5jähriger Arbeit eine ausgezeichnete Publikation vorgelegt haben. Die wissenschaftlichen Abbildungen von 2814 Pflanzenarten repräsentieren mit Ausnahme ganz weniger Sippen alle im Band 2 der Exkursionsflora enthaltenen Arten, darüber hinaus einige wenige im Band 2 nicht aufgeführte Arten. Die Abbildungen sind so angelegt, daß die wesentlichen Merkmale trotz der notwendigen Verkleinerung deutlich sichtbar bleiben. Der größte Teil wurde nach Frischmaterial gezeichnet, die übrigen nach Herbarbelegen aus dem Gebiet. Das Ergebnis ist wirklich erstaunlich. Die Abbildungen sind durchweg klar, übersichtlich, sachgerecht und aussagekräftig. Fast immer ist es gelungen, einen klaren Eindruck vom Habitus (auch bei Bäumen!) zu vermitteln. Wenn einzelne Abbildungen nicht ganz dem allgemein hohen Standard entsprechen, weil sie drucktechnisch nicht gelungen sind (z. B. Seite 546, 580), so schmälert es nicht den hervorragenden Eindruck, den der ganze Band macht. Die Darstellung diagnostisch wichtiger, vor allem vegetativer Merkmale unterstützt die Bestimmungsschlüssel wirksam. Literaturverzeichnis, Register der deutschen und wissenschaftlichen Pflanzennamen sowie ein Nachweis der Bearbeiter und Graphiker vervollständigen das ausgezeichnet gelungene Werk, das über die Darstellung der Pflanzen hinaus zu einer wesentlichen Bereicherung der morphologischen Kenntnisse der heimischen Flora führen wird. Mit dem Atlas dürfte sich die Lebensfähigkeit des ROTHMALER erneut erwiesen haben. Text- und Bildautoren gilt die volle Anerkennung der Benutzer.

S. J. CASPER, Jena

J. J. SYMOENS (ed.): **Vegetation of Inland Waters** (Handbook of Vegetation Science **15/1**). Dordrecht: Dr. W. Junk Publishers. XIV, 385 S., 40 Abb., 34 Tab. Geb. Dfl. 250.— /US $140.—/ UK £ 76.50

Dieses, in Englisch geschriebene Buch ist ein Teilband des in den Niederlanden erscheinenden, von R. TÜXEN begründeten, von H. LIETH herausgegebenen „Handbuchs der Vegetationskunde". Es enthält 11 Beiträge von insgesamt 20 Autoren, von denen 6 in den Niederlanden, 3 in Südafrika, je 2 in Belgien, Frankreich, England und den USA und je einer in Österreich, der BRD und Dänemark tätig sind. Dargestellt werden „Wasser als Umwelt für Pflanzliches Leben" (WETZEL), „Methoden der Erforschung und Analyse der Umwelt aquatischer Vegetation" (GOLTERMAN, CLYMO, BEST & LAUGA), „Photosynthese von Wasserpflanzen unter natürlichen Bedingungen" (SØNDERGAARD), „Strukturaspekte von aquatischen Pflanzengesellschaften" (DEN HARTOG & VAN DER VELDE), „Phytosoziologische Methoden zur Beschreibung und Klassifikation der aquatischen Makrophytenvegetation" (BEST), „Algengesellschaften kontinentaler Gewässer" (SYMOENS, KUSEL-FETZMANN & DESCY), „Vegetationsprozesse in Sümpfen und Überflutungsgebieten" (BREEN, ROGERS & ASHTON). „Die Vegetation von Niedermooren in Abhängigkeit von ihrer Hydrologie und Nährstoffdynamik — eine Fallstudie" (VERHOEWEN, KOERSELMAN & BELTMAN), „Wasserfluß und die Vegetation von Fließgewässern" (DAWSON), „Analyse von Flora und Vegetation in Flüssen: Konzepte und Anwendungen" (WIEGLEB), „Wasserpflanzen in Extremstandorten " (MELACK).

Wie man sieht, ist der Inhalt dieses Bandes etwas heterogen; neben Beiträgen, welche einen Gesamtüberblick über Wissensstand und Methoden der betreffenden Teilbereiche anstreben, gibt es solche, die sich auf die Darlegung der eigenen Konzeptionen und Arbeitsweisen beschränken oder die Ergebnisse spezieller Studien vorstellen. Trotzdem und trotz mancher Subjektivitäten und vorzugsweiser Berücksichtigung der westeuropäischen und amerikanischen Literatur vermittelt das vorliegende Buch eine Fülle von Ergebnissen und Methoden der Vegetationsforschung in Binnengewässern und kann somit jedem Limnologen empfohlen werden.

H.-D. KRAUSCH, Potsdam

J. POKORNÝ, O. LHOTSKÝ, P. DENNY & G. TURNER (eds.): **Waterplants and Wetland Processes** (Arch. Hydrobiol., Beih. Ergebn. Limnol. 27). Stuttgart: E. Schweizerbartsche Verlagsbuchhandlung (Nägele u. Obermiller) 1987. VIII, 265 S., 71 Abb., 24 Tab. Brosch. DM 110,—

Vom 13.—23. Juni 1984 fand in der inmitten des südböhmischen Teichgebietes gelegenen Stadt Třeboň die vom Institut für Botanik der Tschechoslowakischen Akademie der Wissenschaften organisierte 2. Intecol-Feuchtgebiets-Konferenz statt. Die 4 Hauptthemen dieser Tagung waren: Böden und mikroklimatische Diagnosen von Feuchtgebieten, Anpassung der Feuchtgebietspflanzen und ihre Rolle im Haushalt von Feuchtgebieten, Feuchtgebiets-Inventarisation und ihre Veränderungen durch Bewirtschaftungseinflüsse sowie die Modellierung von Feuchtgebiets-Ökosystemen. Von den auf der Konferenz gehaltenen Vorträgen enthält der vorliegende Band 22. Die Skala der behandelten Pflanzensippen reicht dabei vom Phytoplankton und von litoralen Algen über höhere Wasserpflanzen bis hin zu Arten der Röhrichte und Moore. Aus der großen Zahl der Einzelbeiträge können hier nur einige besonders bemerkenswerte herausgegriffen werden.

Einen interessanten Überblick über den Mineralstoffhaushalt der Feuchtgebietspflanzen gibt P. DENNY (London), S. 1—25. M. ZOBEL & V. MASING (Tartu) machen grundsätzliche Bemerkungen zu Veränderungen von Mooren in Raum und Zeit (S. 41—55). Die Dynamik der Wasservegetation in 2 Buchten des Ladoga-Sees im Verlauf von 20 Jahren schildern I. M. RASPOPOV, V. V. MENŠUTKIN & O. N. DOCENKO (Leningrad), S. 75—82. H. BRUX, D. TODESKINO & G. WIEGLEB (Oldenburg) beschreiben Wachstum und Reproduktion von *Potamogeton alpinus* in norddeutschen Gewässern (S. 115—127). Der Taxonomie und der ökologischen Signifikanz von *Typha*-Hybriden ist der Beitrag von S. G. SMITH (Whitewater, Wisconsin) gewidmet (S. 129—138). Eine Übersicht über verschiedene bei Wasserpflanzen in seichten Gewässern vorkommende Adaptionsmorphosen geben S. HEJNÝ & Z. HROUDOVÁ (Průhonice), S. 157—166. Fragen der Allelopathie bei Wasserpflanzen bzw. bei Helophyten behandeln die Beiträge von S. WIUM-ANDERSEN (Hillerod), S. 167—172, bzw. W. SZCZEPANSKA (Nikolaiken), S. 173—179. Eine besondere Aufmerksamkeit fanden auf dieser Konferenz *Phragmites australis* und das weltweite Problem des „Schilfsterbens", was in diesem Band durch 5 der Aut- und Synökologie, der Morphologie und dem Stoffhaushalt dieser Art gewidmeten Beiträgen zum Ausdruck kommt.

In den genannten wie auch in den übrigen Beiträgen dieses Bandes werden nicht nur eine Fülle von Beobachtungs- und Meßdaten geboten, sondern auch zahlreiche gute Gedanken, Hinweise und Anregungen vorgelegt, welche die weiteren Arbeiten über Feuchtgebiete und Feuchtgebietspflanzen in vieler Hinsicht fördern und voranbringen werden. Vorangestellt ist dem Band — neben einem kurzen Vorwort — eine von der Konferenz verabschiedete Resolution zum Schutz von Feuchtgebieten.

H.-D. KRAUSCH, Potsdam

T. PLATT & W. K. W. LI (eds.): **Photosynthetic Picoplankton** (Can. Bull. Fish. Aquat. Sci. **214**). Department of Fisheries and Oceans Ottawa 1986. V, 583 S., 207 Abb. 55 Tab. Brosch. US $ 24.00

Im Oktober 1985 fand in San Miniato (Italien) eine Tagung statt, auf der Beiträge über die gegenwärtige Kenntnis von den photosynthetischen Picoplanktern der Ozeane vorgestellt wurden. Die Vortragenden, die allesamt zu dieser Tagung eingeladen worden waren, waren angehalten, in den Fällen, in denen nicht genügend Fakten zur Verfügung standen über die Eigenschaften dieser winzigen Organismen, auf der Basis ihrer Erfahrungen in anderen Algengruppen zu referieren. Das vorliegende Buch ist eine Zusammenstellung von 16 Artikeln, die sowohl prokaryotische als auch eukaryotische Pico- und Ultraplankter betreffen. Die Hauptthemen sind: Physiologische und ökologische Begrenzungen der extremen Kleinheit des Zellkörpers; Verbreitung; Häufigkeit und taxonomischer Status der Picoplankter; Stoffwechsel der Blaualgen (Cyanobacterien); Beziehung zwischen Ernährung und Kultur der Picoplankter; Physiologie des Phytoplanktons im Freiland und im Laboratorium; Unterschiede zwischen autotrophen und heterotrophen Picoplanktern; optimale Eigenschaften und Pigmente des Picoplanktons und Rolle des Picoplanktons im pelagischen Ökosystem.

Alle Beiträge tragen Übersichtscharakter und haben einen hohen Informationswert. Besonderes Interesse können die von THOMSEN (Übersicht über die kleinsten eukaryotischen Organismen des marinen Phytoplanktons), von WATERBURY et al. (Biologische und ökologische Kennzeichnung der einzelligen marinen Blaualge *Synechococcus*) von BRYANT (Der photosynthetische Apparat der Blaualgen) sowie von SILVER et al. (Die Vergesellschaftung des photosynthetischen Pico- und Ultraplanktons mit dem pelagischen Detritus in der Wassersäule zwischen 0 und 2000 m) beanspruchen. Offensichtlich haben die Herausgeber und Autoren nicht die Absicht gehabt, den Spezialisten Fesseln dadurch anzulegen, daß sie den Begriff des Picoplanktons nur für solche Organismen gelten ließen, deren maximaler Durchmesser 1 µm nicht überschreitet. Sie haben es für richtig gehalten, auch „größere" Organismen zu berücksichtigen, die gemeinhin seit JORGENSEN als Ultraplankter bezeichnet werden. Da es sich hier im wesentlichen um ein methodisches Problem handelt, ist der Beitrag von LI über die Methoden, mit denen das Vorkommen und die Häufigkeit des photosynthetischen Picoplanktons ermittelt werden können, besonders aufschlußreich.

Insgesamt gibt der Band einen umfassenden Einblick in den Forschungsstand über Organismen, von denen noch vor nicht langer Zeit angenommen wurde, daß es sich bei ihnen lediglich um Fragmente von Algenzellen handelte, die durch den Filtrationsprozeß zerbrochen worden seien. Jetzt muß man feststellen, daß das Phytoplankton unter 1 µm Größe, eben das Picoplankton, von größerer Bedeutung für die Produktion der Ozeane sein kann als Kieselalgen oder Dinoflagellaten.

S. J. CASPER, Jena

KURT KRAMMER & HORST LANGE-BERTALOT: **Bacillariophyceae. 2. Teil. Bacillariaceae, Epithemiaceae, Surirellaceae.** In: H. ETTL, J. GERLOFF, H. HEYNIG & D. MOLLENHAUER (Hrsg.): Süßwasserflora von Mitteleuropa. Band **2/2**. Begründet von A. PASCHER. Stuttgart, New York: Gustav Fischer Verlag 1988. XII, 596 S., 184 Tafeln mit 1914 Fig. Geb. DM 168,—

Zwei Jahre nach Erscheinen des ersten Bandes der Bacillariophyceen, in dem die Naviculaceae bearbeitet wurden, wird jetzt der bereits lang erwartete, weil dringend benötigte 2. Teil vorgelegt. Er enthält mit den Bacillariaceae, Epithemiaceae und Surirellaceae drei weitere Familien der pennalen Diatomeen, so daß nun der Großteil der Raphidineae überarbeitet ist. Die Eunotiaceae und Achnanthaceae, die Araphidineae und die Centrales werden im letzten Teilband erscheinen.

Der allgemeine Teil dieses Bandes fällt mit 5 Seiten kurz aus, da über die zur Bestimmung wesentlichen Strukturen, die Morphologie, Fortpflanzung und Ökologie schon ausführlich im Band 1 informiert wurde. Im ersten Abschnitt dieses Teiles stehen die Merkmale der behandelten Familien und ihre Abgrenzung zu anderen im Mittelpunkt. Auf den folgenden 215 Seiten und 181 Tafeln werden dann in ausführlicher Form alle Arten mit Kanalraphen vorgestellt. Dabei erheben die Autoren keineswegs den Anspruch, das in einigen Fällen vorhandene „taxonomische Chaos", vor allem bezüglich der Gattung *Nitzschia*, vollständig zu beseitigen, tragen aber durch ihre kritische Darstellung wesentlich zur Ordnung und besseren Überschaubarkeit von Sektionen, Arten und Sippen bei. Wie schon im ersten Band erfolgt die Abbildung der Arten nicht als Zeichnung sondern als fotografische Darstellung des lichtmikroskopischen Bildes. Dies vermittelt einen realistischeren Eindruck vom Aussehen der Alge im Mikroskop. Da ohnehin in der Regel nach den lichtoptischen Erscheinungen bestimmt wird, wird so das „Wiedererkennen" der untersuchten Arten im Buch beträchtlich erleichtert. Die vielen zusätzlichen REM-Aufnahmen ermöglichen den Vergleich zum lichtmikroskopischen Bild und führen zum besseren Verständnis der beschriebenen Strukturen. Die Bestimmungsschlüssel enthalten zahlreiche Abbildungshinweise, wodurch eine eindeutige Zuordnung zu den Merkmalskomplexen erleichtert wird, was besonders dem Nichtspezialisten helfen wird.

Der Band benötigt, ebenso wie der 1. Teil, keine Empfehlung. Wer Kieselalgen zu bestimmen hat, ist gut beraten, wenn er auf diese handliche, benutzerfreundliche und aktuelle Bestimmungsliteratur zurückgreift. Hoffentlich läßt auch der letzte Teil nicht mehr lange auf sich warten.

H. HORN, Neunzehnhain

HANUŠ ETTL & GEORG GÄRTNER: **Chlorophyta II. Tetrasporales, Chlorococcales, Gloeodendrales.** In: H. ETTL, J. GERLOFF, H. HEYNIG & D. MOLLENHAUER (Hrsg.): Süßwasserflora von Mitteleuropa. Band **10**. Begründet von A. PASCHER. Stuttgart, New York: Gustav Fischer Verlag 1988. 436 S., 311 Abb. Geb. DM 158,—

Vorliegender Band bearbeitet die gloeomorphen (capsalen) Chlamydophyceen (Tetrasporales), die zystomorphen (coccalen) Chlamydophyceen (Chlorococcales) und als Anhang die gloeomorphen (capsalen) Chlorophyceen (Gloeodendrales). Damit finden die von ETTL (1981) im Rahmen der Begründung der neuen Klasse Chlamydophyceae vorgeschlagenen Einteilungskriterien konsequente Anwendung bei der Abfolge der ersten Grünalgenbände der Süßwasserflora.

Die Chlamydophyceae sind durch Flagellaten des *Chlamydomonas*-Typus charakterisiert, die im Prinzip im Verlaufe der gesamten Ontogenese von einer Zellwand (Chlamys) umgeben sind. Die beweglichen Stadien der Chlorophyceae sind hingegen nackt.

Die monadoiden Chlamydophyceae mit den Ordnungen Chlamydomonadales und Volvocales sowie die monadoiden Chlorophyceae mit der Ordnung Dunaliellales wurden im Band 9 (Chlorophyta I, ETTL 1983) behandelt. Die coccalen und protosiphonalen Chlorophyceae sind für den Band 11 (Chlorophyta III) vorgesehen. Der hier vorgelegte Band 10 stellt also den Übergang zwischen beiden Bänden her.

Jeder der in diesem Buch bearbeiteten 3 Ordnungen wird eine ausführliche allgemeine Charakteristik vorangestellt. Bei den Tetrasporales ist die palmelloide Ausbildung der Chlamydomonadales zur trophischen Phase im Lebenszyklus geworden. Die vegetativen Zellen haben den gleichen Bau wie die Chlamydomonaden, sind jedoch unbeweglich geworden und durch Gallertbildungen charakterisiert. Sie sind ohne vorangehende Teilung fähig, sich in vegetative Schwärmer umzuwandeln. Der überwiegende Teil der Tetrasporales lebt benthisch, tychoplanktisch, planktisch oder epibiontisch im Wasser. Die Taxonomie dieser Gruppe ist noch ziemlich unklar und wird dadurch erschwert, daß über die meisten dieser Organismen nur spärliche Freilandbeobachtungen bekannt sind. ETTL & GÄRTNER greifen die Einteilung nach FOTT (Das Phytoplankton des Süßwassers, 3. Teil; Stuttgart 1972) auf, unterziehen einige Sippen jedoch einer stärkeren Bearbeitung. Die Tetrasporales werden in die drei Familien Chlorangiellaceae (Sippen mit beschränkter Gallertbildung mit Befestigungsfunktion), Palmellopsidaceae (Sippen mit starker Gallertlagerbildung) und Tetrasporaceae (Sippen mit Gallertgeißeln) gegliedert; insgesamt werden 26 Gattungen mit knapp 90 Arten vorgestellt.

Die Chlorococcales entsprechen den gallertlosen Ruhezellen (Akineten) der Chlamydomonadales. Während aber bei *Chlamydomonas* die Ruhezelle nur ein vorübergehendes Stadium der Ontogenese ist, stellt sie bei den Chlorococcales die Hauptphase der vegetativen Entwicklung dar. Die vegetativen Zellen sind, im Gegensatz zu den Tetrasporales, nicht mehr in der Lage, sich direkt in Schwärmer umzuwandeln.

Die Bearbeitung der Ordnung Chlorococcales s. str. war ein besonders hartes Brot für die Autoren. Vor ihnen stand die schwierige Aufgabe, diese natürliche Gruppe in der neuen Auffassung einer gründlichen Analyse hinsichtlich ihrer allgemeinen Merkmale sowie ihrer Taxonomie zu unterziehen. Dabei war die alte Auffassung von der weiten Umgrenzung der Chlorococcales zu überwinden. Die meisten Sippen der Chlorococcales wurden auf der Basis von Kulturmaterial beschrieben und gehen auf die amerikanische Schule von BOLD zurück. Da von BOLD und seinen Schülern physiologische Merkmale bevorzugt wurden, mußten, um der Aufgabe der „Süßwasserflora" als Identifikationshilfsmittel bei Routinearbeiten zu dienen, Rechnung zu tragen, alle verfügbaren Typuskulturen von den Autoren eingehend hinsichtlich zytomorphologischer und ontogenetischer Charakterisierung untersucht werden. Auf diese eigenen Untersuchungen von ETTL & GÄRTNER gehen die meisten Zeichnungen und Beschreibungen im vorliegenden Bestimmungsbuch zurück.

Die Autoren verfügen über eine langjährige Praxis mit den coccalen Chlamydophyceen. So hat ihr intensives Studium der Typusstämme, verbunden mit der kompletten Auswertung der einschlägigen Literatur, eine echte Pionierleistung hervorgebracht; erstmals in der Geschichte dieser schwierigen Algen (die „grünen Kugeln" sind für die meisten Algologen ein ausgesprochenes Schreckgespenst) liegt ein Werk vor, das eine reale Basis zur Determination der Sippen der Chlorococcales s. str. schafft und darüber hinaus ein Gerüst zum gezielteren Eindringen in die Fortpflanzungsvorgänge dieser Organismen liefert.

Die allgemeine Charakteristik der Chlorococcales wird auf 28 Seiten besonders gründlich dargelegt. Dieser Abschnitt enthält nicht nur Angaben über den Zellbau, zur Entwicklung und zur Fortpflanzung, sondern auch zur Ökologie, Kultivation, Untersuchung und Determination. Die meisten Chlorococcales sind Bodenalgen; die Ausführungen über die Ökologie und Kultur der Bodenalgen sind von hohem Interesse. Doch auch limnologisch sind sie vor allem auch für kleinere Gewässer, in die sie eingespült werden und in denen sie zur Entwicklung kommen können, relevant. Viele der behandelten Arten wurden allerdings erst einmal gefunden und zwar oft nur in Kulturen, ohne daß irgendwelches Freilandmaterial analysiert werden konnte.

Der Spezielle Teil der Chlorococcales wird nach neuesten Erkenntnissen von ETTL & GÄRTNER in die drei Familien Chlorococcaceae (einkernige, freilebende Sippen), Characiochloridaceae (einkernige, festsitzende Sippen) und Actinochloridaceae (mehrkernige Sippen) gegliedert. Es werden ca. 150 Arten in 22 Gattungen besprochen. Wenn Typuskulturen vorhanden sind, wird auch die Nummer der Isolate aus den einzelnen Sammlungen mit angegeben.

Die capsalen Chlorophyceen (Gloeodendrales) umfassen Vertreter, deren Zellen den Bau nackter Flagellaten aufweisen, sie sind jedoch geißellos und daher im vegetativen Zustand unbeweglich. Innerhalb der Familien Gloeodendraceae (einkernige Sippen ohne Gallertgeißeln), Chaetopeltidaceae (einkernige Sippen mit Gallertgeißeln) und Characiosiphonaceae (polyenergide Sippen) werden 12 Gattungen mit 18 Arten bearbeitet. Auch diese Sippen sind in der Regel wenig erforscht.

Der Band 10 der „Süßwasserflora" ist ein äußerst wichtiges Buch, das von den aktiven Erfahrungen der Autoren und der von ihnen eingebrachten Fülle an eigenständigen Beiträgen lebt. Es vermittelt kräftige Impulse für das Studium dieser so bedeutsamen, aber auch noch so geheimnisvollen Algengruppe.

L. KRIENITZ, Neuglobsow

Veröffentlichungen des Geobotanischen Instituts der ETH, Stiftung Rübel (Heft 95). E. LANDOLT & R. KANDELER: The Family of Lemnaceae — a Monographic Study. Vol. 2. Phytochemistry, Physiology, Application, Bibliography [Biosystematic Investigations in the Family of Duckweeds (Lemnaceae) (Vol. 4.)]. Zürich 1987. 638 S., 60 Abb., 53 Tab. Brosch. US $ 48.00

Der vorliegende zweite Band der Monographie der Lemnaceae — der erste Band erschien als Heft 71 der „Veröffentlichungen ... Stiftung Rübel" im Jahre 1986 und brachte Taxonomie, Ökologie, Geographie und Morphologie — enthält die Kapitel Phytochemie (S. 9—57), Physiologie (S. 58—368), Anwendung und ökonomische Bedeutung (S. 369—409) sowie die Bibliographie (S. 413—580), außerdem eine Übersicht über die akzeptierten Arten und supraspezifischen Taxa, ein Verzeichnis der Abkürzungen, Listen der Abbildungen und Tabellen sowie den Sachindex. Es ist überraschend, welche Fülle an Informationen und Fakten die Autoren zusammengetragen, ausgewertet und selbst erarbeitet haben (das Literaturverzeichnis umfaßt rund 3650 Titel!!). Im einzelnen werden die biochemischen Charakteristika insbesondere die organischen Komponenten ausführlich hinsichtlich ihres Auftretens und ihrer Verteilung in verschiedenen Sippen der Familie diskutiert. Das besonders umfangreiche Kapitel über die Physiologie beinhaltet die Kulturtechnik, die Bedingtheiten des vegetativen Wachstums, die Fragen der Entwicklung und des Stoffwechsels (hier finden sich auch Angaben über DNS-Struktur und -Funktion in Chloroplasten, Mitochondrien und im Kern). Das dritte Kapitel befaßt sich u. a. mit den Fragen der Produktivität, der Nutzung als Nahrungspflanzen, der Eignung als „Filter" in Abwässern und der Bedeutung als Test- und Indikatorpflanzen. Diese grobe Inhaltsangabe gibt nur einen ganz blassen Eindruck von dem, was der Leser wirklich erwarten kann. Offensichtlich gehören die Wasserlinsen zu den am meisten bekannten und untersuchten höheren Pflanzen überhaupt, was sie nicht zuletzt ihrer Kleinheit und daher guten Handhabbarkeit im Laboratorium zu verdanken haben. Die Autoren haben mit diesem zweiten Band nicht nur eine wichtige und unentbehrliche Ergänzung zu dem oben erwähnten ersten Band erarbeitet, sondern gezeigt, wie fruchtbar die Zusammenarbeit von Taxonomen und Physiologen sein kann.

Taxonomen, Morphologen, Ökologen, Physiologen, Biochemiker und Zytologen werden die Darstellung mit Gewinn lesen.

S. J. CASPER, Jena

Veröffentlichungen des Geobotanischen Institutes der ETH, Stiftung Rübel, Zürich (Heft 96). HANNA MEISTERHANS-KRONENBERG: Auswirkungen des Skibetriebes auf subalpine Heuwiesen bei Davos. Zürich 1988. 79 S., 18 Abb., 16 Tab. Brosch. SFr 36.— **Heft 97.** EDWIN MEISTERHANS: Vegetationsentwicklung auf Skipistenplanierungen in der alpinen Stufe bei Davos. Zürich 1988. 169 S., 34 Abb., 10 Tab., 12 Beilagen. Brosch. SFr 45.—

Zu den Aktivitäten des geobotanischen Instituts gehört auch die Untersuchung der von dem Wintersportbetrieb in den Alpen verursachten Umweltschäden. Die erstgenannte Arbeit über die Auswirkungen des Abfahrtskibetriebes auf eine subalpine Fettwiese (Trisetion) auf der Clavadeler Alp bei Davos vermag die aufgetretenen negativen Veränderungen durch eine Fülle von Daten zu belegen. Zu den wesentlichsten Folgen gehören eine verspätete Crocus-Blüte, eine nur zwergenhafte Entwicklung der blühenden Löwenzahnpflanzen, eine deutliche Zunahme von *Alchemilla* und *Trifolium repens*, die als trittfeste Arten verschiedene trittempfindliche Wiesenkräuter verdrängen, eine in allen Phasen niedrigere Krautschicht und ein verminderter Ertrag gegenüber ungestörten Wiesen.

Die zweitgenannte Arbeit untersucht die Folgen der in der Nähe von Davos oberhalb von 2200 m durchgeführten Planierungen von Skipisten aus pflanzenökologischer und standortskundlicher Sicht. Die sorgfältige Untersuchung erbrachte den Nachweis so vieler und schwerwiegender Auswirkungen, u. a. die Vernichtung zahlreicher, z. T. unter Naturschutz stehender Alpenpflanzen und die Beeinträchtigung des Landschaftsbildes, daß der Autor als Fazit seiner Studie von Skipistenplanierungen in der alpinen Stufe dringend abrät.

H.-D. KRAUSCH, Potsdam

Veröffentlichungen des Geobotanischen Instituts der ETH, Stiftung Rübel (Heft 98). A. MIYAWAKI & E. LANDOLT (Hrsg.): Contributions to the Knowledge of Flora and Vegetation of Japan. Proceedings of the 18th International Phytogeographic Excursion (IPE), 1984, through Central Japan. Zürich 1988. 385 S., 79 Abb., 27 Tab. Brosch. US $ 48.00

In bekannter und bewährter Weise legen die Herausgeber die Ergebnisse der 18. Internationalen Pflanzengeographischen Exkursion nach Mittel-Honshu vor. Es ist erstaunlich, daß nur 10 Jahre nach der ersten großen Exkursion durch Japan von Kiushu bis Hokkaido bereits wieder eine derartige Veranstaltung stattfinden konnte. Im Vorwort betont E. LANDOLT die besonderen Verdienste von A. MIYAWAKI um die Durchführung der Exkursion, aber auch die Schwierigkeiten, die sich der aktiven Feldarbeit entgegenstellten. Das Programm war sehr gedrängt. Trotzdem fanden verschiedene Teilnehmer noch Zeit, ihre Wahrnehmungen zu einer Originalarbeit zu verarbeiten. Diese Beiträge sind in den Kapiteln 3 (Vergleich der Vegetation Japan mit anderen Teilen der Welt) und 4 (Beiträge zur Kenntnis einiger japanischer Pflanzenarten in Beziehung auf Europa) zusammengestellt. Das Kapitel 2 enthält eine Übersicht über die Vegetation der Flora von Japan einschließlich je 1 Karte über die potentielle und aktuelle Vegetation (Beilage). Von den insgesamt 12 Beiträgen sind 7 in englischer, der Rest in deutscher Sprache verfaßt. Das Teilnehmerverzeichnis weist aus, daß über 70 Wissenschaftler an den Exkursionen und Veranstaltungen teilgenommen haben. Druck und Ausstattung des informativen Bandes entsprechen dem gewohnten Niveau. Ein auch für alle Nichtteilnehmer informativer Band.

S. J. CASPER, Jena

Veröffentlichungen des Geobotanischen Institutes der ETH, Stiftung Rübel, Zürich (Heft 100). WALTER HUBER: Natürliche Bastardierungen zwischen weißblühenden *Ranunculus*-Arten in den Alpen. Zürich 1988. 160 S., 24 Abb., 26 Tab. Brosch. SFr. 48.—

In der vorliegenden Arbeit wird die Bastardnatur rätselhafter, bis in die Gegenwart umstrittener *Ranunculus*-Taxa anhand von umfangreichem Material nachgewiesen. Dabei handelt es sich jedoch nicht um die weißblühenden Wasserhahnenfüße des Subgenus *Batrachium*, vielmehr um Arten der Sektionen Aconitifolii und Ranuncella. Zur Klärung der Sachverhalte wurden pflanzensoziologische, ökologische und zytologische Untersuchungen, ferner Pollenuntersuchungen, Bestäubungsexperimente und Keimungsversuche durchgeführt. Es werden 5 Bastardtaxa beschrieben und zusätzlich 2 nomenklatorische Änderungen vorgenommen. Die Arbeit besticht durch ihre subtile Arbeitsweise, die klaren Fotos, die instruktiven Verbreitungskarten und den eingehenden Bestimmungsschlüssel. Ihre Ergebnisse kommen in erster Linie der Taxonomie und Ökologie der Gattung *Ranunculus* zugute.

H.-D. KRAUSCH, Potsdam

DIETHER ENNET: **Heilpflanzen und Drogen** (BI-Lexikon). Leipzig: Bibl. Institut 1988. 368 S., 512 Abb. auf 128 Farbt. Lw. M 32,00

In der Reihe der vom Bibliographischen Institut in Leipzig herausgegebenen Lexika darf das vorliegende besonderes Interesse beanspruchen. Die Wahrnehmung der Verantwortung des Menschen für seine Gesundheit fördert sein Streben nach Anwendung natürlicher Mittel und Methoden zur Behandlung von Störungen des Wohlbefindens und leichten Erkrankungen ohne ärztliche Verordnung. In diesem Sinne stellt das Buch Arzneipflanzen und Pflanzen, die in der Volksmedizin und der Homöopathie Bedeutung besitzen, ausführlich vor, wobei streng darauf geachtet wird, Hinweise in den Fällen zu geben, in denen auf eine ärztliche Hilfeleistung keinesfalls verzichtet werden darf.

Besonderer Wert wird auf eine ausgewogene und oft kritische Darstellung der Wirkung und der Gewinnungsmöglichkeiten der Drogen und pflanzlichen Zubereitungen gelegt. Die Artikel über die wichtigsten Arzneipflanzen werden durch Informationen über Erkrankungen, die für eine Behandlung mit Arzneimitteln pflanzlicher Herkunft in Betracht kommen, ergänzt. Insgesamt werden 256 heimische und fremdländische Arzneipflanzen, deren Wirksamkeit wissenschaftlich bestätigt ist, beschrieben, auf 512 Farbfotos abgebildet und deren Drogen gezeigt. Das soll die Identifizierung der Pflanzen in der Natur und der Drogen im Haushalt erleichtern. Bei den Teedrogen werden die Vorschriften zur Teebereitung und die übliche Dosierung angegeben. Jeder Artikel beginnt mit dem deutschen Pflanzennamen, gefolgt von dem wissenschaftlichen Namen und der Beschreibung. Angaben über Blütezeit und Vorkommen schließen sich an, außerdem Abschnitte über Drogengewinnung, Drogenbeschreibung und Inhaltsstoffe. Wirkung, Verwendung, auch Nebenwirkungen, werden erörtert. Am Schluß eines jeden Artikels finden sich Anmerkungen über Geschichtliches. Die Darstellung ist klar und einfach, sachgerecht und informativ. Es ist allerdings schade, daß der Autor sich die Gelegenheit hat entgehen lassen, noch etwas mehr über den Zusammenhang von Volksnamen und Volksmedizin zu bringen. Bei dem Abschnitt Geschichtliches stört etwas die stereotype Wiederholung von Wendungen „und wurden z. B. von HILDEGARD VON BINGEN in ihrer Physica genannt", wobei diese Formel auch noch mannigfach verändert wird. Zu begrüßen ist, daß außer den Artikeln über Drogen oder Stoffe selbst auch solche über Behandlungsformen oder Erkältungsphänomene gebracht werden. Der Leser kann dadurch die bei den einzelnen Pflanzen erwähnten Wirkungen und Verwendungen besser einordnen. Besondere Sorgfalt ist auch darauf gelegt worden, die in den Artikeln genannten Autoritäten gesondert aufzuführen. Im Falle von SEBASTIAN KNEIPP ist allerdings der Lapsus passiert, daß hier nur ausführlich über dessen Wassertherapie berichtet wird, nicht aber über seine Aktivitäten auf dem Gebiete der Heilpflanzen und Drogen.

Einzelne mißverständliche Formulierungen — z. B. wird beim Wolligen Fingerhut angegeben, daß er vereinzelt in Österreich wildwachsend vorkommt, im Abschnitt Geschichtliches aber steht, daß er in Mitteleuropa nicht wild vorkommt — schmälern nicht den Wert des Buches, das ausgezeichnet gedruckt und sehr gut illustriert ist.

Eine erfreuliche Publikation.

S. J. CASPER, Jena

BRUNO P. KREMER: **Welche Heilpflanze ist das?** Kennen — Verwenden — Anbauen (Kosmos-Naturführer). Stuttgart: Franck/ Kosmos-Verlagsgruppe 1987. 128 S., 121 Farbfotos, 15 Farbzeichn., 13 Symbolzeichn. Karton. DM 16,80

In diesem Taschenbuch werden mehr als 100 Pflanzenarten vorgestellt, die als Heilkräuter Verwendung finden oder als Würzkräuter unseren Küchenzettel bereichern können.

Jedem, der selbst diese Pflanzen sammeln oder anbauen will, wird mit der vorliegenden kleinen Kräuterkunde ein geradezu idealer Naturführer in die Hand gegeben. Hervorragende Farbfotos ermöglichen ein Bestimmen der Pflanzen, ein kurzer Text informiert über die wichtigsten Merkmale, das Vorkommen, die Inhaltsstoffe, die verwendeten Pflanzenteile und die Anwendungsmöglichkeiten der einzelnen Kräuter.

All das geschieht in so knapper und übersichtlicher Form, daß es geradezu Spaß macht, mit diesem Büchlein mehr über Heilkräuter in der Natur zu erfahren. Besonders nützlich dürfte auch das Kapitel „Was hilft bei ..." sein oder der Sammelkalender, der gleich auf den inneren Umschlagseiten zu finden ist.

Es ist zu begrüßen, daß auch die Belange des Naturschutzes nicht vergessen worden sind. Viele Pflanzen sind schon viel zu selten geworden, als daß sie unserer Sammelleidenschaft zum Opfer fallen sollten.

A. LANGE, Jena

HEINRICH DÖRFELT: **Die Erdsterne.** Geastraceae und Astraeaceae (Die Neue Brehmbücherei **573**). Wittenberg-Lutherstadt: A. Ziemsen Verlag 1985. 108 S., 71 Abb. (z. T. farb.). M 10,80

Von den annähernd 50 bekannten Arten von Erdsternen kommt ungefähr wie Hälfte in Mitteleuropa vor. Obwohl sie nicht in dem Maße wie andere Gift- oder Speisepilze das Interesse der Pilzsammler erregt haben, sind sie doch wegen ihrer ungewöhnlichen Vielfalt bekannt. Es handelt sich um Bauchpilze, deren Außenhülle sternförmig aufreißt. Im 17. Jahrhundert zeichnete man bei der Abbildung der reifen Fruchtkörper ein menschliches Gesicht ein, um die absonderliche menschenähnliche Gestalt zu betonen. Der Verfasser hat sich intensiv mit den formschönen Pilzen befaßt und stellt in Wort und Bild 25 Arten von ihnen vor. Die Ausführungen betreffen Morphologie, Anatomie, Lebensweise, Verbreitung und stammesgeschichtliche Entwicklung, außerdem aber auch die Ökologie, da Erdsterne als Zeigerpilze für spezielle Standortverhältnisse herangezogen werden können. Ein Schlüssel ermöglicht die Bestimmung aller mitteleuropäischen Arten.

Im allgemeinen Teil werden die Geschichte und die aktuellen Probleme der Erdsternsystematik ausführlich vorgestellt. Die Wiedergabe historischer Abbildungen ist sehr zu begrüßen. Sie leidet aber unter drucktechnischen Unzulänglichkeiten, was übrigens für den gesamten Band gilt. Ein ausführliches Literaturverzeichnis und ein zureichendes Namen- und Sachwortregister unterstützen die kompetente Darstellung, die übersichtlich und informativ ist. Die Beschreibungen enthalten einige sprachliche Unschärfen (z. B. willkürliche Verwendung von Begriffen wie kugelig, rund, eiförmig, ellipsoid, nicht hygrometrisch, ahygrometrisch usw.). Die in der Fußnote 2 auf Seite 65 geübte Kritik am ICBN wirkt vor dem Hintergrund der eigenen nomenklatorischen Bemühungen etwas deplaziert. Rezensent versteht manche Wiederholungen von gültigen Artnamen unter den Synonymen nicht, so z. B. die erneute Anführung von *Geastrum lageniforme* VITTADINI 1842 unter den Synonymen.

Unabhängig davon ist die Darstellung sehr wohl geeignet, neue Freunde für die attraktiven Erdsterne zu gewinnen.

S. J. CASPER, Jena

Archiv für Hydrobiologie. Hrsgeg. von H.-J. ELSTER & W. OHLE. **Supplementband 76, Heft 1/2. Monographische Beiträge.** Stuttgart: E. Schweizerbart'sche Verlagsbuchhandlung (Nägele u. Obermiller) 1987. 196 S., 85 Abb., 19 Tab. Brosch. DM 156,—

Der vorliegende Supplementband enthält 7 Beiträge, in gewohnter Weise ausgezeichnet von den Herausgebern betreut und vom Verlag gedruckt. Im einzelnen handelt es sich um:

- CUFFNEY, T. F. & WALLACE, J. B.: Leaf litter processing in Coastal Plain streams and floodplains of southeastern Georgia, U.S.A. S. 1—24.
- BOURNAUD, M.; TACHET, H. & ROUX, A. L.: The effects of seasonal and hydrological influences on the macroinvertebrates of the Rhone River, France. II. Ecological aspects. S. 25—51.

— SCHRÖDER, R.: Das Schilfsterben am Bodensee-Untersee. Beobachtungen, Untersuchungen, Gegenmaßnahmen. S. 53—99.
— SEN, B.: Fungal parasitism of planktonic algae in Shearwater. I. Occurrence of *Zygorhizidium affluens* CANTER on *Asterionella formosa* HASS. in relation to the seasonal periodicity of the alga. S. 101—127.
— SEN, B.: Fungal parasitism of planktonic algae in Shearwater. II. A study of the chytrid parasites of the diatom *Fragilaria crotonensis* KITTON. S. 129—144.
— TILLY, L. J. & GARCIA-SAIS, J. R.: Limnological features of a Puerto Rican reservoir. S. 145—167.
— UZUNOV, Y. & KOVACHEV, S.: The macrozoobenthos of Struma River: An example of a recovered community after the elimination of a heavy industrial impact with suspended materials. S. 169—196.

Obwohl alle Beiträge das Interesse der Limnologen verdienen, möchte Rezensent besonders auf die von SCHRÖDER und SEN hinweisen. ROLAND SCHRÖDER weist nach, daß die Eutrophierung des Bodensees eine Zunahme der Populationsdichte von *Phragmites* und gleichzeitig eine Schwächung der Stengelfestigkeit bewirkte. Resultat war die Entstehung eines Schilfdickichts mit vermindertem Wasseraustausch zwischen See und Bestandsgürtel. Zur selben Zeit entwickelten sich im Litoral große Massen von Fadenalgen, vor allem von *Cladophora glomerata*, die im Herbst mit fallendem Wasserspiegel abstarben. Biotests zeigten, daß die *Cladophora*-Abbauprodukte in der Lage waren, die Entwicklung der Sprosse und die Wasseraufnahme durch die Wurzeln zu behindern. Diese unbekannten bioaktiven Substanzen verlieren bei Belüftung ihre Wirkung. Als Therapiemaßnahme schlägt der Autor die Mahd der Schilfbestände und die Entfernung des Materials vor. Die sich entwickelnde neue Generation bringt kürzere und kräftigere Sprosse hervor. Kurzfristige Erfolge der Maßnahmen sind nicht zu erwarten, da mit einer Anreicherung der schädlichen bioaktiven Substanzen in den tiefen Schichten der anoxischen Rhizosphäre gerechnet werden muß.

BÜLENT SEN beleuchtet in zwei Arbeiten die Auswirkungen des Pilzbefalls auf die Massenentwicklung und das jahreszeitliche Auftreten von zwei Kieselalgenarten, *Asterionella formosa* und *Fragilaria crotonensis*. Es ist aus der Literatur bekannt, daß Parasitismus einer der Faktoren ist, der für die Abnahme der Zellzahl in *Asterionella*-Populationen verantwortlich ist. Die Untersuchungen SENS zeigen, daß der Chytridiomyzet *Zygorhizidium affluens* zu jeder Jahreszeit und epidemisch vor allem bei sehr niedrigen Wassertemperaturen auftreten kann. Die höchste Infektionsrate betrug 63%. Die Zuwachsrate von *Asterionella* scheint mit dem Infektionsgrad abzunehmen. Der Niedergang der *Asterionella*-Population während starken Pilzbefalls förderte die Entwicklung anderer Kieselalgen. Der Pilzbefall eliminiert die *Asterionella*-Population nicht; *Asterionella* erholte sich stets wieder. Die normale jahreszeitliche Periodizität der Kieselalge wurde nicht verändert. *Fragilaria crotonensis* wurde von zwei morphologisch ähnlichen Chytridiomyzeten, *Rhizophydium fragilariae* und der sog. Spezies 3 (nach CANTER & JAWORSKI 1982, 1983) befallen, und zwar gewöhnlich während der Periode des aktiven Wachstums der Kieselalge. Die Pilzinfektion findet meist durch enzystierte Zoosporen und reife Sporangien statt. Wenigstens 50 Algenfäden per ml scheinen nötig zu sein, damit der Pilzbefall epidemisch wird (die höchste Infektionsrate betrug 55%). Der Pilzbefall bewirkte keine merkliche Änderung des Wachstumsmusters der Kieselalge. Das Herbstmaximum von *Fragilaria* war nach Pilzbefall niedriger und trat später ein. Beide Arbeiten bestätigen erneut die Tatsache, daß der Einfluß des Parasitismus auf den Entwicklungszyklus der phytoplanktischen Population komplex und nicht leicht durchschaubar ist.

S. J. CASPER, Jena

PETER LEINS, SHIRLEY C. TUCKER & PETER K. ENDRESS (eds.): **Aspects in Floral Development** — (With the assistance of CLAUDIA ERBAR). Berlin, Stuttgart: J. Cramer in der Gebrüder Borntraeger Verlagsbuchhandlung 1988. VII, 239 S., 399 Abb., 11 Tab. Kunststoffeinb. DM 120,—/US-$ 68.80

Die 19 Beiträge des Sammelbandes wurden am 29. Juli 1987 in Berlin (West) auf dem Symposium „Floral development: Evolutionary aspects and special topics“ im Rahmen des XIV. Internationalen Botanischen Kongresses präsentiert. Sie befassen sich mit morphologischen, systematischen und phylogenetischen Aspekten der Blütenentwicklung. Die Autoren — 24 ausgewiesene Blütenmorphologen aus Canada, den USA, der Bundesrepublik Deutschland, den Niederlanden, Südafrika, Argentinien und Japan — verstehen Blütenentwicklung als wesentlichstes Element einer Morphologie der Blüten und begreifen sie als Versuch, eine Brücke von funktioneller Morphologie zur Physiologie zu schlagen. In den überwiegend theoretisch orientierten Beiträgen (SATTLER, ERBAR, TUCKER, VAN HEEL) wird deutlich, daß die moderne Blütenmorphologie das überkommene „statische“ Konzept durch ein „dynamisches“ ablösen will. Morphologische Kategorien wie Kaulom, Phyllom oder Stamen werden nicht länger als a priori gegeben akzeptiert, sondern als „geworden“ betrachtet, d. h. als a posteriori Strukturen erkannt. In den sogenannten typischen Fällen bleiben strukturelle Kategorien brauchbar. Das Festhalten an ihnen wird dort zweifelhaft, wo es sich um „Übergänge“ handelt. Form wird nicht nur als Ergebnis eines Prozesses, sondern als Prozeß selbst verstanden. Letztendlich münden diese „biophilosophischen“ Grundgedanken doch ein in direkte Arbeit am Objekt mit konkreter, also praktikabler Terminologie. Die Bedeutung der frühen Phasen der Blütenentwicklung für Systematik und Phylogenie wird an zahlreichen Beispielen demonstriert: Cactaceae (LEINS & SCHWITALLA), Loasaceae (HUFFORD), Ericaceae (NISHINO), Sterculiaceae (JENNY), Rhizophoraceae (JUNCOSA), Vitaceae (GERRATH & POSLUSZNY), Rhamnaceae (MEDAN), Polygalaceae (KRÜGER & ROBBERTSE), Leguminosae (RAMIREZ-DOMENÈCH & TUCKER), Scrophulariaceae (KAMPNY & CANNE-HILLIKER), Solanaceae (DIGGLE), Palmae (UHL), Zingiberales (KIRCHOFF), Bromeliaceae (BROWN & GILMARTIN) und Cycadales (STEVENSON). Die klassifikatorischen Bemühungen folgen meist dem kladistischen Konzept von HENNIG. Der enorme Anteil der Rasterelektronenmikroskopie an der Entwicklung der Blütenmorphologie in den letzten 15 Jahren wird eindrucksvoll durch die zahlreichen informativen Abbildungen demonstriert. Der Informationsgehalt des vorzüglich ausgestatteten und gedruckten Bandes ist sehr hoch. Eine Autorenliste und ein Gattungsindex ergänzen die Darstellung. Auch wenn viele der Ergebnisse nicht unbedingt revolutionär für Systematik und Evolution der Blütenpflanzen sind, so belegen sie doch die Lebenskraft einer so alten wissenschaftlichen Disziplin wie der Morphologie in der modernen Botanik.

S. J. CASPER, Jena

G. W. LAWSON & D. M. JOHN: **The Marine Algae and Coastal Environment of Tropical West Africa** (2. edition) (Beiheft zur Nova Hedwigia, Heft 93). Berlin, Stuttgart: J. Cramer in der Gebrüder Borntraeger Verlagsbuchhandlung 1987. VI, 416 S., 65 Taf., 16 Fig. Brosch. DM 220,—

Die erste Auflage der vorliegenden Meeresalgenflora erschien 1982. Seitdem haben die Kenntnisse auf den Gebieten der Taxonomie, Ökologie und Verbreitung derart zugenommen, daß kaum eine Seite des Textes unverändert in die neue Ausgabe übernommen werden konnte. Die Bibliographie schwoll um 15% an. Weitgehend unverändert konnten die meist originalen Abbildungen bleiben. Das Gesamtkonzept wurde beibehalten: Dem systematischen Hauptteil, der die Abteilungen Chloro-, Phaeo-, Rhodo- und Cyanophyta umfaßt (S. 49—350) stehen die kürzeren Abschnitte Einleitung (S. 1—15, mit einem historischen Überblick und einer Kennzeichnung der physikalischen Umwelt), Marine Phytogeographie (S. 16—21) und Marine Ökologie (S. 22—48, mit Angaben über Zonierung und Habitate) gegenüber. Bestimmungsschlüssel führen zu den Gattungen und Arten. Die Sippen werden ausreichend beschrieben und die meisten von ihnen abgebildet. Die Gattung *Callithamnion* wurde von J. H. PRICE revidiert. Spezialisten finden Anmerkungen über Verbreitung, Ökologie, Taxonomie und Morphologie, die oft auf unveröffentlichten Beobachtungen der Autoren beruhen. Nützlich sind ein umfangreiches Glossar und ein systematischer Index. Im Anhang werden u. a. Sammeln, Untersuchen und Aufbewahren von Proben beschrieben. Druck und Ausstattung sind gut. Eine Pioniertat.

S. J. CASPER, Jena

M. A. RAGAN & CAROLYN BIRD (eds.): **Twelfth International Seaweed Symposium.** Proceeding of the Twelfth International Seaweed Symposium held in Sao Paulo, Brazil, July 27—August 1, 1986 (Developm. in Hydrobiol. **41**). Dordrecht, Boston, Lancaster: Dr. W. Junk Publishers 1987. XXVIII, 590 S., 339 Abb., 182 Tab. Geb. Dfl. 375.00/US $ 182.50/UK £ 116.00

Der umfangreiche Sammelband (Nachdruck aus Hydrobiologia 151 und 152) umfaßt über achtzig (!) Einzelbeiträge von rund 200 Autoren (Liste der registrierten Teilnehmer). Der zum ersten Male in Südamerika abgehaltene Phykologenkongreß hatte sich zum Ziel gesetzt, einen repräsentativen Überblick über die weltweiten Aktivitäten auf dem Gebiete der theoretischen und angewandten Forschung der Meeresalgen zu geben und insbesondere die Wege aufzuzeigen, auf denen die Prinzipien und Methoden der Biotechnologie für die Meeresalgenforschung und -nutzung umgesetzt werden können. Plenarvorträge, Minisymposia, Workshops und Einzelbeiträge befassen sich mit Algengenetik, Produktion und Kultur von *Spirulina* und anderen Mikroalgen (*Dunaliella*), Anwendung der Biotechnologie auf Algen (*Fucus, Gracilaria, Porphyra, Pterocladia, Agardhiella*), ökonomischen Aspekten der Meeresalgenkultur, Algenpolysacchariden, Morphologie und Taxonomie (*Dictyota, Gracilaria, Callithamnion, Gelidium, Pterocladia*), Ökologie und Verbreitung (*Chondrus, Laminaria, Gracilaria, Sargassum*), Management, Kulturtechniken (*Eucheuma, Gracilaria, Laminaria, Sargassum*), Physiologie, Biochemie, Chemie und Molekulargenetik (*Klebsiella*). Der technische Apparat umfaßt Sach-, Sippen- und Autorenindex sowie ein Verzeichnis der chemischen Verbindungen bzw. Bestandteile. Der wichtige Band ist wie immer vom Verleger gut betreut und ausgestattet worden.

S. J. CASPER, Jena

Umweltprobenbank, Bericht und Bewertung der Pilotphase. — Hrsgeg. vom Bundesministerium für Forschung und Technologie. Berlin, Heidelberg, New York, London, Paris, Tokyo: Springer-Verlag 1988. 158 S., 33 Abb., 18 Tab. Brosch. DM 34,—

1983 waren ca. 6 Millionen chemische Verbindungen bekannt, von denen ungefähr 100 000 kommerziell wichtig sind. Zu diesen kommen jährlich etwa 1000 neue Chemikalien hinzu. Es ist unvermeidbar und zum Teil auch gewünscht, daß ein mehr oder weniger großer Teil dieser Stoffe direkt auf den Menschen einwirkt, oder indirekt über vielfältige Wege durch die Umwelt an ihn gelangt. Angesichts einer solchen Situation wurde in verschiedenen Industriestaaten der Erde schon Anfang der 70er Jahre der Wunsch nach biologischen Proben als Referenzmaterial für den Nachweis der Umweltbelastung früherer Zeiten deutlich. Ursprünglich hoffte man, dafür Material aus Museumssammlungen verwenden zu können. Aber die Untauglichkeit dieser Idee stellte sich schon bald heraus, weil die Einlagerung niemals unter dem Gesichtspunkt der Konservierung von Umweltschadstoffen erfolgte. Das Bundesministerium für Forschung und Technologie förderte deswegen ab 1976 einen 3-Stufenplan, an dessen Ende gegebenenfalls die Einrichtung einer speziellen Umweltprobenbank stehen sollte, in der ausgewählte Umweltproben für viele Jahre praktisch ohne Veränderung ihrer chemischen Zusammensetzung aufbewahrt werden können.

Der vorliegende Band beinhaltet den Bericht und die Bewertung der Pilotphase.

Den ersten Teil bildet die Darstellung und die Bewertung der Pilotphase durch den Projektträger, das Umweltbundesamt. Hier werden die Aufgaben klar definiert: Archivierung von Umweltproben für noch nicht als Schadstoffe bekannte Verbindungen, Neuanalysen bei verbesserter Analysentechnik, Nachprüfung früherer Monitoring-Ergebnisse, Trendabschätzungen, Erfolgskontrolle eingeleiteter Maßnahmen, flächendeckende Umweltbeobachtung. Die Pilotphase sollte vor allem zeigen, ob das anspruchsvolle Projekt praktisch durchführbar ist, wobei technische und analytische Fragestellungen im Vordergrund standen.

Insgesamt beteiligten sich 10 Institutionen, die im zweiten Teil des Buches in Form von 11 Kurzberichten ihre Erkenntnisse und Erfahrungen vorlegen.

Dabei geht es vor allem um die Probenauswahl, die Probennahme, die Lagerung, die Analytik und die Dokumentation der Ergebnisse. Hinter diesen Schlagworten verbirgt sich eine Vielzahl von Einzelfragen, die alle sorgfältig geprüft werden mußten: Welche Organismen eingelagert werden sollen, da sie beispielhaft und repräsentativ zugleich sein müssen. In welchen Gefäßen die Proben aufbewahrt werden sollen, denn sie dürfen das Material nicht kontaminieren und müssen Temperaturen von —85 °C bzw. —140 °C auf lange Zeit widerstehen. Welche Sicherheitssysteme installiert sein müssen, um die Tiefstkühlung unter allen Umständen zu sichern. Wie die Analysenverfahren auszusehen haben, mit denen die z. T. äußerst geringen Substanzmengen sicher nachgewiesen werden können und vieles andere mehr.

Im dritten Teil, der zusammenfassenden Beurteilung, wird festgestellt, daß die grundlegenden Probleme während der Pilotphase gelöst werden konnten. Aus wissenschaftlich-technischer Sicht ist der Dauerbetrieb einer Umweltprobenbank möglich und im Hinblick auf die eingangs genannten Fragestellungen dringend notwendig.

Die während des Pilotprojektes gesammelten Erfahrungen veranlaßten die Bundesregierung, den Betrieb einer Umweltprobenbank unter Leitung des Umweltbundesamtes ab 1985 aufzunehmen.

P. KASPRZAK, Neuglobsow

Wasser und Abwasser, Supplementband I/88. KÁLMAN BÍRÓ: Kleiner Bestimmungsschlüssel für Zuckmückenlarven (Diptera, Chironomide). Übersetzt aus dem Ungarischen von E. DANECKER. Hrsg. u. Verlag: Bundesanstalt für Wassergüte Wien, 1986. Schriftl. Hofrat Univ.-Prof. Dr. WERNER KOHL. 329 S., 27 Abb., 100 Tafeln. Brosch. öS 255,—

Angesichts der großen Bedeutung, die die Chironomiden in aquatischen Ökosystemen haben, sowie der nach wie vor erheblichen taxonomischen Schwierigkeiten, die diese Familie bereitet, ist jeder Versuch einer zusammenfassenden Bearbeitung zu begrüßen.

Das vorliegende Werk gibt einen Überblick über die Larven der in Ungarn nachgewiesenen bzw. zu erwartenden Zuckmückenarten und stellt, erstmals in deutscher Sprache, einen alle Unterfamilien berücksichtigenden Bestimmungsschlüssel vor. Es beginnt mit einem Familienschlüssel, der die Zuordnung aller im Gebiet vorkommenden aquatischen Dipterenlarven erlaubt. Dem schließt sich eine „Allgemeine Charakteristik der Zuckmücken" an, die grundlegende Kenntnisse zur Larvenmorphologie und zu Methoden der Aufsammlung und Bearbeitung vermittelt.

Den Hauptteil des Buches nimmt der Bestimmungsschlüssel ein. Er führt über Unterfamilien zu Arten bzw. Artengruppen. Dieser Abschnitt ist gut illustriert, zahlreiche Strichzeichnungen auf 100 Tafeln zeigen die wichtigsten Merkmale. Folgende Informationen sind in der Regel bei den Arten zu finden: Synonyme, ökologische Kurzcharakteristik, Verbreitung in Europa, Vorkommen in Ungarn.

Das Literaturverzeichnis umfaßt 71, vorwiegend ungarische Arbeiten. Einige grundlegende Publikationen, die nach 1980 erschienen sind und somit im Original nicht berücksichtigt werden konnten, wurden von der Übersetzerin ergänzend aufgeführt. Den Abschluß bildet ein Arten- und Stichwortverzeichnis. Das von Herausgeber und Autor gleichermaßen formulierte Ziel der Arbeit besteht darin, Praktikern der Wasserwirtschaft, Lehrern und Laienforschern ein Hilfsmittel in die Hand zu geben, welches es ermöglicht, Chironomidenlarven sicher bis zur Art zu determinieren. Eine genauere Betrachtung der Ausführungen zeigt jedoch, wie weit der aktuelle Wissensstand noch von diesem Ziel entfernt ist. So muß man in vielen Fällen lesen: „Die Arten können anhand der Larven nicht sicher unterschieden werden." Da vorliegende Bearbeitung mit 287 Arten nur knapp 20% des in Europa nachgewiesenen Bestandes berücksichtigt, wird der Wert der mit ihrer Hilfe erzielten Ergebnisse erheblich beeinträchtigt. Darüberhinaus ist zu bedauern, daß das 1983 erschienene, grundlegende Werk „Chironomidae of the Holarctic region-keys and diagnosis, Part 1 — Larvae" (Hrsg. T. WIEDERHOLM, Ent. Scand., Suppl. **19**) im Text keine Berücksichtigung findet und lediglich in der Literatur-Ergänzung darauf verwiesen wird.

Im Gegensatz zum besprochenen Buch wird heute im allgemeinen die Unterfamilie Corynoneurinae nicht mehr vertreten, ihre zwei Gattungen werden den Orthocladiinae zugeordnet.

RAINER SAMIETZ, Gotha

W. F. VINCENT (ed.): **Dominance of Bloom-forming Cyanobacteria (Blue-green Algae)** (New Zealand Journal of Marine and Freshwater Research Vol. **21**, Issue 3). Wellington: Science Information Publishing Centre 1987. 150 S. NZ $ 35.75/US $ 20.00

In einem Sonderheft werden Ergebnisse eines internationalen Forums über Blaualgen-Dominanz vorgestellt, das 1987 in Neuseeland stattfand. Eine Reihe namhafter Spezialisten stellt in Form von längeren Übersichtsartikeln oder von kürzeren Orginalmitteilungen abiotische und biotische Steuergrößen der Massenentwicklung vor.

Die Autoren versuchen, kausale Einsichten über eine holistische Gesamtbetrachtung zu erlangen. Dazu werden sowohl Erkenntnisse auf dem niedrigsten Niveau der strukturellen Organisation (z. B. photosynthetische Primärprozesse) als auch auf höheren trophischen Ebenen (z. B. Zooplankton-Interaktion) herangezogen. Wechselwirkungen mit dem physiko-chemischen Milieu stehen im Vordergrund.

Einleitend klassifizieren und quantifizieren R. H. SPIGEL & J. IMBERGER Durchmischungsprozesse. Es werden Vorgänge untersucht, die die Wechselwirkung der schwimmfähigen Blaualgen mit ihrer Umwelt beeinflussen (Lichtdurchlässigkeit, Durchmischung, Beweglichkeit, Nährstoffaufnahme, Produktivität). Ein Mechanismus, der die Schwimmfähigkeit der Blaualgen reguliert, wurde von C. S. REYNOLDS und A. E. WALSBY schon in den 70-iger Jahren beschrieben. Gemeinsam mit R. L. OLIVER stellen sie nun der turgorbeeinflußten Gasvesikeldynamik einen weiteren Typ der Dichteregulation gegenüber (Ballastkohlenhydratsynthese und -abbau). In diesen Kontext fügt sich die Übersicht über die Bildung und funktionelle Bedeutung von Speicherprodukten (Polyphosphatkörper, Cyanophycingranula und Polysaccharide) und ihrer Rolle als interne Nährstoff- bzw. Energiequelle von J. KROMKAMP ein. R. D. ROBARTS & T. ZOHARY fassen Befunde zur Wirkung der Temperatur auf die Photosynthese, Atmung und das Wachstum von Blaualgen zusammen. M. TILZER untersucht das Phänomen, daß Blaualgen während der Stratifikation die höchstmöglichen flächenspezifischen Photosyntheseraten bei effizienter Umsetzung der Lichtenergie und minimaler Lichtinhibition bzw. Photooxidation realisieren. Es werden physiologische Adaptationsprozesse des Photosyntheseapparates in einer fluktuierenden Umwelt berücksichtigt. Die Bedeutung der Makronährstoffe, insbesondere von Fe, für die N_2-Fixation planktischer Blaualgenpopulationen arbeiten A. J. HORNE & M. L. COMMINS heraus. Ausgehend von der Hypothese, daß das Wachstum im Freiland entscheidend von der Effizienz des Ammoniumassimilationssystems abhängig ist, stellen KERBY et al. Orginalbefunde zur Aufnahme, zum Transport und zur Assimilation von Ammoniumanaloga (Methyl-, Ethylendiamin) sowie zur Nitrogenaseregulation vor. F. R. PICK & D. R. S. LEAN diskutieren auf der Grundlage von Korrelationsanalysen phänomenologisch die Beziehung zwischen der Dominanz von Blaualgen und einem niedrigen N/P-Verhältnis (Bedeutung der Temperatur, der Durchmischungsverhältnisse, der Transparenz sowie der Fe- und C-Verfügbarkeit). Die bisher wenig beachtete Rolle der Spurenmetalle für die Nährstoffversorgung der Blaualgen wird von J. G. RUETTER & R. R. PETERSEN dargestellt. Ausgehend von der Schlüsselfunktion von Fe und Mo im Biochemismus der Photosynthese bzw. Nitratreduktion und N_2-Fixation wird eine hypothetische Trennung von Phytoplanktongesellschaften vorgenommen, die bei einem niedrigen $Fe:NO_3$-Verhältnis von Eukaryoten und bei einem hohen von Cyanobakterien beherrscht werden. Die umfassende und abgestimmte Darstellung der Zooplankton-Cyanobakteria-Wechselwirkungen (Inhibition von Blaualgen, Verwertbarkeit, Hemmung des Wachstums und der Reproduktion des Zooplanktons) erfuhr entsprechend dem experimentellen Ansatz eine Dreiteilung: J. F. HANEY widmete sich den Freilanduntersuchungen, C. W. BURNS den Mesocosmen („enclosures") und W. LAMPERT den Laborexperimenten.

Abschließend berichten die Teilnehmer des Forums in geraffter Form über ihre gemeinsamen Untersuchungen von Umweltfaktoren und Mechanismen, die mit der Dominanz von Blaualgen in Verbindung stehen (Durchmischung im Tagesverlauf, Koexistenz und Ausschluß von Zooplankton, Effekte unterschiedlicher Verfügbarkeit von N und P bzw. ihres Verhältnisses, Dichteregulation, Kohlenhydrat-Proteinverhältnis und Glutaminsynthetaseaktivität als Indikator der Wasserbewegung bzw. N-Versorgung). Es wurden ein P-reicher See mit hoher Biomasse N_2-fixierender *Anabaena*-Arten und ein nährstoffärmerer, hydrothermal gespeister Explosionskrater-See mit dominanter Grünalgenpopulation und *Microcystis*-Blüte untersucht.

Das Heft sei allen allgemein am Thema Interessierten empfohlen, denn viele Beiträge sind „lehrbuchgerecht" abgefaßt und die Bibliographie vermittelt weiterführende Literatur. Physiologisch-biochemische Grundlagenerkenntnisse wurden aus unterschiedlichsten Gebieten (z. B. Adaptation des Pigmentapparates, Regulation der Nitrogenase und Glutaminsynthetase, Energiespeicherung bei Prokaryoten usw.) unter dem Aspekt der angewandten Limnologie einbezogen. Auch der Spezialist wird fündig und erhält viele Anregungen.

G. DUDEL, Berlin

ISBN 3-05-500643-7
ISSN 0075-9511

Erschienen im Akademie-Verlag Berlin, Leipziger Straße 3—4, Berlin, DDR-1086

Printed in the German Democratic Republic
Gesamtherstellung: Druckhaus „Thomas Müntzer" GmbH, DDR-5820 Bad Langensalza
Umschlaggestaltung: Wolfgang Janisch
LSV 1305
Bestellnummer: 7639493 (2111/20/2)

Inhalt/Contents

Preface 215
WILLÉN, E., HADJU, S., & PEJLER, Y.: Summer Phytoplankton in 73 Nutrient-poor Swedish Lakes. Classification, Ordination and Choice of Long-term Monitory Objects. With 8 Figures 217
ILMAVIRTA, V., & HUTTUNEN, P.: Characterization of 151 Small Lakes in Eastern Finland Based on their Water Chemistry, Land-use Catchment and Phytoplankton Communities. With 12 Figures 229
ARVOLA, L., SALONEN, K., & RASK, M.: Chemical Budgets for a Small Dystrophic Lake in Southern Finland. With 5 Figures 243
WALLS, M., RAJASILTA, M., SARVALA, J., & SALO, J.: Diel Changes in Horizontal Microdistribution of Littoral Cladocera. With 2 Figures 253
JÓNASSON, P. M., LINDEGAARD, C., DALL, P. C., HAMBURGER, K., & ADALSTEINSSON, H.: Ecosystem Studies on Temperate Lake Esrom and the Subarctic Lakes Mývatn and Thingvallavatn. With 3 Figures 259
STENSON, J. A. E.: Interspecific and Intertrophic Relations in Oligotrophic Water Systems. With one Figure 267
LINDHOLM, T., & ERIKSSON, J.: Limnological Changes in Two Recently Isolated Lakes. With 8 Figures 273
BONSDORFF, E., & STORBERG, K.-E.: Ecological Changes in a Formerly Meromictic Coastal Lake. With 7 Figures 279
HOLTAN, H.: Lake Mjøsa, Recipient and Ecosystem. With 8 Figures 285
MORLING, G., & WILLÉN, T.: Acidification and Phytoplankton Development in Some West-Swedish Lakes 1966—1983. With 9 Figures 291
MORLING, G., & PEJLER, B.: Acidification and Zooplankton Development in Some West-Swedish Lakes 1966—1983. With 16 Figures 307
APPELBERG, M.: Population Regulation of the Crayfish *Astacus astacus* L. after Liming an Oligotrophic, Low-alkaline Forest Lake. With 8 Figures 319
BLOMQVIST, P., & HEYMAN, U.: Effects of Arsenate Additions to Pelagic Mesocosms on Phytoplankton Biomass and Composition. With 5 Figures 329
HÅKANSON, L.: Environmental Hazard Analysis — Contamination of Nutrients, Mercury and Cesium-137 in Natural Waters. With 4 Figures 339
ANDERSSON, I., PARKMAN, H., & JERNELÖV, A.: The Role of Sediments as Sink or Source for Environmental Contaminants — a Case Study of Mercury and Chlorinated Hydrocarbons. With 3 Figures 347
Rezensionen 361

Band 21 der LIMNOLOGICA umfaßt zwei Hefte (Volume 21 of LIMNOLOGICA contains two numbers).
Band 21, Heft 1, wird folgende Arbeiten enthalten (Volume 21, number 1) will contain the following contributions):

RIPL, W., M. MOTTER, E. WESSLER, & W. FISCHER: Regional-ökologische Studien zum Plankton und Benthos Berliner Gewässer. — KREY, L., & K. KLOSS: Geographische und pollenanalytische Untersuchungen des Kleinen Barsch-Sees (Bez. Potsdam, DDR). — FLÖSSNER, D.: Die Geschichte der Cladocerenfauna des Kleinen Barsch-Sees, eines sauren kalkarmen Moorweihers im mitteleuropäischen Flachland. — SCHÖNBORN, W.: Analyse subfossiler Protozoenschalen der Sedimente eines kleinen sauren Waldsees (Kleiner Barsch-See, nördliche DDR). — CASPER, S. J., & W. SCHEFFLER: Diatom Analysis, Late-glacial and Post-glacial Development of Lake Kleiner Barsch-See (GDR). — PROFT, G., & L. KREY: Die chemische Zusammensetzung der Spät- und Postglacialsedimente des Kleinen Barsch-Sees (Fe, Mn, P, C, N, H und BSi), eines dystrophen Moorweihers im Norden der DDR. — KRIENITZ, L.: Coccale Grünalgen der mittleren Elbe. — ARNDT, H., C. SCHRÖDER & W. SCHNESE: Rotifers of the Genus Synchaeta — an Important Component in the Coastal Waters of the Southern Baltic. — BERG, G.: Saprobiologische Untersuchungen an der Benthosflora der Recknitz im Norden der Mecklenburger Seenplatte (DDR). — OPRAVILOVA, V.: Microzoobenthos of the River Jihlava after the Construction of Dalešice Waterworks. — PENTECOST, A.: The Identity of Suspended Particles in a Calcite-Depositing Stream and their Significance in Trapping and Binding Phenomena. — BRINKMANN, R., & K. BÖTTGER: Zur Fischfauna einiger stehender Gewässer bei Concepción (Paraguy). — NWADIARO, C.: A Hydrobiological Survey of the Chanomi Creek System, Lower Niger Delta, Nigeria. — SABRI, A. W.: Local and Seasonal Variation of the Epipelic Algae in Samarra Impoundment, Iraq. — JANA, B. B., & G. P. PAL: Production of Chironomid Larvae in Culturing Media of Various Organic Wastes. — YADAVA, Y. S., & S. C. DEY: Impact of Physico-chemical Complexes on Plankton Density in Dhir Beel of Assam. — SINGHAL, S., G. M. A. RAMANI, & U. S. GUPTA: Seasonal Variations and Relationships of Different Physico-chemical Characteristics in Newly Made Tawa Reservoir. — Rezensionen.

ISBN 3-05-500643-7
ISSN 0075-9511

Bei Fragen zur Produktsicherheit wenden Sie sich bitte an:
If you have any questions regarding product safety,
please contact:

Walter de Gruyter GmbH
Genthiner Straße 13
10785 Berlin
productsafety@degruyterbrill.com